AF887176

Pedestrian and Evacuation Dynamics

Springer
Berlin
Heidelberg
New York
Barcelona
Hong Kong
London
Milan
Paris
Tokyo

Michael Schreckenberg
Som Deo Sharma
Editors

Pedestrian and Evacuation Dynamics

With 175 Figures and 28 Tables

Springer

Editors

Michael Schreckenberg
Physik von Transport und Verkehr
Gerhard-Mercator-Universität Duisburg
Lotharstraße 1
47048 Duisburg, Germany
e-mail: schreckenberg@uni-duisburg.de

Som Deo Sharma
Institut für Schiffstechnik Duisburg (ISD)
Gerhard-Mercator-Universität Duisburg
Bismarckstraße 69
47048 Duisburg, Germany
e-mail: sharma@nav.uni-duisburg.de

Library of Congress Cataloging-in-Publication Data applied for

Die Deutsche Bibliothek - CIP-Einheitsaufnahme
Pedestrian and evacuation dynamics : with 28 tables /
Michael Schreckenberg ; Som Deo Sharma ed.. -
Berlin ; Heidelberg ; New York ; Barcelona ; Hong Kong ;
London ; Milan ; Paris ; Tokyo : Springer, 2002
ISBN 3-540-42690-6

Mathematics Subject Classification (2000): 49-XX, 49-06, 65-XX, 65C99, 68-XX, 68U20

ISBN 3-540-42690-6 Springer-Verlag Berlin Heidelberg New York

This work is subject to copyright. All rights are reserved, whether the whole or part of the material is concerned, specifically the rights of translation, reprinting, reuse of illustrations, recitation, broadcasting, reproduction on microfilm or in any other way, and storage in data banks. Duplication of this publication or parts thereof is permitted only under the provisions of the German Copyright Law of September 9, 1965, in its current version, and permission for use must always be obtained from Springer-Verlag. Violations are liable for prosecution under the German Copyright Law.

Springer-Verlag Berlin Heidelberg New York
a member of BertelsmannSpringer Science+Business Media GmbH

http://www.springer.de

© Springer-Verlag Berlin Heidelberg 2002
Printed in Germany

The use of general descriptive names, registered names, trademarks, etc. in this publication does not imply, even in the absence of a specific statement, that such names are exempt from the relevant protective laws and regulations and therefore free for general use.

Typeset by the authors
Cover design: *design & production* GmbH, Heidelberg

SPIN: 10853382 46/3142LK - 5 4 3 2 1 0 – Printed on acid-free paper

Opening Speech at the University

Dear Professor Wolff,
dear Professor Schreckenberg,
Professor Usadel,
dear Madams and Sirs,

it's my pleasure to welcome you – such an important international group of scientists – in the name of the city of Duisburg.

The Lord Mayor, Mrs. Zieling asked me to greet you also in her name. Tomorrow at 5 p.m. she will welcome you in the town-hall personally.

You came together here in Duisburg to discuss current questions about topics like pedestrian dynamics and simulation of evacuation. As a politician of the green party I generally have a liking for pedestrian motion, but I won't make a secret out of it that your special topic is what people call a closedbook for me. If I got involved in your expert discussion, I would possibly trip myself up… and I need to be evacuated myself… For this reason I want to tell you about the virtues of Duisburg and its people as many are not familiar with 'Duisburg' as they are with 'London' or 'Berlin'.

Our university carries the name of Gerhard Mercator, a name which occurs here in Duisburg more than once. The spectrum goes from Mercator-Pharmacy over Mercator-Hall, from Mercator-Center to Mercator-Restaurant. This led a former director of the university to the assertion that 'Mercator would reside in a state of continual beatification'. Mercator lived and worked as one of the most important scholars in the 16th century here in Duisburg and created his well known globes, atlases, and maps. Moreover he developed the so called 'Mercator-Projection' without which cartography and navigation would not be what they are today. In a way Mercator stands as an example for the tradition of science in Duisburg, a city which was first mentioned in documents in 883. When you – up to now – didn't know that Duisburg is such an old city with lots of traditions, you are in good company. What is usually associated with the city of Duisburg, has rarely anything to do with its traditions, but with its industrial development – to cut a long story short: With coal and steel. Especially in the perception of people not from this area, Duisburg has been the symbol of smokestack industries for decades. Up to the 1960s Duisburg was perceived as a synonym for an industrial metropolis maybe like Manchester in England or Pittsburg in the United States. For generations people showed an empathy with the environment of coal mines and iron- and steal-industries. Losing this sector of industry, not only the factor of identification got lost, but also the most dominant employer in this area. As a result we find high unemployment in our city. You can imagine, how difficult it is to find a way into future, to create a structural change and offer new jobs in a modern city on the one hand, and to find an appropriate way to deal with the past on the other. I don't betray a secret when I tell you that Duisburg had a hard time dealing with this. The memories about this time were in danger of becoming those with a somewhat tarnished history where people were more likely to be ashamed of instead of being proud. The question of Duisburg's future profile was unanswered for a long time. Today the city has overcome its hurdles to present a new profile that has been carefully and diligently formed. You can see this not only at the port which is being remodelled under the motto "Life, Work, Culture and Recreation along the water", but also at the Landschaftspark Duisburg Nord which you will have the chance to visit tonight - an excursion that you shouldn't miss. Where we don't try to hide the disused steel mills we show, that the handling with the smokestack industries of our past gained more self-confidence. We no longer condemn the relicts of the past, but we identified them as the potential we need to design the future of our city. This also makes this area distinctive.

But enough for now – I will stop with this little sketch and I wish you all an effective conference.

I hope that in addition to the conference you get the chance to see the delights of Duisburg. In this sense to all of you: A successful conference and a cheerful Glueckauf!

Thank you for your attention.

April 04, 2001 Doris Janicki

Mayor of Duisburg

Welcome Speech in the Town Hall

Dear Professor Wolff,
dear Professor Schreckenberg,
Professor Sharma,
dear Madams and Sirs,

it's a pleasure for me to welcome you here in the Town Hall today in the name of the city of Duisburg. Special greetings to the guests from abroad. You have come to Duisburg to participate in the international conference "Pedestrian and Evacuation Dynamics".

I must admit that the topic you deal with is in needs to be explained to me. When I got it right, it is a topic that might be of extraordinary importance for anybody who is a part of the dynamical processes, that you investigate as pedestrians or vessels. Moreover, one is usually not even aware of this fact.

Therefore we all benefit from your work and of course hope, that every one of you will profit from this international conference and enjoy your stay in Duisburg.

The room we are in right now is the assembly hall of the city council which meets here about every other month to discuss essential questions for Duisburg. The seats you are sitting on right now, are usually those of 74 councillors; Social Democrats, Christian Democrats, members of the green, socialistic and liberal parties. For about 100 years the political meetings have taken place in this room. When this Town Hall was built in the beginning of the 20th century for 1.5 million Goldmark, which was an enormous amount in that time, a lot of money was spent for the design of the rooms. You can still notice some of the magnificence. At that time nobody really thought about questions like fire prevention, a fact that we really have to deal with today. So it is an interesting, but of course coincidental fact, that international experts who discuss about emergency questions and evacuation, come together in this Town Hall. This summer we will have to reconstruct the building to meet the current safety standards. Of course we try carefully not to destroy any of the historic environment. The assembly hall will have to be remodelled. We will probably turn the seats by 180 degrees to be able to use the door next to me as an emergency exit as well. If you have a look at the painting in front of you gives the impression as if it was placed there just for this meeting today. On the right hand sight you see, that Duisburg has a long scientific tradition. The University of Duisburg was established as far back as 1655. You can see the inaugural ceremony on the painting right above me. On your left you may notice the confusion and tangle that happened when the archbishop of Cologne wanted to conquer this city. The behaviour of the escaping crowd on a narrow bridge can also be observed. Even without specialised knowledge one can realize that the manner of evacuation in panic situations didn't really change from 1445 up to today.

At this point I want to thank the Department of Physics as well as the Institute of Ship Technology of the University of Duisburg for organizing this conference. In this way they enable the city of Duisburg to greet such an reputable group of scientists.

I wish you all a successful conference and a pleasant stay in our city.

But for now I wish you a nice evening in the Town Hall.

Thank you for your attention!

April 05, 2001

Bärbel Zieling

Lord Mayor of Duisburg

Preface

Pedestrian dynamics is a field of growing interest for researchers from various disciplines. Improvements in the area of complex dynamics and the availability of fast computation facilitate the performance of large scale simulations. Moreover the inclusion of psychological and social aspects has boosted the scope of potential applications. Evacuation assessment is one of the major applications in this area.

The analysis of escape routes and safety standards for buildings and passenger vessels has seen a tremendous development over the last decades. However, there are still many open questions and challenges, especially around passenger ships. The aim of the conference was bringing together researchers and experts from the fields of modelling and simulation of pedestrian motion, safety assessment, especially with regard to ship design and operation, regulatory action, safety consultancy, simulation of evacuation processes, etc. and developers and users of software tools. The range of topics included:

- Pedestrian motion
- Multimodal and intermodal traffic
- Evacuation simulation
- Video detection methods
- Cellular Automata
- Multi-Agent Models
- Operations Research and Optimization
- Offshore Vessel Evacuation
- Evacuation from Ships
- Requirements and Regulations
- Directional sound

We would like to thank the sponsors: Duisburger Universitäts-Gesellschaft, Flensburger Schiffbau-Gesellschaft, Intraplan Consult, Westdeutsche Landesbank, Stadt Duisburg, Duisburger Verkehrs-Gesellschaft, Gerhard-Mercator-Universität, Institut für Schiffstechnik, and Institut für Physik.

Last but not least we are grateful to the conference secretaries: Birgit Dahm-Courths, Peter Pfeifer, and Hubert Klüpfel, as well as to the technical and organisational staff: Tim Meyer-König, Joachim Wahle, Roland Chrobok, Thomas Grunewald, Peer Wetterling, Rupert Henn, Robert Barlovic, Torsten Huisinga, Wolfgang Knospe, Andreas Keßel, Andreas Pottmeier, Bernd Pluntke, and finally Andreas Keßel for the poster and the cover picture.

Duisburg, July 2001

Michael Schreckenberg

Som D. Sharma

Contents

Ship Evacuation

List of Participants

1. Guido Adam, BKP Brandschutzplanung Klingsch, Flughafen Düsseldorf, ZG-Ost, Raum 6040, 40474 Düssseldorf, Germany; g.adam@bpk-mail.de
2. Robert Barlovic, Physik von Transport und Verkehr, Gerhard-Mercator-Universität, Lotharstr. 1, 47048 Duisburg, Germany; barlovic@uni-duisburg.de
3. Andreas Baumgart, Germanischer Lloyd AG, Vorsetzen 32/35, 20459 Hamburg, Germany; abau@germanlloyd.org
4. Martin Berends, Global Maritime, Scheepmakershaven 27/28, 3011 VA Rotterdam, The Netherlands
5. Jean Louis Berrou, Monte Carlo Sun, 74 Bd D'Italie, MC98000 Monaco, Monaco; jlberrou@jdc.monaco.mc
6. Willem Bles, TNO Human Factors, Kampweg 5, PO Box 23, 3769 ZG Soesterberg, The Netherlands; bles@tm.tno.nl
7. Victor Blue, New York State Department of Transportation, Poughkeepsie, NY 12603, USA; vblue@gw.dot.state.nyus
8. Piet Bovy, SubFaculty of Civil Engineering, Delft University of Technology, Stevinweg 1, 2628 CN, Delft, The Netherlands; bovy@ct.tudelft.nl
9. Elmar Brockfeld, Deutsches Zentrum für Luft- und Raumfahrt e.V. (DLR), Linder Höhe, 51147 Köln, Germany; elmar.brockfeld@dlr.de
10. Carsten Burstedde, Institut für Theoretische Physik, Universität zu Köln, Habsburgerring 5, 50674 Köln, Germany; cb@thp.uni-koeln.de
11. Thomas Christensen, Blohm + Voss GmbH, Hermann-Blohm-Str. 3, 20457 Hamburg, Germany; christensent@blohmvoss.thyssen.com
12. Guro Christiansen, Ship Stability Research Centres, University of Strathclyde, 801 Colville BLD, 48 North Portland ST, Glasgow, G1 IXN, Scotland, UK; guro.christiansen@strath.ac.uk
13. Roland Chrobok, Physik von Transport und Verkehr, Gerhard-Mercator-Universität, Lotharstr. 1, 47048 Duisburg, Germany; chrobok@traffic.uni-duisburg.de
14. Winnie Daamen, TU Delft, Stevinweg 1, 2600 Ga Delft, The Netherlands; w.daamen@ct.delft.nl
15. Birgit Dahm, Physik von Transport und Verkehr, Gerhard-Mercator-Universität, Lotharstr. 1, 47048 Duisburg, Germany; dahm@uni-duisburg.de
16. Cemalettin Demirel, BKP Brandschutzplanung Klingsch, Flughafen Düsseldorf, ZG-Ost, Raum 6040, 40474 Düssseldorf, Germany; c.demirel@bpk-mail.de
17. Jan Dijkstra, Department of Architecture, OS, Eindhoven University of Technology, P.O. Box 513, 5600 MB Eindhoven, The Netherlands; j.dijkstra@tue.nl
18. Mario Dogliani, RINA s.p.a., Via Corsica 12, 16128 Genova, Italy; mario.dogliani@rina.org

19. Harald Drager, A/S Quasar Consultants, P.O. Box 388 Skøyen, 0212 Oslo, Norway; harald@quasar.no
20. Michiel Flier, Global Maritime, Scheepmakershaven 27/28, 3011 VA Rotterdam, The Netherlands; mfl@globalmaritime.nl
21. Edwin Galea, School of Computing and Mathematical Sciences, University of Greenwich, Wellington Street Woolwich, London SE18 6PF, UK; e.r.galea@gre.ak.uk
22. Cees Glansdorp, Global Maritime, Scheepmakershaven 27/28, 3011 VA Rotterdam, The Netherlands
23. Tore Grünert, Operations Research, RWTH Aachen, Templergraben 64, 52056 Aachen, Germany; tore@or.rwth-aachen.de
24. Thomas Grunewald, Physik von Transport und Verkehr, Gerhard-Mercator-Universität, Lotharstr. 1, 47048 Duisburg, Germany; grune@traffic.uni-duisburg.de
25. Stefan Haarhoff, Institut für Schiffstechnik Duisburg (ISD), Gerhard-Mercator-Universität, Bismarckstr. 69, 47048 Duisburg, Germany; haarhoff@uni-duisburg.de
26. Jenny Hall, Maritime & Coastguard Agency, Spring Place, Commercial Rd, Southampton, SO151EG, UK; jhall@mcga.gov.uk
27. Horst W. Hamacher, Fachbereich Mathematik, Universität Kaiserslautern, Postfach 3049, 67653 Kaiserslautern, Germany; hamacher@itwm.uni-kl.de
28. Dirk Helbing, Institut für Wirtschaft und Verkehr, TU Dresden, Andreas-Schubert-Str. 23, 01062 Dresden, Germany; helbing@trafficforum.de
29. Rupert Henn, Institut für Schiffstechnik Duisburg (ISD), Gerhard-Mercator-Universität, Bismarckstr. 69, 47048 Duisburg, Germany; rupert@uni-duisburg.de
30. Christoph Hermanns, RWTH Aachen, Im Esch 9, 46354 Südlohn-Oeding, Germany; bkempkes@sparkasse-vreden.de
31. Julien Hine, Transport Research Institute, 66 Spylaw Road, EH10 5BR Edinburgh, UK; j.hine@napier.ac.uk
32. Serge Hoogendoorn, SubFaculty of Civil Engineering, Delft University of Technology, Stevinweg 1, 2628 CN, Delft, The Netherlands; s.hoogendoorn@ct.tudelft.nl
33. Torsten Huisinga, Physik von Transport und Verkehr, Gerhard-Mercator-Universität, Lotharstr. 1, 47048 Duisburg, Germany; huisinga@uni-duisburg.de
34. Mikko Ilus, Kvaerner Masa-Yards, P.O. Box 666, 20101 Turku, Finland; mikko.ilus@masa-yards.fi
35. Nejat Ince, Istanbul Technical University Foundation, Ferit Tek Sok. 4/8 Moda, 81300 Istanbul, Turkey; ituvsam@ibm.net
36. Marco Irzik, Institut für Verkehrswirtschaft, Straßenwesen und Städtebau, Uni Hannover, Appelstr. 9A, 30167 Hannover, Germany; irzik@ivh.uni-hannover.de
37. Armin Jesse, Fr. Lürssen Werft, Friedrich-Klippert-Str. 1, 28759 Bremen, Germany; t@luerssen.de

38. Frank Josten, Blohm + Voss GmbH, Hermann-Blohm-Str. 3, 20457 Hamburg, Germany; josten@blohmvoss.thyssen.com
39. Markku Kauriala, Kauriala Ltd Fire Engineering and Fire Safety Design, Satakunnantie 164, 20320 Turku, Finland; markku.kauriala@inststokauriala.fi
40. Jon Kerridge, Transport Research Institute, 66 Spylaw Road, EH10 5BR Edinburgh, UK; j.kerridge@napier.ac.uk
41. Andreas Keßel, Physik von Transport und Verkehr, Gerhard-Mercator-Universität, Lotharstr. 1, 47048 Duisburg, Germany; kessel@traffic.uni-duisburg.de
42. Kevin Kiefer, USCG, Washington, 2100 Second St., S.W., Washington, DC 20593, USA; kkiefer@comdt.uscg.mil
43. Ansgar Kirchner, Institut für Theoretische Physik, Universität zu Köln, Zülpicher Str. 77, 50937 Köln, Germany; aki@thp.uni-koeln.de
44. Hubert Klüpfel, Physik von Transport und Verkehr, Gerhard-Mercator-Universität, Lotharstr. 1, 47048 Duisburg, Germany; kluepfel@traffic.uni-duisburg.de
45. Wolfgang Knospe, Physik von Transport und Verkehr, Gerhard-Mercator-Universität, Lotharstr. 1, 47048 Duisburg, Germany; knospe@uni-duisburg.de
46. Andreas Könnecke, intraplan consult GmbH, Orleansplatz 5a, 81667 München, Germany; andreas.koennecke@intraplan.de
47. Rainer Könnecke, I.S.T. Integrierte Sicherheits-Technik GmbH, 60325 Frankfurt / Main, Feuerbachstr. 19, Germany
48. Robert Kukla, Transport Research Institute, 66 Spylaw Road, EH10 5BR Edinburgh, UK; r.kukla@dcs.napier.ac.uk
49. Dongkon Lee, Korea Research Institute of Ships and Ocean Eng. (KRISO); YuSungKu, P.O. Box 23, 305-600 Taejeon, Korea; dklee@kriso.re.kr
50. Ville Lehmuskoski, LT-Consultants Ltd., Melkonkatu 9, 00210 Helsinki, Finland; ville.lehmuskoski@ltcon.fi
51. Bengt Lyderson, Swedish Maritime Administration, Slottsgatan 82, 60278 Norrkoping, Sweden; bengt.lyderson@sjofartsverket.se
52. Hani S. Mahmassani, Department of Civil Engineering, Ernest Cockrell Jr. Hall, Suite 6.2, University of Texas at Austin, Austin, Texas 78712-1076, USA; masmah@mail.utexas.edu
53. Sara Manzoni, Dipartimento di Informatica, Sistemistica e Università di Milano – Bicocca, Via Bicocca degli Arcimboldi, 8, 20126 Milano, Italy; manzoni@disco.unimib.it
54. Paul Martin, 26 Albert Embankment, SE1 7TE London, UK; cherrich@jdc.monaco.mc
55. Merdan Metin, Istanbul Technical University Foundation, Ferit Tek Sok. 4/8 Moda, 81300 Istanbul, Turkey; ituvsam@ibm.net
56. Tim Meyer-König, Physik von Transport und Verkehr, Gerhard-Mercator-Universität, Lotharstr. 1, 47048 Duisburg, Germany
57. A.A.J. Mulder, Netherlands Shipping Inspectorate, Pr. Willem Alexanderlaan 49, 3273 AS Westmaas, The Netherlands; amulder@sir.dgg.minvenw.nl

58. Kai Nagel, ETH Zürich, ETH Zentrum IFW B27.1, 8092 Zürich, Switzerland; nagel@inf.ethz.ch
59. Matthijs Noordegraaf, van der Giessen de Noord, Juweelstraat 81, 2403 BK Alphen a/d Rijn, The Netherlands; m_b_nordegraaf@hotmail.com
60. Paul Norda, Jos. L. Meyer GmbH, Postfach 1555, 26855 Papenburg, Germany; nordap@meyerwerft.de
61. Thomas Ohm, Technische Universität Hamburg-Harburg, Am Langenzug 7, 22085 Hamburg, Germany; ohm@tu-harburg.de
62. Peter Olsson, BMVBW, Ref. LS 23, Postfach 200100, 53170 Bonn, Germany; peter.olsson@bmvbw.bund.de
63. Carsten Ortloff, Flensburger Schiffbaugesellschaft, Batteriestr. 52, 24939 Flensburg, Germany; ortloff@fsg-ship.de
64. Edwin Pang, Three Quays Marine Services Ltd, 12-20 Camomile Street, EC3A 7AS London, UK; edwin.pang@threequays.com
65. Alan Penn, UCL, Gower St, WC1E 6BT London, UK: a.penn@ucl.ac.uk
66. Peter Pfeifer, Institut für Schiffstechnik Duisburg (ISD), Gerhard-Mercator-Universität, Bismarckstr. 69, 47048 Duisburg, Germany; pfeifer@nav.uni-duisburg.de
67. Andreas Pottmeier, Physik von Transport und Verkehr, Gerhard-Mercator-Universität, Lotharstr. 1, 47048 Duisburg, Germany; pott@traffic.uni-duisburg.de
68. Martin Rose, Institut für Bauinformatik, Universität Hannover, Am kleinen Felde 30, 30167 Hannover, Germany; rose@bauinf.uni-hannover.de
69. Andreas Schadschneider, Institut für Theoretische Physik, Universität zu Köln, 50937 Köln, Germany; as@thp.uni-koeln.de
70. Hans-Joachim Schlüter, Institut für Schiffstechnik Duisburg (ISD), Gerhard-Mercator-Universität, Bismarckstr. 69, 47048 Duisburg, Germany; schlueter@nav.uni-duisburg.de
71. R. Schouten, van der Giessen de Noord, Schaardijk 23, 2921 LG Krimpen a/d Yssel, The Netherlands; rschouten@gn.nl
72. Michael Schreckenberg, Physik von Transport und Verkehr, Gerhard-Mercator-Universität, Lotharstr. 1, 47048 Duisburg, Germany; schreckenberg@uni-duisburg.de
73. Daniela Shala, Institut für Schiffstechnik Duisburg (ISD), Gerhard-Mercator-Universität, Bismarckstr. 69, 47048 Duisburg, Germany; d.shala@uni-duisburg.de
74. Som Deo Sharma, Institut für Schiffstechnik Duisburg (ISD), Gerhard-Mercator-Universität, Bismarckstr. 69, 47048 Duisburg, Germany; sharma@nav.uni-duisburg.de
75. Norbert Stuntz, Institut für Schiffstechnik Duisburg (ISD), Gerhard-Mercator-Universität, Bismarckstr. 69, 47048 Duisburg, Germany; norbert@vug.uni-duisburg.de
76. Yuki Sugiyama, Division of Mathematical Science, City College of Mie, Ishinden Nakano 157, Tsu, Mie 514-01, Japan; genbey@eken.phys.nagoya-u.ac.jp

77. Josef Thaler, Graf-Anton-Günther-Str. 22, 26931 Elsfleth, Germany; qw.ecoplan@t-online.de
78. Tomáš Tichý, CTU Faculty of Transportation Sciences, Konviktská 20, 11000 Prague 1, Czech Republic; tichy@fd.cvut.cz; tichyt@eltodo.cz
79. Stevanus A. Tjandra, Fraunhofer Institut Techno- und Wirtschaftsmathematik (ITWM), Gottlieb-Daimler-Str., Gebäude 49, 67663 Kaiserslautern, Germany; tjandra@itwm.fhg.de
80. Margarete Tonecker, Aker MTW-Werft GmbH, Wendorfer Weg 5, 23966 Wismar, Germany; m.tonecker@akermtw.de
81. Alasdair Turner, CASA, University College London, Gower Street, WC1E 6BT London, UK; a.turner@ucl.ac.uk
82. Fulvio Vaccarezza, RINA s.p.a., Via Corsica 12, 16128 Genova, Italy; Fulvio.Vaccarezza@rina.org
83. S.L. van der Voort, Netherlands Shipping Inspectorate, Egmonderstraat 35, 2201 RK Noordwijk, The Netherlands; vdvoort@sir.dgg.minvenw.nl
84. Dracos Vassalos, Department of Ship and Marine Technology, University of Strathclyde, 100 Montrose ST, Glasgow, G4 OLZ, Scotland, UK; d.vassalos@strath.ac.uk
85. Antonino Vitetta, DIMET – Università di Reggio Calabria – Fac. Ing., Feo di Vito, 89100 Reggio Calabria, Italy; vitetta@unirc.it
86. Joachim Wahle, Physik von Transport und Verkehr, Gerhard-Mercator-Universität, Lotharstr. 1, 47048 Duisburg, Germany; wahle@traffic.uni-duisburg.de
87. Thomas Weigend, Jos. L. Meyer GmbH, Postfach 1555, 26855 Papenburg, Germany; weigend@meyerwerft.de
88. Frank Weßel, BKP Brandschutzplanung Klingsch, Flughafen Düsseldorf, ZG-Ost, Raum 6040, 40474 Düsseldorf, Germany; f.wessel@bpk-mail.de
89. Peer Wetterling, Physik von Transport und Verkehr, Gerhard-Mercator-Universität, Lotharstr. 1, 47048 Duisburg, Germany
90. Carsten-S. Wibel, Forschung und Entwicklung, Transport & Service GmbH & Co., Geo-Plate-Str. 1, 27568 Bremerhaven, Germany; carsten.wibel@gmx.de
91. Rainer Wiebusch-Wothge, Ruhr-Universität Bochum, Universitätsstr. 150, 44780 Bochum, Germany; rainer.wiebusch-wothge@ruhr-uni-bochum.de
92. Alexandra Willis, Transport Research Institute, 66 Spylaw Road, EH10 5BR Edinburgh, UK; a.willis@napier.ac.uk
93. Deborah Withington, Department of Biomedical Science, University of Leeds, LS2 9NQ Leeds, UK; d.j.withington@leeds.ac.uk

Pedestrian Dynamics

A Speed-Concentration Relation for Bi-Directional Crowd Movements with Strong Interaction

S.A.H. AlGadhi[1], H.S. Mahmassani[2], and R. Herman[2†]

[1] Department of Civil Engineering, King Saud University, Saudi Arabia

[2] Department of Civil Engineering, The University of Texas at Austin, USA

A bi-directional speed concentration relation is developed for crowd movements in opposite directions with strong interactions. Two types of formulations of this relation are investigated: static and dynamic. The investigation is based on observations of crowd movement in the "Jamarat" system, in connection with the pilgrimage rites in Makkah. The estimated models revealed that the impedance to one's movement due to pedestrians going in the same direction is more than twice that due to those going in the opposite direction. Furthermore, the analysis shows that there is no significant interaction term, between the concentrations of the two streams, in the estimated speed-concentration models.

1 Introduction

1.1 Background

Uni-directional Speed-Concentration Relation Studies.

These studies deal primarily with uni-directional pedestrian flows encountered on a daily basis in high traffic areas such as passageways, stairways, plazas, sidewalks, etc.

The Institute of Architecture of the Russian Academy of Arts (VAKH) appears to have been the first to study pedestrian traffic flow, in 1937 [24]. The principal scientific achievement of that study was establishing that the speed of pedestrian flow, u, is inversely proportional to the prevailing concentration of pedestrians, k, though the specific quantitative results obtained were unreliable because of the relatively small number of actual observations, and other inadequacies discussed in Predtechenskii and Milinski [24].

The Central Scientific Research Institute of the Russian Fire Protection Service (VNIIPO) conducted a study, between 1946 and 1948, to obtain reliable quantita-

† This paper was completed after the regrettable passing away of Professor Robert Herman, whose contribution to the research was immense.

tive expressions characterizing pedestrian traffic flows. Average sizes of the horizontal projection of the human body (approximated by an ellipse) for different age groups, in both winter and summer clothing, were established, based on a large number of observations made in public buildings such as theaters, and industrial, educational and transportation facilities. The study introduced the measurement of the concentration of pedestrian traffic as the sum of the horizontal projections of the individuals occupying a unit area, or the occupancy ratio. This work observationally verified that the pedestrian flow rate, q, equals the product of the stream's speed times its concentration, that is $q = u\ k$, which is a well known identity in traffic flow theory [24].

Hankin and Wright [8] investigated the flow of passengers in London subways (passageways), and observationally established relations between speed, concentration and flow for uni-directional pedestrian flow. More recent studies of the relationships among these variables include those by Oeding [17] on mixed traffic (e.g., shoppers, commuters, sports spectators), Predtechenskii [23] on mixed mass[1], Older [18] on shoppers, Navin and Wheeler [16] on students, and Fruin [7] on commuters. A synthesis of the above speed-concentration relations is shown in Fig. 1, which depicts a family of curves constructed from the measurements reported by the investigators cited previously and converted to common units. It should be noted that all of these studies applied only to streamlined and orderly pedestrian movements under approximately steady-state conditions.

While most of these studies have dealt with uni-directional pedestrian flows, Older [18], Navin and Wheeler [16] and Fruin [7] studied pedestrian flows in two opposing directions, in a sidewalk environment. These studies found that the two groups of pedestrians split into two distinct streams. Pedestrians travelling in the same direction tend to follow one another in files which interweave with those from the opposite direction. Thus, the interaction between the two flows is reduced and limited to the interface line of the two streams [18].

A linear relation was specified between speed and concentration in all the reported studies, with the exception of those by Hankin and Wright [8] and Predtechenskii [23], (Fig. 1), according to the following form:

$$u = u_f(1 - k/k_{jam}) \qquad (1)$$

where u_f represents the theoretical speed attained by a traffic stream under conditions of completely free flow, with a practically unlimited amount of space per pedestrian, and k_{jam} is the jam concentration, at which all movement in a traffic stream grinds to a halt and speed becomes zero.

---

[1] Predtechenskii and Milinski [24] defined mass movement as that characterized by simultaneous displacement of a large number of people within a relatively limited area.

Of course, the linearity of the speed-concentration relation has long been questioned for both vehicular [9] and pedestrian flows [25], because observations generally depart from a constant slope both in the high- and low-concentration ranges, for both vehicles and pedestrians. At very high concentrations, flow resists grinding to a complete halt and attempts to maintain some speed (i.e. shuffling) even with minimal allocation of space [8], while at very low concentrations, speeds may not increase proportionately to rising space allocations because of imposed or inherent speed limits [25]. However, a thorough statistical investigation of seven different hypotheses concerning the shape of the speed-concentration relationship for vehicular flow indicated that the differences among them are rather small [5]. No such testing appears to have been performed for pedestrian flows.

1.2 Jamarat Model Framework

The Jamarat system of the Hajj is composed of three stone monuments symbolizing the devil, which are supposed to be stoned (i.e. pebbles thrown at each of them) by pilgrims in a given sequence, starting with the first, "Small", second, "Middle", then the third, "Big" Jamarah. Each monument (called Jamarah in the Arabic language) is enclosed by a ring of about 16 meters in diameter, referred to hereafter as the Jamarah ring, to collect pebbles. Pilgrims squeeze through the crowd to get their pebbles into the ring, and preferably, to hit the monument. At the same time, others who have finished stoning squeeze and push their way out. On each of the three days of stoning[2], pilgrims arrive at about the same time, creating extremely crowded conditions.

In a previous paper, the first two authors developed a general crowd behavior and movement model, which consists of a set of simultaneous partial differential equations describing the principal processes that govern the dynamics of the system [1]. The system of governing differential equations can be solved numerically by discretizing in time and space. Such solution yields a profile of the system's evolution and allows the computation of various performance measures and figures of merit. The details of the application of this general model to the Jamarat system of the Hajj, which focuses on the processes taking place around an individual Jamarah ring, are given elsewhere [3].

A fundamental concept in the Jamarat model is the classification of the pilgrims in the Jamarat system into two major classes: class 1 users, who have the intention of performing stoning, and class 2 users, who intend to leave the Jamarah after having completed the rite. For each class of users, two types of movements around the Jamarah can be identified: in the radial direction (forward or backward), and in the circumferential direction (clockwise or counter-clockwise). In the model, mathematical relations are defined for three fundamental processes: 1) radial

---

[2] On the first day of stoning, only the Big Jamarah is stoned by pilgrims over a period extending from Sunrise to Sunset. However, on each of the other days of stoning all three Jamarat are stoned in sequence, over a period extending from Noon till Sunset.

movement, 2) lateral movement, and 3) stoning process. These relations were developed and calibrated using actual measurements taken at the site. The overall model was applied to the evaluation of possible design and control strategies aimed at improving the efficiency and throughput of the system.

The radial movements of pilgrims at the Jamarah were modeled by the mathematical product of the prevailing radial speed and the corresponding concentration. The average radial speed of a given class of users at a given location around the Jamarah ring is a function not only of the concentration of users of the same class in that location, but the concentration of both classes of users.

The speed-concentration model, for pedestrian traffic in a bi-directional movement setting, is one of the fundamental relations used in the Jamarat model. A discussion of the theoretical aspects of the bi-directional speed-concentration relation specification and calibration is given next, followed by a description of the data collection and reduction procedures employed in the observational study. Data analysis and results are then presented.

2 Theoretical Setting

As stated earlier, previous research on pedestrian movement has dealt mainly with orderly and streamlined flows; hence, the speed-concentration relations developed by such studies were for uni-directional pedestrian streams. A linear speed-concentration formulation was used by most of these studies, but with different parameter values for different types of pedestrians (Eq. 1; Fig. 1).

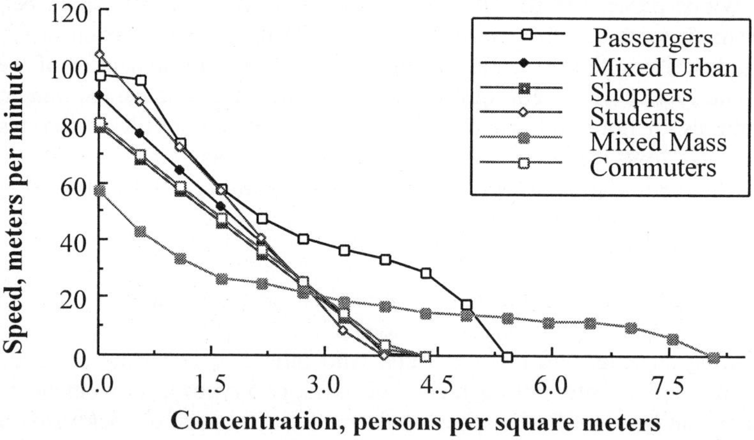

Figure 1: Speed-concentration relations for normal steady-state uni-directional pedestrian flows.

Although previous research on pedestrian movement has considered such areas as stairways, escalators, and merging-crossing situations, no formal models of pedestrian traffic flow in multi-directional movement settings have been developed [2]. It is the objective of this paper to develop such models for pedestrian movement in two opposite directions with strong interactions. This section starts by a presentation of the relevant traffic stream models of vehicular traffic flow, then analogies are drawn to develop pedestrian speed-concentration models.

2.1 Model Specification

In modeling vehicular traffic flow, the most widely known dynamic formulations can be classified as either simple continuum or high order continuum, e.g., [10-12, 14, 19-21]. The simple continuum formulation is based on the conservation equation supplemented by a quasi static equation of state. In the higher order continuum models, a so-called "momentum equation" (actually an acceleration equation) is added to the conservation in order to account for purely dynamic effects [14]. By analogy, pedestrian traffic flow can be modeled using both types of formulations. The simple continuum formulation is considered first, where a "static" speed-concentration relation is specified. The static model is most naturally applicable under equilibrium conditions, where speeds and concentrations are only slowly-varying over time. However, in the simulation of both vehicular and pedestrian traffic, this type of model has also been considered as an instantaneous relation between prevailing speeds and concentrations. Such interpretation may be questionable. Thus, the discussion is expanded to the higher order continuum formulation and a "dynamic" speed-concentration model specification.

2.1.1 Static Model

It was postulated, in the previous section, that the radial speed of either class of users (i.e. pilgrims seeking to perform and those who have completed the stoning process at the Jamarah) is a function of the total concentration of both classes:

$$u_{1F} = \mathrm{Max}(u_{min}, f_1(k_1, k_2)) \tag{2a}$$

$$u_{2B} = \mathrm{Max}(u_{min}, f_2(k_1, k_2)) \tag{2b}$$

where u_{1F} is the speed of class 1 users moving towards the Jamarah ring (i.e. forward radial direction), u_{2B} is the speed of class 2 users moving away from the Jamarah ring (i.e. backward radial direction), u_{min} is the shuffling speed, and k_m is the class m users concentration (m=1, 2). The presence of the laterally moving pedestrians will also impede forward and backward radial movement. Thus, the radial speed of a class m user (m=1, 2) is a function not just of the radial concentrations, k_{mr} but of the total concentrations (i.e. radial plus lateral), k_m. The functional forms, $f_1(.)$ and $f_2(.)$, relating pedestrians' radial speed to the two types of concentrations, are specified below.

First, we consider a linear speed-concentration model with speed as a function only of the concentrations of both user types:

$$u_{1F} = u_{f1} [1 - (\alpha_{11} k_1 + \alpha_{21} k_2)/k_{jam}], \text{ for } u_{1F} > u_{min} \tag{3a}$$

$$u_{2B} = u_{f2}\,[1 - (\alpha_{12}\,k_2 + \alpha_{22}\,k_1)/k_{jam}], \text{ for } u_{2B} > u_{min} \tag{3b}$$

where u_{fm} is the class m users' free-flow speed (m=1, 2), and α_{1m} and α_{2m} are parameters (expected to be positive) that reflect the relative impedance, on one's speed, of the concentration of the pedestrian's own stream vs. that of the opposing stream, respectively. A key question that arises is: which concentration type exerts greater impedance on an individual's speed, his/her own stream concentration or that of the opposite stream? It seems reasonable to argue that the impedance, on an individual's speed, of pedestrians moving in same direction is greater than that of those moving in the opposite one, i.e., $\alpha_{1m} > \alpha_{2m}$, (m=1, 2). The speed of an individual moving within a group of people in one direction is controlled by the aggregate speed of the group, leaving the pilgrim little control over his/her own speed. On the other hand, an individual going against the main flow can often exert some degree of will in maneuvering his/her way in the crowd. It seems that a counter-argument (i.e., $\alpha_{2m} > \alpha_{1m}$) might be more plausible for pedestrian flows under low levels of crowding, and hence higher speed of movements, than under crowded conditions. This question is investigated observationally in Section 4.

Another important question is whether the speed-concentration relation for class 1 users is different from that for class 2 users, or both exhibit the same behavior. In other words, do these two equations (Eqs. 3a,b) share the same underlying parameters? If so, then the calibrated speed-concentration model would not be limited to conditions prevailing for a specific direction of movement (i.e. forward or backward), but would have more general applicability.

A more general linear speed-concentration formulation includes the interaction between the two opposing streams, captured by a multiplicative interaction term in the specification, namely:

$$u_{1F} = u_{f1}\,[1-(\alpha_{11}\,k_1 + \alpha_{21}k_2 + \alpha_{31}k_1k_2/k_{jam})], \text{ for } u_{1F} > u_{min} \tag{4a}$$

$$u_{2B} = u_{f2}\,[1-(\alpha_{12}\,k_2 + \alpha_{22}\,k_1 + \alpha_{32}k_1k_2/k_{jam})], \text{ for } u_{2B} > u_{min} \tag{4b}$$

The inclusion of the interaction term is useful for investigating if the effect of the concentration of class m users (m=1, 2) on the speed of movement is independent of the concentration of the other class of users. For example, consider the impact of a change in k_1 on its forward radial speed, u_{1F}. Without the interaction term, the effect ($\partial u_{1F}/\partial k_1$) would be ($-\alpha_{11}$)($u_{f1}/k_{jam}$). However, with the interaction, the effect would be ($-\alpha_{11} - \alpha_{31}\,k_2$)($u_{f1}/k_{jam}$). Thus, the effect of k_1 on u_{1F} would depend on the level of k_2. If α_{31} is positive, then the negative effect of k_1 on u_{1F} will increase as the value of k_2 increases. The two static models, with and without the interaction term, are calibrated in Section 4.

2.1.2 Dynamic Model

The simple continuum model does not include acceleration and inertia effects; these can be taken into account by the momentum equation in the higher order models. For example, Payne's model for vehicular traffic [19] employs the usual continuity equation and the following "momentum equation" :

$$\partial u/\partial t = -u\,(\partial u/\partial x) - 1/T\,(u - ue(k)) - (\upsilon/T)\,(1/k)\,(\partial k/\partial x) \tag{5a}$$

where u=u(x,t) is the vehicle speed at time t at road location x, T represents a constant reaction time, $u_e(k)$ is the equilibrium speed (which would be compatible with the concentration level, k, at steady state) and $\upsilon = -1/2[du_e(k)/dk]$. Equation 5a can be written in the following difference form [19]:

$$\begin{aligned} u_j^{t+1} = u_j^t - \Delta t\, \{ & u^t_j\, [(u^t_j - u^t_{j-1})/\Delta x_j] \\ & \text{(convection term)} \\ & + 1/\,T_j\, [(u^t_j - u_e(k^t_j)] \\ & \text{(relaxation to equilibrium term)} \\ & + (\upsilon_j/T_j)\,(1/k^t_j)\,(k^t_{j+1} - k^t_j/\Delta x_j)\},\ j=1, 2, \ldots, N \\ & \text{(anticipation term)} \end{aligned} \tag{5b}$$

where j is the roadway section index, t is the time, Δx_j is the length of section j, Δt is the numerical solution time step, N is the total number of sections in the roadway and other terms are as previously defined. The three groups of terms express three physical processes. The first of these is convection, i.e., the fact that vehicles traveling at speed u_{j-1} in the upstream section (section j-1) will tend to continue to travel at that speed as they enter section j. The second represents the tendency of drivers to adjust their speeds to the equilibrium one. The third term is a model of anticipation of travel conditions ahead; i.e. drivers tend to slow down if the concentration is seen to be increasing.

However, in their review of continuum models of freeway vehicular flow, Michalopoulos and Beskos [14] concluded that Payne's model is valid only at low concentrations, and they proposed that an improved practical treatment would be to drop the "momentum equation" at high concentrations; they suggested a concentration threshold of $k_{jam}/3$, beyond which the "momentum equation" can be dropped.

By analogy to vehicular traffic flow, we propose here to explore a dynamic pedestrian speed-concentration relationship. However, data collected in connection with this study, as described in the next sub-section, does not allow the investigation of all the three processes proposed in connection with Eq. 5. Only the relaxation to equilibrium term is applicable since the data are restricted to a single section of the physical space. By dropping the convection and anticipation terms, and changing the notation to fit ours, Eq. 5b can be written as:

$$u_{1F,t+1} = u_{1F,t} - \Delta t/T\ [(u_{1F,t} - u_e(k_{1,t}, k_{2,t})] \tag{6a}$$

$$u_{2B,t+1} = u_{2B,t} - \Delta t/T\ [(u_{2B,t} - u_e(k_{2,t}, k_{1,t})] \tag{6b}$$

where $u_e(k_{1,t}, k_{2,t})$ is the equilibrium speed-concentration function, which could take either form of Eqs. 3 or 4. One useful aspect of this formulation is that it enables one to calibrate the parameters of the equilibrium speed-concentration model, $u_e(k_1, k_2)$, and test whether they are equal to those obtained from a "direct" calibration of the static model.

3 Observational Work

The objective of the observational component of this study is to obtain the data needed to calibrate the speed-concentration models described in the previous section and to quantify the interactions between pilgrims approaching the Jamarah ring and those leaving it. This can be accomplished by measuring the speeds and the corresponding concentrations of both classes of pilgrims: class 1 users, who have the intention of performing stoning, and class 2 users, who are leaving the Jamarah ring after having completed stoning. Field procedures employed in this observational study are described next, followed by an outline of the data reduction method used.

3.1 Field Procedures

The study was conducted at the "Big" Jamarah during the 1989 Hajj season. Finding a vantage point from which to view the entire scene to be studied is of critical importance to providing sufficient field length and width for analysis. The observation station of the Hajj Research Center (HRC) of Umm-Al-Qura University, Makkah, Saudi Arabia, was used as the study's observation post. It is located on top of a (small) mountain overseeing the Big Jamarah, about 120 meters across from it, and at an elevation of 20 meters.

Time-lapse photography was the principal method used in this study. This method has been effectively used for pedestrian movement studies, especially for complex heavy flows that preclude manual recording [15]. It is useful in measurements of pedestrian stream speed, concentration, and flow, identification of conflicts, interactions, and interferences, as well as tracking specific movements within pedestrian streams.

The camera system, mounted on a tripod for stability and using a telephoto lens (800 mm focal length), was focused on a relatively small area (5 x 18 = 90 square meters) upstream of the Big Jamarah. This area constitutes the observation frame of this study. The selection of its location with respect to the Jamarah ring is critical in the sense that if it is adjacent to the ring, the movement of class 1 pilgrims towards the ring will be difficult to detect because the majority will be in a stationary condition, busy casting their pebbles. On the other hand, an observation frame far from the Jamarah ring would not serve the purpose of this study, because class

2 pilgrims who have finished stoning would have already chosen to go to the sides of the structure to avoid interaction with oncoming pilgrims from the middle Jamarah. The center of the selected observation frame is about 6 meters from the rim of the ring, and along its center-line.

After the initial setup, the camera was focused on that observation frame and ground measurements were conducted by two teams, one at the observation post and the other on the Jamarat bridge. Coordination was achieved through a two-way wireless communication system (walkie-talkie), with the ground team physically measuring, using a measurement tape, the dimensions of the observation frame as it appeared in the camera field of view, in addition to measuring its position relative to the Jamarah ring.

Using the time-lapse photography, with a sequence of one exposure every 2 seconds, filming was conducted during the two peak periods of the first and second days of stoning (July 13 and 14,1989). On the first day, only the Big Jamarah was stoned by pilgrims, with the peak taking place in the period of 6:00-9:00 A.M. On that morning, photography started at 6:20 A.M. for one hour, during which 15 rolls of 36-exposure colored films were used, and the average time taken for unloading the exposed film and loading the new one was about 3 minutes. The same procedure was repeated on the next day, during which all three Jamarat were stoned, with photography taking place during the different peak period of 1:00 to 4:00 P.M. (photography started at 1:20 P.M. for one hour).

3.2 Data Reduction

Using a standard slide projector, each of the above 1100 exposures were projected on a grid sheet mounted on a wall at a fixed distance from the projector. As mentioned earlier, the observation frame covers an actual area of 90 square meters, which can accommodate up to 630 pilgrims (at the observed maximum concentration of 7 p/m^2). To manually count the total number of pilgrims within this area for each slide would have been cumbersome and time consuming. A more efficient approach was employed to facilitate the data reduction process, by restricting it to a smaller frame of (2.5 m x 4 m) 10 m^2 in size at the center of the larger one. Its size (length) is such that no pilgrim can cross it in one two second interval.

For each slide, the number of class 1, class 2 and the sum of both classes of pilgrims were counted (heads only). A person was considered of class 1 if his/her face was pointed towards the Jamarah ring region, that is if his/her speed vector was directed towards the center of the Jamarah ring or within 90 degrees from it, and vice versa. In case of ambiguity, the class was identified simply by advancing or backing the slide projector by a few slides to ascertain the pilgrim's intention. The corresponding concentrations, k_1, k_2, and k, were obtained by dividing the above numbers of pilgrims by the ground area of the observation frame.

Due to the limitations of the manual data reduction method employed, it was not possible to measure the speed of movement for all pilgrims; a representative sample of each class of users was obtained instead. Each 36-exposure film was broken into four groups of 9 slides each, such that the frame length (2.5 meters) could be traversed during that period by a pilgrim moving at a shuffling speed of 500 meters per hour. Within each group of slides, and starting with the first one, five representative pilgrims of each class were selected as they entered the frame. Each of them was traced as he/she traversed the frame until one of the following three conditions occurred: (1) the pilgrim left the frame, (2) the end of the slides group was reached, or (3) he/she stopped to perform stoning (class 1 only). The pilgrim's time of travel and distance traversed were then recorded.

The selection of the representative pilgrims for speed measurement was based mainly on their identifiability and traceability, by tagging each of them with a special reference mark, such as one's clothes type and color, height, hair shape and color, and the presence of eye glasses. On the first day of stoning, it was more difficult to identify and tag individual pilgrims, because all male pilgrims were required to wear the same clothes, consisting of two pieces of seamless, stitchless white cloth, one covering the body from waist to ankle and the other worn over the pilgrim's shoulder. The difficulties associated with identifying and tracing the same pilgrim, the size of the observation frame, and the period of tracing, 18 seconds, limited the total number traced to five of each class of pilgrims. Using this approach it was possible to measure the speeds of 1200 pilgrims, of both classes.

For each class of pilgrims, within each group of slides, the space-mean speeds were then calculated:

$$u_m = \sum_{i=1}^{5} D_{im} \Big/ \sum_{i=1}^{5} t_{im} \;, \; m = 1,2$$

where u_m is the space-mean speed of class m users (m=1, 2), in meters per hour, and D_{im} is the distance, in meters, traversed by pilgrim i of class m in time t_{im}, hours. The above speed-concentration measurements are then used to calibrate the models specified earlier, as discussed next.

4 Data Analysis and Results

Measurements of each class of pilgrims' radial speeds and the corresponding concentrations, which were collected during the first two days of stoning around the Big Jamarah, are now used to estimate the parameters of the static and dynamic speed-concentration models specified in Section 2.

The speeds of movement and concentration levels, for each class of pilgrims, were measured over a time interval of 18 seconds (i.e. 9 slides) and 2 seconds, respectively. Thus, for each speed measurement, the corresponding concentrations were obtained by averaging the respective individual measurements over the corresponding period of 18 seconds. Table 1 present summary statistics for the principal variables in the sample of observations used for estimation.

Variable	Day	Average	St. Dev.	Maximum	Minimum
Speed of Class 1 Users	1	727	197	1306	300
u_{1F}(m/hr)	2	1200	410	2284	214
	both	965	400	2284	214
Speed of Class 2 Users	1	1021	267	1667	548
u_{2B}(m/hr)	2	1519	477	2606	587
	both	1272	460	2606	548
Concentration of	1	2.91	0.72	4.72	1.40
Class 1 Users	2	1.88	0.35	2.58	1.12
$k_1(p/m^2)$	both	2.39	0.76	4.72	1.12
Correlation of	1	1.04	0.38	1.98	0.38
Class 2 Users	2	0.74	0.21	1.27	0.37
$k_2(p/m^2)$	both	0.89	0.34	1.98	0.37
Total Concentration	1	3.95	0.67	5.86	2.70
$k\ (p/m^2)$	2	2.62	0.32	3.29	1.92
	both	3.28	0.85	5.86	1.92

Table 1: Statistical Summary of Observed Speed and Concentration Variables (n = 119 groups of slides).

4.1 Static Model

For estimation purposes, Eq. 3 can be rewritten as:

$$u_{1Fi} = \beta_{01} + \beta_{11}k_{1i} + \beta_{21}\,k_{2i} + \varepsilon_{1i} \tag{7a}$$

$$u_{2Bi} = \beta_{02} + \beta_{12}\,k_{2i} + \beta_{22}\,k_{1i} + \varepsilon_{2i} \tag{7b}$$

where the i subscript denotes the i-th observation, i =1, ..., n (number of observations). The terms ε_{1i} and ε_{2i} are random error terms reflecting the discrepancies between observed and predicted values; β_{0m}, β_{1m} and β_{2m} (m=1,2) are the parameters to be estimated.

More generally, let u_i denote the average speed of users of a given class going in a particular direction, k_i the corresponding concentration of users of the same class, and $k_{opp,i}$ the concentration of users in the opposite direction. The corresponding speed - concentration relation can be expressed as:

$$u_i = \beta_0 + \beta_1\,k_i + \beta_2\,k_{opp,i} + \varepsilon_i \tag{7c}$$

The above specification can be calibrated using a pooled sample of class 1 and class 2 users, and then used to test if the two classes of users share the same underlying coefficients (i.e. that $\beta_{j1} = \beta_{j2} = \beta_j$, j = 0, 1, 2).

Calibration is performed using OLS estimation. One of the assumptions of OLS is that the random error terms are independently and identically normally distributed, i.e. that $\varepsilon_i \sim N(0, \sigma^2)$. However, because the field measurements were obtained from a time-series study, the above assumption may not hold as the errors associated with consecutive observations may be serially correlated. It is therefore necessary to test for the presence of serial correlation and correct for it if present. The standard Durbin-Watson (DW) test is used this purpose [6, 22].

By employing OLS estimation method and using the functional form specified by Eq. 7, the following static speed-concentration relations for class 1, class 2, and both classes of pilgrims were obtained, respectively:

$$\begin{aligned} u_{1F} &= 1775 - 289\, k_1 - 135\, k_2 \\ &\quad (14.1)\ \ (-7.2)\quad\ (-1.5) \\ &\quad R^2 = 0.32,\ DW = 1.3,\ n = 119 \end{aligned} \tag{8a}$$

$$\begin{aligned} u_{2B} &= 2194 - 341\, k_2 - 259\, k_1 \\ &\quad (14.4)\ (-3.1)\quad\ (-5.4) \\ &\quad R^2 = 0.26,\ DW = 1.57,\ n = 119 \end{aligned} \tag{8b}$$

$$\begin{aligned} u &= 1996 - 353\, k - 182\, k_{opp} \\ &\quad (20.7)\ (-11.1)\quad (-5.7) \\ R^2 &= 0.35,\ DW = 1.43,\ n = 238 \end{aligned} \tag{8c}$$

where speeds are in meter/hr, concentrations are in p/m^2, R^2 is the coefficient of determination, DW is the Durbin-Watson serial correlation test statistic, n is the sample size and the numbers in parentheses are the t-statistics for the null hypothesis that the corresponding parameter is equal to zero.

The DW test statistic calculated for each of the above models reveals that positive serial correlation is present in all three models (d_l=1.63 and d_u=1.72, at the 5 percent level of significance). These models were then corrected for serial correlation by utilizing the Durbin procedure [6, 22][1] resulting in the following model parameter estimates[2]:

---

[1] The Durbin procedure assumes a first-order auto-correlation model of the form $\varepsilon_i = \rho\, \varepsilon_{i-1} + \upsilon_i$, where $1 < \rho \leq 0$ and $\upsilon_i \sim N(0, \sigma^2)$ is independent of other errors over time as well as of the ε's. Any standard econometrics textbook can be consulted for a review of serial correlation correction methods [13,22].

[2] Note the lower values of R^2 after correcting for positive serial correlation, where its presence resulted in underestimation of the true residual variance, hence overestimating R^2.

$$u_{1F} = 1708 - 275\ k_1 - 80\ k_2^3$$
$$(8.6)\quad(-4.7)\quad(-0.7)$$
$$R^2 = 0.2,\ DW = 2.26,\ n = 89 \qquad (9a)$$

$$u_{2B} = 2278 - 450\ k_2 - 239\ k_1$$
$$(8.5)\quad(-2.9)\quad(-2.7)$$
$$R^2 = 0.14,\ DW = 1.91,\ n = 89 \qquad (9b)$$

$$u = 1907 - 336\ k - 128\ k_{opp}$$
$$(12.0)\quad(-6.4)\quad(-2.5)$$
$$R^2 = 0.2,\ DW = 2.0,\ n = 178 \qquad (9c)$$

We then test if the two classes of pilgrims share the same underlying coefficient values. The Chow test is used for this purpose [13]. Under the null hypothesis that the restrictions are valid, the test statistic is F-distributed with (r,N-GP) degrees of freedom, where r denotes the number of restricted parameters values, N the number of observations, G is the number of groups and P is the numbers of parameters to be estimated in the model.[3] For our case, the calculated F^* statistic equals 1.51, while the theoretical value of $F_{3,172}$ at the 5 percent level of significance is 2.68; therefore, we can conclude that the restrictions are valid, i.e., coefficients of the speed concentration relation are the same for the two classes of pilgrims and are best estimated using the pooled model (Eq. 9c).

Under uni-directional pedestrian flow (i.e. k_{opp}=0), and assuming that the intercept of Eq. 9c represents the free-flow speed (u_f=1907 m/hr), the jam concentration k_{jam}, can be found to equal 5.68 p/m^2 (=1907/336). Using this jam concentration, the general static bi-directional speed concentration model can be rewritten as:

$$u = 1907\ [1 - (k + 0.38\ k_{opp})/k_{jam}] \qquad (10)$$

An interesting aspect of this relation is that the impedance exerted on the average speed of a stream of pilgrims by pilgrims moving in the same direction is more than twice that of those moving in the opposing one.

The other linear speed-concentration formulation, with multiplicative interaction terms (Eq. 4,) was calibrated, using the OLS estimation method, and the following models were obtained for class 1, class 2, and both classes of pilgrims, respectively, after correcting for (positive) serial correlation that was present in all models:

---

[3] The test statistic is calculated as $F^* = [(Q^R - Q^U)/r]/[Q^U/(N\text{-}GP)]$ where Q^R is the sum of squared errors in the pooled (restricted model), Q^U is the total sum of squared errors taken over all the separate models, and the other terms are defined in the text.

$$u_{1F} = 1960 - 380\ k_1 - 345\ k_2 + 111\ k_1 k_2$$
$$(4.8)\ (-2.3)\quad (-0.9)\quad (0.66)$$
$$R^2 = 0.22,\ DW = 2.27,\quad n = 89 \tag{11a}$$

$$u_{2B} = 2834 - 1174\ k_2 - 469\ k_1 + 302\ k_1 k_2$$
$$(5.2)\quad (-2.1)\quad (-2.1)\quad (1.3)$$
$$R^2 = 0.16,\ DW = 1.89,\ n = 89 \tag{11b}$$

$$u = 1927 - 348\ k - 141\ k_{opp} + 9.2\ k\, k_{opp}$$
$$(10.2)\ (-3.9)\quad (-1.6)\quad (0.14)$$
$$R^2 = 0.21,\ DW = 2.0,\ n = 178 \tag{11c}$$

The null hypothesis that the coefficient of the interaction term is zero cannot be rejected at the 5 percent level of significance, for all the three models. Hence, the original formulation without the interaction term (Eq. 10) is retained as the static speed-concentration relationship for pedestrian movement in bi-directional movement setting.

4.2 Dynamic Model

Consider the dynamic speed-concentration formulation of Eq. 6. Assuming that the equilibrium speed-concentration function, $u_e(k_1, k_2)$, takes the form of Eq. 3, the following models were estimated for class 1, class 2, and both classes of users, respectively:

$$u_{1F,t+1} = 849 + 0.47\ u_{1F,t} - 96\ k_{1,t} - 108\ k_{2,t}$$
$$(4.1)\quad (4.8)\quad (-2.0)\quad (-1.2)$$
$$R^2 = 0.39,\ DW = 2.17,\ n = 89 \tag{12a}$$

$$u_{2B,t+1} = 1327 + 0.36\ u_{2B,\,t} - 165\ k_{2,t} - 141\ k_{1,t}$$
$$(4.6)\quad (3.3)\quad (-1.3)\quad (-2.3)$$
$$R^2 = 0.3,\ DW = 1.85,\ n = 89 \tag{12b}$$

$$u_{t+1} = 1042 + 0.43\ u_t - 166\ k_t - 67\ k_{opp,t}$$
$$(6.0)\quad (6.0)\quad (-4.0)\quad (-1.8)$$
$$R^2 = 0.4,\ DW = 1.93,\ n = 178 \tag{12c}$$

Note that the DW test statistic calculated for each of the above models indicates that, as expected, serial correlation is not significant. This is due to the inclusion of the lagged speed variable in the specification.

The Chow test is then performed to determine if the two classes share the same parameter values. The calculated F^* statistic is 1.13, while the theoretical value of $F_{4,170}$ at the 5 percent level of significance is 2.45; thus, it can be concluded that the coefficients of the dynamic model are the same for the two classes, and are best estimated by the values shown in Eq. 12c. By comparing Eqs. 5b and 12c, the implied value of the relaxation reaction time, T, can be found to be $\simeq$ 32 seconds, by substituting the value of Δt=18 seconds in the expression of the coefficient of u_t [$(1-\Delta t/T)$=0.43]. Hence the implied equilibrium speed-concentration model takes the form:

$$u_e(k, k_{opp}) = 1828 - 291\,k - 118\,k_{opp} \tag{13}$$

When the equilibrium speed-concentration model was assumed to take the form of Eq. 4, which involves an interaction term, the following dynamic models were estimated, for class 1, class 2, and both classes of users:

$$\begin{aligned} u_{1F,t+1} &= \underset{(3.2)}{2165} + \underset{(4.5)}{0.44}\,u_{1F,t} - \underset{(-1.8)}{234}\,k_{1,t} - \underset{(-1.4)}{455}\,k_{2,t} + \underset{(1.1)}{141}\,k_{1,t}\,k_{2,t} \\ R^2 &= 0.4,\ DW = 2.22,\ n = 89 \end{aligned} \tag{14a}$$

$$\begin{aligned} u_{2B,t+1} &= \underset{(3.1)}{2448} + \underset{(3.0)}{0.34}\,u_{2B,t} - \underset{(-1.0)}{445}\,k_{2,t} - \underset{(-1.4)}{251}\,k_{1,t} + \underset{(0.7)}{112}k_{1,t}\,k_{2,t} \\ R^2 &= 0.3,\ DW = 1.9,\ n = 89 \end{aligned} \tag{14b}$$

$$\begin{aligned} u_{t+1} &= \underset{(5.3)}{1879} + \underset{(5.8)}{0.42}\,u_t - \underset{(-2.5)}{194}\,k_t - \underset{(-1.3)}{94}\,k_{opp} + \underset{(0.4)}{22}\,k_t\,k_{opp,t} \\ R^2 &= 0.4,\ DW = 1.93,\ n = 178 \end{aligned} \tag{14c}$$

As was the case with the corresponding static model, the null hypothesis of zero interaction term coefficient cannot be rejected at the 5 percent level of significance. Hence the other dynamic model, without the interaction term (Eq. 12c), is retained for further investigation. The dynamic speed-concentration model can then be rewritten in a form similar to Equation 6 as follows:

$$u_{t+1} = u_t - \Delta t/32\,[(u_t - (1828 - 291\,k_t - 118\,k_{opp,t})] \tag{15}$$

We can test if the differences between the "equilibrium" speed-concentration model parameters are different from those of the previously calibrated "static" model (Eq. 10). To do so, the static model parameters were restricted to those of the equilibrium one (Eq. 13), and two regression models were considered: with and without the restriction. The sum of squared errors for the restricted model, Q^R, and the unrestricted one (corrected for serial correlation), Q^U, were calculated as follows, respectively:

$$Q^R = \sum_{i=1}^{n}\left[u_i - \left(1828 - 291k_i - 118k_{opp,i}\right)\right]^2 \tag{16a}$$

$$Q^U = \sum_{i=1}^{n}\left[u_i - \left(1907 - 336k_i - 128k_{opp,i}\right)\right]^2 \tag{16b}$$

Thus, the calculated F* statistic is equal to 1.1, while the theoretical value of $F_{3,175}$ at the 5 percent level of significance is 2.68; thus, we can conclude that the restrictions are valid, i.e., the calibrated static speed-concentration model parameters (Eq. 10) are not significantly different from those of the equilibrium model.

5 Conclusions

In this paper, two types of speed-concentration model specifications were investigated for pedestrian movement in two opposite directions: static and dynamic. The static formulation assumes that the average speed of a stream of pedestrians is determined solely by the prevailing concentration of those moving in the same direction as well as those moving in the opposite one. On the other hand, the dynamic formulation assumes that this average speed at a given time further depends also on the speed at the previous time step.

Both types of models were estimated, using data collected for this purpose. The estimated models revealed some interesting features of the process. The speed of a group of pedestrians moving in one direction, in a bi-directional movement setting, was found, as expected, to depend on the concentration levels of each of the two opposing pedestrian streams present. Furthermore, these models showed that the impedance, on users' speed, of people moving in the same direction is more than twice that of those moving in the opposite direction. Another important aspect of these calibrated models is that the marginal impedance of each of the two types of concentrations involved, on the speed of movement, is independent of the level of the other stream concentration. There is no interaction term between the two types of concentration in the estimated speed-concentration model.

The estimation results of the dynamic model, suggested that it takes 32 seconds, on average, for a stream of pedestrians to adjust its average speed of movement to the equilibrium one. This value was inferred from the calibrated results, though not directly identified in the calibration. However, this value may be somewhat higher than one would expect given the flexibility of human beings and their ability to change speed quickly in response to stimulus. Thus, more investigation of the dynamic formulation, which incorporates all three processes proposed in connection with Eq. 5, is needed, before a more definitive conclusion is reached.

References

1. S.A.H. AlGadhi and H.S. Mahmassani, *Modelling Crowd Behavior and Movement: Application to Makkah Pilgrimage*, in: *Proc. of the 11th Int'l Symposium on Transportation and Traffic Theory*, M. Koshi, (Ed.), pp. 59-78 (Yokohama, Japan, July 1990).

2. S.A.H. AlGadhi, *Characterization of Crowd Behavior and Movement*, Ph.D. Dissertation, (Department of Civil Engineering, The University of Texas at Austin, 1990).

3. S.A.H. AlGadhi and H. S. Mahmassani, *Simulation of Crowd Behavior and Movement: Fundamental Relations and Application*, Transportation Research Record **1320**, pp. 260-268 (1991).

4. G.C. Chow, *Tests of Equality between sets of Coefficients in two Linear Regressions*, Econometrica **28**, no.3, (1960).

5. J.S. Drake, J.L. Schofer, and A.D. May, *A Statistical Analysis of Speed Density Hypotheses*, Highway Research Record **154**, 53-87 (1967).

6. J. Durbin, *Estimation of Parameters in Time-Series Regression Models*, Journal of Royal Statistical Society, Ser. B, **22**, 139-151 (1960).

7. J.J. Fruin, *Pedestrian Planning and Design*, (Metropolitan Association of Urban Designers and Environmental Planners, New York, N.Y., 1971).

8. B. Hankin and R. Wright, *Passenger Flow in Subways*, Operational Research Quarterly **2**, 81-88 (1958).

9. Highway Research Board, *Highway Capacity Manual*, Special Report no. 87, (National Research Council, Washington, D.C., 1965).

10. R. Kuhne, *Macroscopic Freeway Model for Dense Traffic*, in: *Proc. of the Ninth International Symposium on Transportation and Traffic Theory*, I. Volmuller and R. Hamerslag, (Eds.), pp. 21-42 (VNU Science press, Utrecht, 1984).

11. R. Kuhne, *Freeway Speed Distribution and Acceleration Noise - Calculation from a Stochastic Continuum Theory and Comparison with Measurements*, in: *Proc. of the Tenth International Symposium on Transportation and Traffic Theory*, N.H. Gartner and N.H.M. Wilson, (Eds.), pp. 119-137 (Elsevier Science Publishing Co., Inc., 1987).

12. M.J. Lighthill and G. B. Whitham, Kinematic Waves, II., *A Theory of Traffic Flow on Long Crowded Roads*, Proceedings of the Royal Society (London), **229** A, 317 (1955).

13. G.S. Maddala, *Econometrics*, (McGraw-Hill Book Company, New York 1977).

14. P.G. Michalopoulos and D.E. Beskos, *Improved Continuum Models of freeway flow*, in: *Proc. of the Ninth International Symposium on Transportation and Traffic*

Theory, I. Volmuller and R. Hanersley, (Eds.), pp. 89-111 (VNU Science press, Utrecht, 1984).

15. P.W. Miller and W.H. Kraft, *Data Collection Techniques: Time-Lapse Photography*, in: *Proc. of the seminar on Bicycle/Pedestrian Planning & Design,* pp. 221-236 (Walt Disney, Florida, American Society of Civil Engineers, New York, 1974).

16. F. Navin and R. Wheeler, *Pedestrian Flow Characteristics*, Traffic Engineering **9** (v. 39), 30-33, 36 (1969).

17. D. Oeding, *Verkehrsbelastung und Dimensionierung von Gehwegeon und anderen Anlagen des Fussgangerverkehrs*, [Traffic Loads and Dimensions of Walkways and other Pedestrian Circulation Facilities], Straßenbau und Straßenverkehrstechnik **22**, (Germany, Bonn, 1963).

18. S.J. Older, *Movement of Pedestrians on Footways in Shopping Streets*, Traffic Engineering and Control **4**, (v. 10), 160-163 (1968).

19. H.J. Payne, *Models of freeway traffic and control*, in: *Mathematics of Public Systems*, Vol. 1, No. 1, G.A. Bekey, (Ed.), (Simulation Council, 1971).

20. H.J. Payne, *FREFLO: A macroscopic simulation model of freeway traffic*, Transportation Research Record **772**, 68-75 (1979).

21. W.F. Phillips, *A new Continuum model of traffic flow*, Utah State University, Logan, Utah, Report DoT-RC-82018, (1981).

22. R.S. Pindyack and D.L. Rubinfeld, *Econometric Models and Economic Forecasts*, (2nd ed., McGraw Hill, New York, 1981).

23. V.M. Predtechenskii, *Dvizhenie Lyudskikh Potokovi Operdeknie Razmerov Kommunikatsionnykh Pomeshch enii*, Glov v uchenbnike, Arkhitekura Grashdanskikh i Promyshlennykh Zdanii, [Pedestrian traffic flow and the determination of passage dimensions in buildings. Chapter in the book: Architecture of public and industrial buildings], (Stroiizdat, Moscow, 1966).

24. V.M. Predtechenskii and A.I. Milinski, *Planning for foot traffic flow in buildings*, translated from Russian, (Stroiizdat Publishers, Moscow, 1969). Published for NBS, DoC and NSF Washington, D.C., by Amerind Publishing Co. Pvt. Ltd., New Delhi (1978).

25. B. Pushkarev and J. Zupan, *Urban Space for Pedestrians*, (MIT Press, Cambridge, Mass, 1975).

Simulation of Pedestrian Crowds in Normal and Evacuation Situations

D. Helbing[1], I.J. Farkas[2], P. Molnár[3], and T. Vicsek[2]

[1] Institute for Economics and Traffic, Faculty of Traffic Sciences "Friedrich List", Dresden University of Technology, D-01062 Dresden, Germany
[2] Department of Biological Physics, Eötvös University, Pázmány Péter Sétány 1A, H-1117 Budapest, Hungary
[3] Center for Theoretical Studies of Physical Systems, Clark Atlanta University, James P. Brawley Drive, Atlanta, Georgia 30314, USA

Starting with a short review of the available literature in the field of pedestrian and evacuation research, an overview is given over the observed collective phenomena in pedestrian crowds. This includes lane formation in corridors and oscillations at bottlenecks in normal situations, while different kinds of blocked states are produced in panic situations. By means of molecular-dynamic-like microsimulations based on a generalized force model of interactive pedestrian dynamics, the spatio-temporal patterns in pedestrian crowds are successfully reproduced and interpreted as self-organized phenomena. In contrast to previous socio-psychological approaches, this allows a physical understanding of the observations. Despite the significantly different phenomena occuring in normal and panic situations, the main effects can be described by a unified model containing only well interpretable and plausible terms. The transition between the "rational" normal behavior and the apparently "irrational" panic behavior is controlled by a single parameter, the "nervousness", which influences fluctuation strengths, desired speeds, and the tendency of herding. Thereby, it causes paradoxial effects like "freezing by heating", "faster is slower", and the ignorance of available exits. Nevertheless, there are measures to improve pedestrian flows, both in normal and panic situations. For example, the suitable placement of columns can help, although they reduce the accessible space.

1 Introduction

The various collective phenomena observed in pedestrian crowds have recently attracted the interest of a rapidly increasing number of scientists. In this review, we will always distinguish the dynamics of pedestrians in normal and panic situations. Since both problems are characterized by different characteristic phenomena, they have often been investigated by different scientific communities. However, as we will show in the following, they can be treated in a consistent way by one and the same pedestrian model.

1.1 Research in Normal Pedestrian Behavior

Pedestrian crowds have been empirically studied for more than four decades now [1–8]. The evaluation methods applied were based on direct observation, photographs, and time-lapse films. Apart from behavioral investigations [9,10], the main goal of these studies was to develop a ***level-of-service concept*** [11–13], ***design elements*** of pedestrian facilities [14–19], or ***planning guidelines*** [20–24]. The latter have usually the form of ***regression relations***, which are, however, not very well suited for the prediction of pedestrian flows in pedestrian zones and buildings with an exceptional architecture, or in extreme conditions such as evacuation. Therefore, a number of simulation models have been proposed, e.g., ***queueing models*** [25–27], ***transition matrix models*** [28], and ***stochastic models*** [29,30], which are partly related to each other. In addition, there are models for the ***route choice behavior*** of pedestrians [31–34].

None of these concepts adequately takes into account the self-organization effects occuring in pedestrian crowds. These may, however, lead to unexpected obstructions due to mutual disturbances of pedestrian flows. More promising with regard to this is the approach by Henderson. He conjectured that pedestrian crowds behave similar to gases or fluids ([35–38], see also [39,40]). This could be partly confirmed (see Sec. 2.3). However, a realistic gas-kinetic or fluid-dynamic theory for pedestrians must contain corrections due to their particular interactions (i.e., avoidance and deceleration maneuvers) which, of course, do not obey momentum and energy conservation. Although such a theory can be actually formulated [34,41,42], for practical applications a direct simulation of ***individual*** pedestrian motion is favourable, since this is more flexible. the fluid-dynamic equations is difficult. As a consequence, current research focusses on the ***microsimulation*** of pedestrian crowds, which also allows us to consider incoordination by ***excluded volume effects*** related to the discrete, "granular" structure of pedestrian flows. In this connection, a ***behavioral force model*** of individual pedestrian dynamics has been developed [18,43–53] (see Sec. 3). A discrete and simple forerunner of this model was proposed by Gipps and Marksjö (1985). We also like to mention recent ***cellular automata*** of pedestrian dynamics [54–63], and ***AI-based models*** [64–67].

1.2 Evacuation and Panic Research

Computer models for emergency and evacuation situations have been developed as well [62,68–76]. Most research into panics, however, has been of empirical nature (see, e.g., Refs. [77–80]), carried out by social psychologists and others.

With some exceptions, panics are observed in cases of scarce or dwindling resources [81,77], which are either required for survival or anxiously desired. They are usually distinguished into escape panics ("stampedes", bank or stock market panics) and acquisitive panics ("crazes", speculative manias) [82,83], but in some cases this classification is questionable [84].

It is often stated that panicking people are obsessed by short-term personal interests uncontrolled by social and cultural constraints [77,82]. This is possibly a result of the reduced attention in situations of fear [77], which also causes that options like side exits are mostly ignored [78]. It is, however, frequently attributed to social contagion [77,79,81–89], i.e., a transition from individual to mass psychology, in which individuals transfer control over their actions to others [83], leading to conformity [90]. This "herding behavior" is in some sense irrational, as it often leads to bad overall results like dangerous overcrowding and slower escape [83,84,78]. In this way, herding behavior can increase the fatalities or, more generally, the damage in the crisis faced.

The various socio-psychological theories for this contagion assume hypnotic effects, rapport, mutual excitation of a primordial instinct, circular reactions, social facilitation (see the summary by Brown [88]), or the emergence of normative support for selfish behavior [89]. Brown [88] and Coleman [83] add another explanation related to the prisoner's dilemma [91,92] or common goods dilemma [93], showing that it is reasonable to make one's subsequent actions contingent upon those of others, but the socially favourable behavior of walking orderly is unstable, which normally gives rise to rushing by everyone. These thoughtful considerations are well compatible with many aspects discussed above and with the classical experiments by Mintz [81], which showed that jamming in escape situations depends on the reward structure ("payoff matrix").

Nevertheless and despite of the frequent reports in the media and many published investigations of crowd disasters (see Table 1), 1), a quantitative understanding of the observed phenomena in panic stampedes has been lacking. In this study, we will add another aspect to the explanation of panics by simulating a computer model for the crowd dynamics of pedestrians.

Table 1. Incomplete list of major crowd disasters after J. F. Dickie in Ref. [98], `http://ourworld.compuserve.com/homepages/G_Keith_Still/disaster.htm`, `http://SportsIllustrated.CNN.com/soccer/world/news/2000/07/09/stadium_disasters_ap/`, and other internet sources (from [99]). The number of injured people was usually a multiple of the fatalities.

Date	Place	Venue	Death	Injured	Reason
1863	Santiago, Chile	Church	2000		
1881	Vienna, Austria	Theatre	570		
1883	Sunderland, UK	Theatre	182		
1902	Ibrox, UK	Stadium	26	517	Collapse of West Stand
1903	Chicago, USA	Theatre	602		
1943	London, UK	Subway Station	173		Stampede while air raid
1946	Bolton, UK	Stadium	33	400	Collapse of a wall
1955	Santiago, Chile	Stadium	6		Fans trying to force their way into the stadium
1961	Rio de Janeiro, Brazil	Circus	250		
1964	Lima, Peru	Stadium	318	500	Goal disallowed
1967	Kayseri, Turkey	Stadium	40		
1968	Buenos Aires, Argentina	Stadium	75	150	Fans fleeing from fire
1970	St. Laurent-du-Pont, France	Dance Hall	146		
1971	Ibrox, UK	Stadium	66	140	Collapse of barriers
1971	Salvador, Brazil	Stadium	4	1500	Fight and wild rush
1974	Cairo, Egypt	Stadium	48		Crowds break barriers
1976	Port-au-Prince, Haiti	Stadium	2		Firecracker
1979	Nigeria	Stadium	24	27	Light failure
1979	Cincinatti, USA	Stadium	11		Fans trying to force their way into the stadium
1981	Piraeus, Greece	Stadium	24		Rush of leaving fans
1981	Sheffield, UK	Stadium		38	Crowd surge
1982	Cali, Columbia	Stadium	24	250	Provocation by drunken fans
1982	Moscow, USSR	Stadium	340		Re-entering fans after last minute goal
1985	Bradford, UK	Stadium	56		Fire in wooden terrace section
1985	Mexico City, Mexico	Stadium	10	29	Fans trying to force their way into the stadium
1985	Brussels, Belgium	Stadium	38	> 400	Riots break out
1987	Tripoli, Libya	Stadium	2	16	Collapse of a wall
1988	Katmandu, Nepal	Stadium	93	> 100	Stampede due to hailstorm
1989	Hillsborough, Sheffield, UK	Stadium	96		Fans trying to force their way into the stadium
1990	Mecca, Saudi Arabia	Pedestr. Tunnel	1425		Overcrowding
1991	Orkney, South Africa	Stadium	> 40		Fans trying to escape fighting

Date	Place	Venue	Death	Injured	Reason
1991	New York, USA	Stadium	9		Overcrowding at concert
1992	Rio de Janeiro, Brazil	Stadium		50	Part of the fence giving way
1992	Bastia, Corsica	Stadium	17	1900	
1994	Mecca, Saudi Arabia		270		"Stoning the devil" ritual
1995	New Delhi, India	Tent	≈ 400	> 100	Indian fire
1996	Lusaka, Zambia	Stadium	9	78	Stampede after Zambia's victory over Sudan
1996	Tembisa, South Africa	Railway Station	15	> 20	Electric cattle prods used by security guards
1996	Guatemala City, Guatemala	Stadium	80	180	Fans trying to force their way into the stadium
1997	Las Vegas, USA	Hotel	1	50	Gunshot
1997	Düsseldorf, Germany	Stadium	1	> 300	Overcrowding at concert
1998	Dhaka, Bangladesh	Multi-Storey Building	1	15	Fire stampede
1998	Mecca, Saudi Arabia		107		"Stoning the devil" ritual
1998	Harare, Zimbabwe	Stadium	4	10	Spectators scrambled for seats
1998	Manila, Phillipines	Presid. Action Center	2		Large crowd waiting for jobs and housing
1998	Chervonohrad, Ukraine	Cinema	4		Stampede due to in- and outcoming children
1998	Lima, Peru	Disco	9	7	Tear gas
1999	Minsk, Belarus	Subway Station	51	150	Heavy rain at rock concert
1999	Kerala, India	Hindu Shrine	> 50		Collapse of parts of the shrine
1999	Benin, Nigeria	Religious Place	14		Stampede at a Christian revivalist rally
1999	Innsbruck, Austria	Stadium	5	25	Fans re-entering the stadium?
2000	Kaloroa, Bangladesh	Examination Place	5		Stampede to enter an examination hall
2000	Mecca, Saudi Arabia	Holy Place	2	4	Pilgrim overcrowding
2000	Durban, South Africa	Disco	13	44	Tear gas
2000	Chiaquelane, Mozambique	Chiaquelane Camp	5	10	Aid chaos
2000	Lisbon, Portugal	Nightclub	7	65	Poisonous gas bombs
2000	Seville, Spain			30	Good Friday procession
2000	Monrovia, Liberia	Stadium	3		Fans trying to force their way into the stadium
2000	Lahore, Pakistan	Circus	8	3	Guards used batons
2000	Addis Abeba, Ethiopia	Memorial Place	14		Children trying to cover from a rainstorm
2000	Roskilde, Denmark	Stadium	8	25	Failure of loud speakers
2000	Harare, Zimbabwe	Stadium	12		Tear gas
2000	São Januário, Brazil	Stadium		200	Oversold stadium

2 Observations

2.1 Normal Situations

We have investigated pedestrian motion for several years and evaluated a number of video films. Despite the sometimes more or less "chaotic" appearance of individual pedestrian behavior, one can find regularities, some of which become best visible in time-lapse films like the ones produced by Arns [94]. While describing these, we also summarize results of other pedestrian studies and observations [18,19,45,95]:

1. Pedestrians feel a strong aversion to taking detours or moving opposite to the desired walking direction, even if the direct way is crowded. However, there is also some evidence that pedestrians normally choose the *fastest* route to their next destination, but not the *shortest one* [96]. In general, pedestrians take into account detours as well as the comfort of walking, thereby minimizing the effort to reach their destination [97]. Their ways can be approximated by polygons.
2. Pedestrians prefer to walk with an individual desired speed, which corresponds to the most comfortable (i.e., least energy-consuming) walking speed (see Ref. [8]) as long as it is not necessary to go faster in order to reach the destination in time. The desired speeds within pedestrian crowds are Gaussian distributed with a mean value of approximately 1.34 m/s and a standard deviation of about 0.26 m/s [35]. However, the average speed depends on the situation [21], sex and age, the time of the day, the purpose of the trip, the surrounding, etc. [8].
3. Pedestrians keep a certain distance to other pedestrians and borders (of streets, walls, and obstacles; see [22,24]). This distance is smaller the more a pedestrian is in a hurry, and it decreases with growing pedestrian density. Resting individuals (waiting on a railway platform for a train, sitting in a dining hall, or lying at a beach) are uniformly distributed over the available area if there are no acquaintances among the individuals. Pedestrian density increases (i.e., interpersonal distances lessen) around particularly attractive places. It decreases with growing velocity variance, e.g.,, on a dance floor [41,18]. Individuals knowing each other may form groups which are entities behaving similar to single pedestrians. Group sizes are Poisson distributed [100–102].

2.2 Panic Situations

Panic stampede is one of the most tragic collective behaviors [79,81–83,85–89], as it often leads to the death of people who are either crushed or trampled down by others. While this behavior is comprehensible in life-threatening situations like fires in crowded buildings [77,78], it is hard to understand in cases of a rush for good seats at a pop concert [84] or without any obvious reasons.

Unfortunately, the frequency of such disasters is increasing [84] (see Table 1), as growing population densities combined with easier transportation lead to greater mass events like pop concerts, sporting events, and demonstrations. Nevertheless, systematic studies of panics [81,103] are rare [77,82,84], and there is a scarcity of quantitative theories capable of predicting the dynamics of human crowds [62,68,69,72,73,76]. In spite of this, the following features appear to be typical [52,53]:

1. In situations of escape panics, individuals are getting nervous, i.e., they tend to develop blind actionism.
2. People try to move considerably faster than normal [21].
3. Individuals start pushing, and interactions among people become physical in nature.
4. Moving and, in particular, passing of a bottleneck frequently becomes incoordinated [81].
5. At exits, jams are building up [81]. Sometimes, arching and clogging are observed [21], see Fig. 1.

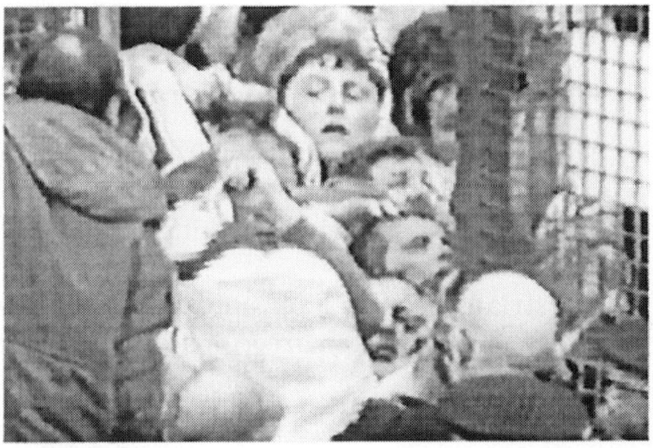

Figure 1. Panicking football fans trying to escape the football stadium in Sheffield. Because of a clogging effect, it is difficult to pass the open door.

6. The physical interactions in jammed crowds add up and can cause dangerous pressures up to 4,500 Newtons per meter [78,98], which can bend steel barriers or tear down brick walls.
7. Escape is slowed down by fallen or injured people turning into "obstacles".
8. People tend to show herding behavior, i.e., to do what other people do [77,86].
9. Alternative exits are often overlooked or not efficiently used in escape situations [77,78].

The following quotations give a more personal impression of the conditions during escape panics:

1. "They just kept pushin' forward and they would just walk right on top of you, just trample over ya like you were a piece of the ground." (After the panic at "The Who Concert Stampede" in Cincinatti.)
2. "People were climbin' over people ta get in ... an' at one point I almost started hittin' 'em, because I could not believe the animal, animalistic ways of the people, you know, nobody cared." (After the panic at "The Who Concert Stampede".)
3. "Smaller people began passing out. I attempted to lift one girl up and above to be passed back ... After several tries I was unsuccessful and near exhaustion." (After the panic at "The Who Concert Stampede".)
4. I "couldn't see the floor because of the thickness of the smoke." (After the "Hilton Hotel Fire" in Las Vegas.)
5. "The club had two exits, but the young people had access to only one, said Narend Singh, provincial minister for agriculture and environmental affairs. However, the club's owner, Rajan Naidoo, said the club had four exits, and that all were open. 'I think the children panicked and headed for the main entrance where they initially came in,' he said." (After the "Durban Disco Stampede".)

2.3 Analogies with Gases, Fluids, and Granular Media

When the density is low, pedestrians can move freely, and crowd dynamics can be compared with the behavior of gases. At medium and high densities, the motion of pedestrian crowds shows some striking analogies with the motion of fluids and granular flows:

1. Footprints of pedestrians in snow look similar to streamlines of fluids [34].
2. At borderlines between opposite directions of walking one can observe "viscous fingering" [104,105].
3. The emergence of pedestrian streams through standing crowds [94,18,19,45] appears analogous to the formation of river beds [106–108] (see Fig. 2).
4. Similar to segregation or stratification phenomena in granular media [109,110], pedestrians spontaneously organize in lanes of uniform walking direction, if the pedestrian density is high enough [2,18,19,43,45] (see Fig. 3).
5. At bottlenecks (e.g., corridors, staircases, or doors), the passing direction of pedestrians oscillates [46,47]. This may be compared to the "saline oscillator" [111] or the granular "ticking hour glass" [112,113].
6. One can find the propagation of shock waves in dense pedestrian crowds pushing forward (see also [114]).
7. The arching and clogging in panicking crowds [53] is similar to the outflow of rough granular media through small openings [115,116].

In summary, one could say that fluid-dynamic analogies work well in normal situations, while granular aspects become important in panic situations.

Figure 2. The long-term photograph of a standing crowd in front of a cinema taken by Thomas Arns shows that crossing pedestrians form a river-like stream (from [18,19,45]).

3 Generalized Force Model of Pedestrian Dynamics

3.1 The Social Force Concept

Human behavior often seems to be “chaotic”, irregular, and unpredictable. So, why and under what conditions can we model it by means of forces? First of all, we need to be confronted with a phenomenon of motion in some (quasi-)continuous space, which may be also an abstract behavioral space or opinion scale [34,117–119]. Moreover, it is favourable to have a system where the fluctuations due to unknown influences are not large compared to the systematic, deterministic part of motion. This is usually the case in pedestrian and vehicle traffic, where people are confronted with standard situations and react automatically rather than taking complicated decisions between various possible alternatives. For example, an experienced driver would not have to think about the detailled actions to be taken when turning, accelerating, or changing lanes.

This automatic behavior can be interpreted as the result of a *learning process* based on trial and error [19], which can be simulated with *evolutionary algorithms* [120–123]. For example, pedestrians have a preferred side of walking [2,4,8], since an asymmetrical avoidance behavior turns out to be profitable [54]. The related *formation of a behavioral convention* can be described by means of *evolutionary game theory* [34,43,119,124–127].

Another requirement is the vectorial additivity of the separate force terms reflecting different environmental influences. This is probably an approxima-

Figure 3. At sufficiently high densities, pedestrians form lanes of uniform walking direction (from [18,19,45]).

tion, but there is some experimental evidence for it. Based on quantitative measurements for animals and test persons subject to separately or simultaneously applied stimuli of different nature and strength, one could show that the behavior in conflict situations can be described by a superposition of forces [128–130]. This fits well into a concept by Lewin [131], according to which behavioral changes are guided by so-called ***social fields*** or ***social forces***, which has been put into mathematical terms by Helbing [34,43,47,117–119]. In some cases, social or ***behavioral forces***, which determine the amount and direction of systematic behavioral changes, can be expressed as gradients of dynamically varying potentials, which reflect the social or behavioral fields resulting from the interactions of individuals. The behavioral force concept was applied to opinion formation [34,117–119] and migration [34,118,119], but it was particularly successful in the description of pedestrian and vehicle traffic [43,46,47,132,133].

For reliable simulations of pedestrian crowds we do not need to know whether a certain pedestrian, say, turns to the right at the next intersection. It is sufficient to have a good estimate what percentage of pedestrians turns to the right. This can be either empirically measured or calculated by means of route choice models like the one by Borgers and Timmermans [31,32]. In some sense, the uncertainty about the individual behaviors is averaged out at the macroscopic level of description, as in fluid dynamics. Nevertheless, instead of a fluid-dynamic model, we will use the more flexible microscopic simulation approach based on the generalized force concept. According to this, the temporal change of the location $\boldsymbol{x}_i(t)$ of pedestrian i obeys the equation of motion

$$\frac{d\boldsymbol{x}_i(t)}{dt} = \boldsymbol{v}_i(t)\,. \tag{1}$$

Moreover, if $\boldsymbol{f}_i(t)$ denotes the sum of forces influencing pedestrian i, m_i is the mass of pedestrian i, and $\boldsymbol{\xi}_i(t)$ are individual fluctuations reflecting unsystematic behavioral variations, the velocity changes are given by the *acceleration equation*

$$m_i \frac{d\boldsymbol{v}_i}{dt} = \boldsymbol{f}_i(t) + \boldsymbol{\xi}_i(t)\,. \tag{2}$$

Particular advantages of this approach are that we can take into account

- the flexible usage of space by pedestrians, requiring a (quasi-)continous treatment of motion, and
- excluded volume effects due to granular properties of panicking pedestrian crowds.

It turns out that these points are essential to reproduce the above mentioned phenomena in a natural way.

3.2 Social Force Model for Normal Pedestrian Dynamics

We will now describe the different motivations of and influences on a pedestrian i by separate force terms. First of all, the desire to adapt the actual velocity $\boldsymbol{v}_i(t)$ to the desired speed v_i^0 and direction $\boldsymbol{e}_i^0(t)$ within a certain "relaxation time" τ_i is reflected by the *acceleration term* $[v_i^0(t)\boldsymbol{e}_i^0(t) - \boldsymbol{v}_i(t)]/\tau_i$. Herein, the contribution $v_i^0(t)\boldsymbol{e}_i^0(t)/\tau_i$ can be interpreted as *driving term*, while $-\boldsymbol{v}_i(t)/\tau_i$ has the meaning of a *friction term* with friction coefficient $1/\tau_i$.

Next, the tendency of pedestrians to keep a certain distance to other pedestrians ("territorial effect") may be described by repulsive social forces

$$\boldsymbol{f}_{ij}^{\text{soc}}(t) = A_i \exp[(r_{ij} - d_{ij})/B_i]\boldsymbol{n}_{ij}\left(\lambda_i + (1-\lambda_i)\frac{1+\cos(\varphi_{ij})}{2}\right)\,. \tag{3}$$

Herein, A_i denotes the interaction strength and B_i the range of the repulsive interactions, which are culture-dependent and individual parameters. $d_{ij}(t) = \|\boldsymbol{x}_i(t) - \boldsymbol{x}_j(t)\|$ is the distance between the centers of mass of pedestrians i and j, $r_{ij} = (r_i + r_j)$ the sum of their radii r_i and r_j, and

$$\boldsymbol{n}_{ij}(t) = \left(n_{ij}^1(t), n_{ij}^2(t)\right) = \frac{\boldsymbol{x}_i(t) - \boldsymbol{x}_j(t)}{d_{ij}(t)} \tag{4}$$

the normalized vector pointing from pedestrian j to i. Finally, with the choice $\lambda_i < 1$, we can reflect the anisotropic character of pedestrian interaction. In other words, with the parameter λ_i we can model that the situation in front of a pedestrian has a larger impact on his or her behavior than things happening behind. The angle $\varphi_{ij}(t)$ denotes the angle between the direction $\boldsymbol{e}_i(t) = \boldsymbol{v}_i(t)/\|\boldsymbol{v}_i(t)\|$ of motion and the direction $-\boldsymbol{n}_{ij}(t)$ of the object exerting the repulsive force, i.e., $\cos\varphi_{ij}(t) = -\boldsymbol{n}_{ij}(t)\cdot\boldsymbol{e}_i(t)$. One may, of course, take into account other details such as a velocity-dependence of the forces and non-circular shaped pedestrian bodies, but this does not have *qualitative* effects on the dynamical phenomena resulting in the simulations. In fact, most observed self-organization phenomena are quite insensitive to the specification of the interaction forces, as different studies have shown [47,49,51–53].

In addition, we may also take into account time-dependent attractive interactions towards window displays, sights, or special attractions k by social forces $\boldsymbol{f}_{ik}^{\text{att}}(t)$ of the type (3). However, compared with repulsive interactions, the corresponding interaction range B_{ik} is usually larger and the strength parameter $A_{ik}(t)$ typically smaller, negative, and time-dependent. Additionally, the joining behavior [134] of families, friends, or tourist groups can be reflected by forces of the type $\boldsymbol{f}_{ij}^{\text{att}}(t) = -C_{ij}\boldsymbol{n}_{ij}(t)$, which guarantee that acquainted individuals join again, after they have accidentally been separated by other pedestrians.

In summary, the force model of pedestrian motion in normal situations corresponds to Eqs. (1) and (2) with

$$\boldsymbol{f}_i(t) = \frac{v_i^0(t)\boldsymbol{e}_i^0(t) - \boldsymbol{v}_i(t)}{\tau_i} + \sum_{j(\neq i)} [\boldsymbol{f}_{ij}^{\text{soc}}(t) + \boldsymbol{f}_{ij}^{\text{att}}(t)] + \sum_b \boldsymbol{f}_{ib}(t) + \sum_k \boldsymbol{f}_{ik}^{\text{att}}(t) \quad (5)$$

In the following, we will use a simplified version of this model by dropping attraction effects and assuming $\lambda_i = 0$, so that the interaction forces become isotropic and conform with Newton's 3rd law.

3.3 Force Model for Panicking Pedestrians

Additional, physical interaction forces $\boldsymbol{f}_{ij}^{\text{ph}}$ come into play when pedestrians get so close to each other that they have physical contact ($r_{ij} \geq d_{ij}$). In this case, which is mainly relevant to panic situations, we assume also a *"body force"* $k(r_{ij} - d_{ij})\,\boldsymbol{n}_{ij}$ counteracting body compression and a *"sliding friction force"* $\kappa(r_{ij} - d_{ij})\,\Delta v_{ji}^t\,\boldsymbol{t}_{ij}$ impeding *relative* tangential motion. Inspired by the formulas for granular interactions [115,116], we assume

$$\boldsymbol{f}_{ij}^{\text{ph}}(t) = k\Theta(r_{ij} - d_{ij})\boldsymbol{n}_{ij} + \kappa\Theta(r_{ij} - d_{ij})\Delta v_{ji}^t\,\boldsymbol{t}_{ij}\,, \quad (6)$$

where the function $\Theta(z)$ is equal to its argument z, if $z \geq 0$, otherwise 0. Moreover, $\boldsymbol{t}_{ij} = (-n_{ij}^2, n_{ij}^1)$ means the tangential direction and $\Delta v_{ji}^t = (\boldsymbol{v}_j - \boldsymbol{v}_i)\cdot\boldsymbol{t}_{ij}$ the tangential velocity difference, while k and κ represent large constants.

Strictly speaking, friction effects already set in before pedestrians touch each other, because of the psychological tendency not to pass other individuals with a high relative velocity, when the distance is small. This is, however, not important for the effects we are going to reproduce later on.

The interactions with the boundaries of walls and other obstacles are treated analogously to pedestrian interactions, i.e., if $d_{ib}(t)$ means the distance to boundary b, $\boldsymbol{n}_{ib}(t)$ denotes the direction perpendicular to it, and $\boldsymbol{t}_{ib}(t)$ the direction tangential to it, the corresponding interaction force with the boundary reads

$$\boldsymbol{f}_{ib} = \{A_i \exp[(r_i - d_{ib})/B_i] + k\Theta(r_i - d_{ib})\}\, \boldsymbol{n}_{ib} - \kappa\Theta(r_i - d_{ib})(\boldsymbol{v}_i \cdot \boldsymbol{t}_{ib})\, \boldsymbol{t}_{ib} \, . \quad (7)$$

Finally, fire fronts are reflected by repulsive social forces similar those describing walls, but they are much stronger. The physical interactions, however, are qualitatively different, as people reached by the fire front become injured and immobile ($v_i = 0$).

4 Simulation Results

The generalized force model of pedestrian dynamics has been simulated on a computer for a large number of interacting pedestrians confronted with different situations. In spite of its simplifications, it describes a lot of observed phenomena quite realistically. Especially, it allows us to explain various self-organized spatio-temporal patterns that are not externally planned, prescribed, or organized, e.g., by traffic signs, laws, or behavioral conventions. Instead, the spatio-temporal patterns discussed below emerge due to the nonlinear interactions of pedestrians even without assuming strategical considerations or communication of pedestrians. Many of these collective patterns of motion are symmetry-breaking phenomena, although the model was formulated completely symmetric with respect to the right-hand and the left-hand side [18,19,45,48].

4.1 Self-organized Pedestrian Dynamics in Normal Situations

Lane Formation. Our microsimulations reproduce the empirically observed *formation of lanes* consisting of pedestrians with the same desired walking direction [18,19,44,46–51] (see Fig. 4). For open boundary conditions, these lanes are dynamically varying. Their number depends on the width of the street [18,47], on pedestrian density, and on the noise level. Interestingly, one finds a ***noise-induced ordering*** [51,135]: Compared to small noise amplitudes, medium ones result in a more pronounced segregation (i.e., a smaller number of lanes), while large noise amplitudes lead to a "freezing by heating" effect (see Fig. 7).

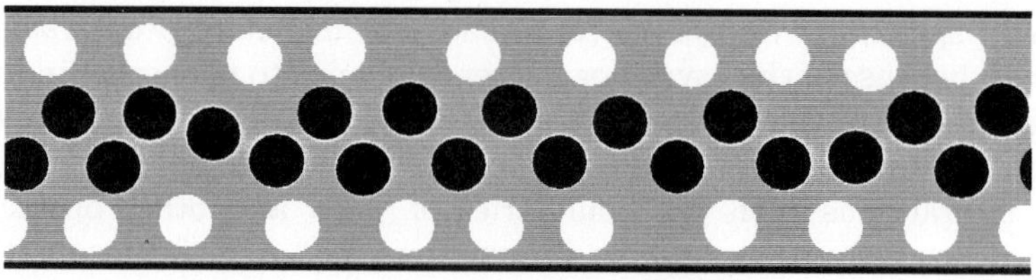

Figure 4. Formation of lanes in initially disordered pedestrian crowds with opposite walking directions (after [52,136]; cf. also [46,47,95,99,136]). White disks represent pedestrians moving from left to right, black ones move the other way round. Lane formation does not require the periodic boundary conditions applied above, see the Java applet `http://www.helbing.org/Pedestrians/Corridor.html`.

The conventional interpretation of lane formation assumes that pedestrians tend to walk on the side which is prescribed in vehicular traffic. However, the above model can explain lane formation even without assuming a preference for *any* side [51,52]. The most relevant point is the higher relative velocity of pedestrians walking in opposite directions. Pedestrians moving against the stream or in areas of mixed directions of motion will have frequent and strong interactions. In each interaction, the encountering pedestrians move a little aside in order to pass each other. This sidewards movement tends to separate oppositely moving pedestrians, which leads to segregation. The resulting collective pattern of motion minimizes the frequency and strength of avoidance maneuvers, if fluctuations are weak. Assuming identical desired velocities $v_i^0 = v_0$, the most stable configuration corresponds to a state with a minimization of the average interaction strength

$$-\frac{1}{N}\sum_{i\neq j} \tau \boldsymbol{f}_{ij} \cdot \boldsymbol{e}_i^0 \approx \frac{1}{N}\sum_i (v_0 - \boldsymbol{v}_i \cdot \boldsymbol{e}_i^0) = v_0(1-E)\,. \qquad (8)$$

It is related with a maximum efficiency

$$E = \frac{1}{N}\sum_i \frac{\boldsymbol{v}_i \cdot \boldsymbol{e}_i}{v_0} \qquad (9)$$

of motion corresponding to ***optimal self-organization*** [51], where the efficiency E with $0 \leq E \leq 1$ describes the average fraction of the desired speed v_0 with which pedestrians actually approach their destinations ($N = \sum_i 1$ is the respective number of pedestrians i). As a consequence, lane formation "globally" maximizes the average velocity into the respectively desired direction of motion, although the model does not even assume that pedestrians would try to optimize their behavior *locally*. This is a consequence of the symmetrical interactions among pedestrians with opposite walking directions. One can even show that a large class of driven many-particle systems, if they self-organize at all, tend to globally optimize their state [51].

Finally, note that lane formation is hard to describe by cellular automata. However, Burstedde *et al.* have recently found a way to reproduce this collective phenomenon by introducing an additional floor field inspired by trail formation models [97,137,138,45], which mimics individual intelligence. In the limit of vanishing diffusion and fast decay of the floor field, this cellular automaton is similar to Helbing and Bolay's implementation of an efficient, discretized version of the social force model [54], where the interaction effects of boundaries and pedestrians are represented by a global potential. Some interactive Java applets based on this model are available at `www.helbing.org`.

Oscillations at Bottlenecks. In simulations of bottlenecks like doors, we observe oscillatory changes of the passing direction, if people do not panic [18,19,44,46–50] (see Fig. 5). Once a pedestrian is able to pass the narrowing, pedestrians with the same walking direction can easily follow. Hence, the number and "pressure" of waiting and pushing pedestrians becomes less than on the other side of the narrowing where, consequently, the chance to occupy the passage grows. This leads to a deadlock situation which is followed by a change in the passing direction.

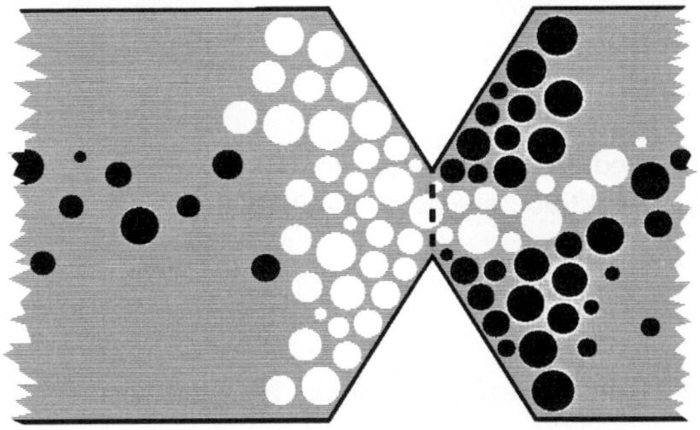

Figure 5. Oscillations of the passing direction at a bottleneck (after [95,99,136]; cf. also [46,47]). Dynamic simulations are available at `http://www.helbing.org/Pedestrians/Door.html`.

Dynamics at Intersections. At intersections one is confronted with various alternating collective patterns of motion which are very short-lived and unstable. For example, phases during which the intersection is crossed in "vertical" or "horizontal" direction alternate with phases of temporary roundabout traffic (see Fig. 6) [46,44,18,49,50,48,19]. This self-organized round-about traffic is similar to the emergent rotation found for self-driven particles [139]. It is connected with small detours but decreases the frequency of necessary deceleration, stopping, and avoidance maneuvers considerably, so that pedestrian motion becomes more efficient on average.

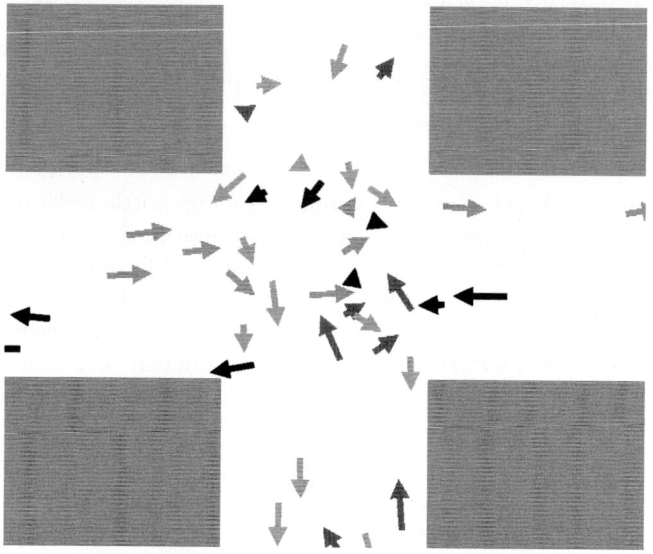

Figure 6. Self-organized, short-lived roundabout traffic in intersecting pedestrian streams (from [18,19,44,48,95]; see also [46,49,50]).

4.2 Collective Phenomena in Panic Situations

In panic situations (e.g., in some cases of emergency evacuation) the following characteristic features of pedestrian behavior are often observed:

1. People are getting nervous, resulting in a higher level of fluctuations.
2. They are trying to escape from the source of panic, which can be reflected by a significantly higher desired velocity.
3. Individuals in complex situations, who do not know what is the right thing to do, orient at the actions of their neighbours, i.e., they tend to do what other people do. We will describe this by an additional herding interaction, but attractive interactions have probably a similar effect.

We will now discuss the fundamental collective effects which fluctuations, increased desired velocities, and herding behavior can have. In contrast to other approaches, we do not assume or imply that individuals in panic or emergency situations would behave relentless and asocial, although they sometimes do.

"Freezing by Heating". The effect of getting nervous has been investigated in Ref. [52]. Let us assume the individual level of fluctuations is given by

$$\eta_i = (1 - n_i)\eta_0 + n_i \eta_{\max} , \tag{10}$$

where n_i with $0 \le n_i \le 1$ measures the nervousness of pedestrian i. The parameter η_0 means the normal and $\eta_{\max}$ the maximum fluctuation strength. It turns out that, at sufficiently high pedestrian densities, lanes are destroyed by increasing the fluctuation strength (which is analogous to the temperature). However, instead of the expected transition from the "fluid" lane state to a disordered, "gaseous" state, a solid state is formed. It is characterized by a blocked situation with a regular (i.e., "crystallized" or "frozen") structure so that we call this paradoxial transition *"freezing by heating"* (see Fig. 7). Notably enough, the blocked state has a *higher* degree of order, although the internal energy is *increased* and the resulting state is *metastable* with respect to structural perturbations such as the exchange of oppositely moving particles [52]. Therefore, "freezing by heating" is just opposite to what one would expect for equilibrium systems, and different from fluctuation-driven ordering phenomena in some granular systems [140–142], where fluctuations lead from a disordered *metastable* to an ordered *stable* state [135].

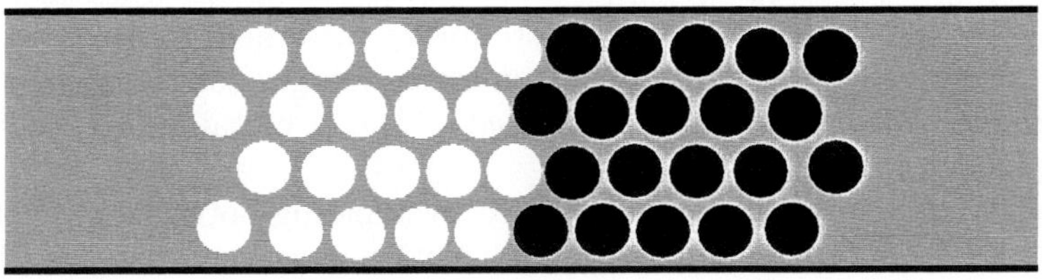

Figure 7. Noise-induced formation of a crystallized, "frozen" state in a periodic corridor used by oppositely moving pedestrians (after [52,95,99,136]).

The precondition for the unusual freezing-by-heating transition are the driving term $v_i^0 \boldsymbol{e}_i^0/\tau_i$ and the dissipative friction $-\boldsymbol{v}_i/\tau_i$, while the sliding friction force is not required. Inhomogeneities in the channel diameter or other impurities which temporarily slow down pedestrians can further this transition at the respective places. Finally note that a transition from fluid to blocked pedestrian counter flows is also observed, when a critical density is exceeded [52,59].

Transition to Incoordination due to Clogging. The simulated outflow from a room is well-coordinated and regular, if the desired velocities $v_i^0 = v_0$ are normal. However, for desired velocities above 1.5 m/s, i.e., for people in a rush, we find an irregular succession of arch-like blockings of the exit and avalanche-like bunches of leaving pedestrians, when the arches break (see Fig. 8a, b). This phenomenon is compatible with the empirical observations mentioned above and comparable to intermittent clogging found in granular flows through funnels or hoppers [115,116] (although this has been attributed to *static* friction between particles without remote interactions, and the transition to clogging has been observed for small enough openings rather than for a variation of the driving force).

"Faster-is-slower Effect" due to Impatience. Since clogging is connected with delays, trying to move faster (i.e., increasing v_i^0) can cause a smaller average speed of leaving, if the friction parameter κ is large enough (see Fig. 8c, d). This "faster-is-slower effect" is particularly tragic in the presence of fires, where fleeing people sometimes reduce their own chances of survival. The related fatalities can be estimated by the number of pedestrians reached by the fire front (see Fig. 9).

Since our friction term has, on average, no deceleration effect in the crowd, if the walls are sufficiently remote, the arching underlying the clogging effect requires a *combination* of several effects:

1. slowing down due to a bottleneck such as a door and
2. strong inter-personal friction, which becomes dominant when pedestrians get too close to each other. It is noteworthy that the faster-is-slower effect also occurs when the sliding friction force changes continuously with the distance rather than being "switched on" at a certain distance r_j as in the model above.

The danger of clogging can be minimized by avoiding bottlenecks in the construction of stadia and public buildings. Notice, however, that jamming can also occur at widenings of escape routes! This surprising result is illustrated in Fig. 10. It originates from disturbances due to pedestrians, who try to overtake each other and expand in the wide area because of their repulsive interactions. They squeeze into the main stream again at the end of the widening, which acts like a bottleneck and leads to jamming. The corresponding drop of efficiency E is more pronounced,

1. if the corridor is narrow,
2. if the pedestrians have different or high desired velocities, and
3. if the pedestrian density in the corridor is high.

"Phantom Panics". Sometimes, panics have occured *without* any comprehensible reasons such as a fire or another threatening event (e.g.,, in Moscow, 1982; Innsbruck, 1999). Due to the "faster-is-slower effect", panics can be triggered by small pedestrian counterflows [78], which cause delays to the crowd intending to leave. Consequently, stopped pedestrians in the back, who do not see the reason for the temporary slowdown, are getting impatient and pushy.

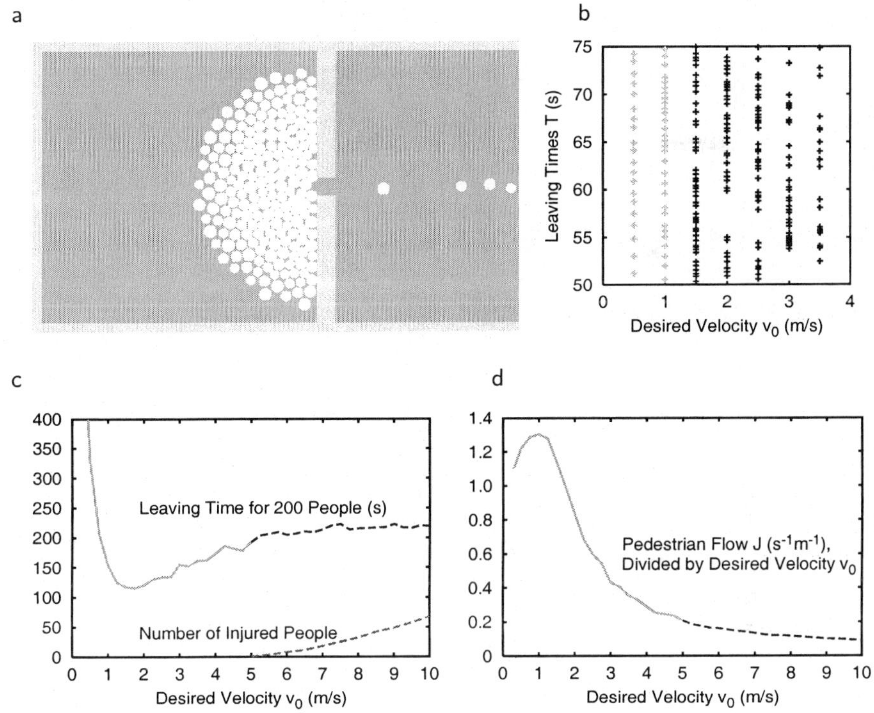

Figure 8. Simulation of pedestrians moving with identical desired velocity $v_i^0 = v_0$ towards the 1 m wide exit of a room of size 15 m × 15 m (from [53], see also [95,99,136]). a Snapshot of the scenario. Dynamic simulations are available at **`http://angel.elte.hu/~panic/`**. b Illustration of leaving times of pedestrians for various desired velocities v_0. Irregular outflow due to clogging is observed for high desired velocities ($v_0 \geq 1.5$ m/s, see dark plusses). c Under conditions of normal walking, the time for 200 pedestrians to leave the room decreases with growing v_0. Desired velocities higher than 1.5 m/s reduce the efficiency of leaving, which becomes particularly clear, when the outflow J is divided by the desired velocity (see d). This is due to pushing, which causes additional friction effects. Moreover, above a desired velocity of about $v_0 = 5$ m/s (– –), people are injured and become non-moving obstacles for others, if the sum of the magnitudes of the radial forces acting on them divided by their circumference exceeds a pressure of 1600 N/m [98].

In accordance with observations [43,18], one may describe this by increasing the desired velocity, for example, by the formula

$$v_i^0(t) = [1 - n_i(t)]v_i^0(0) + n_i(t)v_i^{\max} . \qquad (11)$$

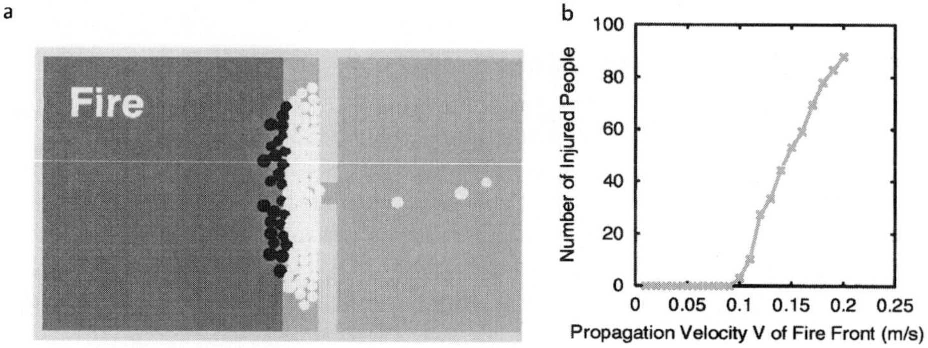

Figure 9. Simulation of $N = 200$ individuals fleeing from a linear fire front, which propagates from the left to the right wall with velocity V, starting at time $t = 5$ s [99] (for a Java simulation applet, see `http://angel.elte.hu/~panic/`). a Snapshot of the scenario for a 15 m × 15 m large room with one door of width 1 m. The fire is indicated by dark grey color. Pedestrians reached by the fire front are injured and symbolized by black disks, while the white ones are still active. The socio-psychological effect of the fire front is assumed 10 times stronger than that of a normal wall ($A_F = 10A_i$). b Number of injured persons (casualities) as a function of the propagation velocity V of the fire front, averaged over 10 simulation runs. Up to a critical propagation velocity $V_{\rm crit}$ (here: about 0.1 m/s), nobody is injured. However, for higher velocities, we find a fast increase of the number of casualties with increasing V. The transition is continuous.

Herein, $v_i^{\max}$ is the maximum desired velocity and $v_i^0(0)$ the initial one, corresponding to the expected velocity of leaving. The time-dependent parameter

$$n_i(t) = 1 - \frac{\overline{v}_i(t)}{v_i^0(0)} \tag{12}$$

reflects the nervousness, where $\overline{v}_i(t)$ denotes the average speed into the desired direction of motion. Altogether, long waiting times increase the desired velocity, which can produce inefficient outflow. This further increases the waiting times, and so on, so that this tragic feedback can eventually trigger so high pressures that people are crushed or falling and trampled. It is, therefore, imperative, to have sufficiently wide exits and to prevent counterflows, when big crowds want to leave [53].

Ignorance of Available Exits. Finally, we investigate a situation in which pedestrians are trying to leave a smoky room, but first have to find one of the invisible exits (see Fig. 11a). Each pedestrian i may either select an individual direction $\boldsymbol{e}_i$ or follow the average direction $\langle \boldsymbol{e}_j^0(t)\rangle_i$ of his neighbours j in a

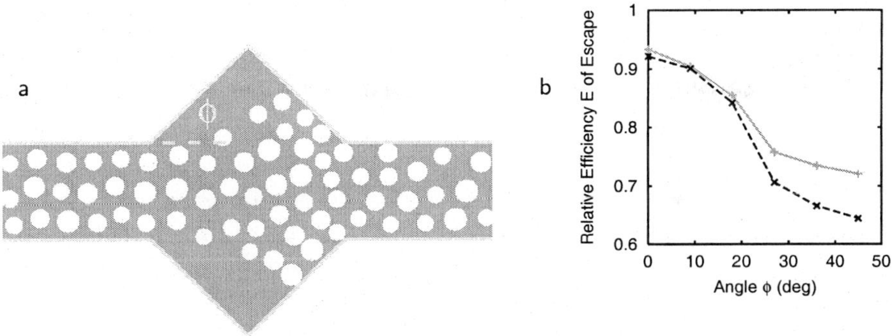

Figure 10. Simulation of an escape route with a wider area (from [53,99], see also the Java applets supplied at `http://angel.elte.hu/~panic/`). a Illustration of the scenario with $v_i^0 = v_0 = 2$ m/s. The corridor is 3 m wide and 15 m long, the length of the triangular pieces in the middle being 2×3 m $= 6$ m. Pedestrians enter the simulation area on the left-hand side with an inflow of $J = 5.5\ \mathrm{s}^{-1}\mathrm{m}^{-1}$ and flee towards the right-hand side. b Efficiency of leaving as a function of the angle ϕ characterizing the width of the central zone, i.e., the difference from a linear corridor. The relative efficiency $E = \langle \boldsymbol{v}_i \cdot \boldsymbol{e}_i^0 \rangle / v_0$ measures the average velocity along the corridor compared to the desired velocity and lies between 0 and 1 (—). While it is almost one (i.e., maximal) for a linear corridor ($\phi = 0$), the efficiency drops by about 20%, if the corridor contains a widening. The decrease of efficiency E is even more pronounced in the area of the widening where pedestrian flow is most irregular (– –).

certain radius R_i [143,144], or try a mixture of both. We assume that both options are weighted with the nervousness n_i:

$$\boldsymbol{e}_i^0(t) = \mathcal{N}\left[(1 - n_i)\, \boldsymbol{e}_i + n_i \,\langle \boldsymbol{e}_j^0(t) \rangle_i\right] , \tag{13}$$

where $\mathcal{N}(\boldsymbol{z}) = \boldsymbol{z}/\|\boldsymbol{z}\|$ denotes the normalization of a vector $\boldsymbol{z}$. As a consequence, we have individualistic behavior if n_i is low, but herding behavior if n_i is high. (In a somewhat different szenario, one may simulate a crowd with a proportion $(1 - n)$ of individualists and a proportion n of herd followers.)

Our model suggests that neither individualistic nor herding behavior performs well (see Fig. 11b). Pure individualistic behavior means that each pedestrian finds an exit only accidentally, while pure herding behavior implies that the complete crowd is eventually moving into the same and probably blocked direction, so that available exits are not efficiently used, in agreement with observations. According to Figs. 11b and c, we expect optimal chances of survival for a certain mixture of individualistic and herding behavior, where individualism allows some people to detect the exits and herding guarantees that successful solutions are imitated by small groups of others. If pedestrians follow the walls instead of "reflecting" at them, we expect that herd following causes jamming and inefficient use of doors as well (see Fig. 8), while individualists moving in opposite directions obstruct each other.

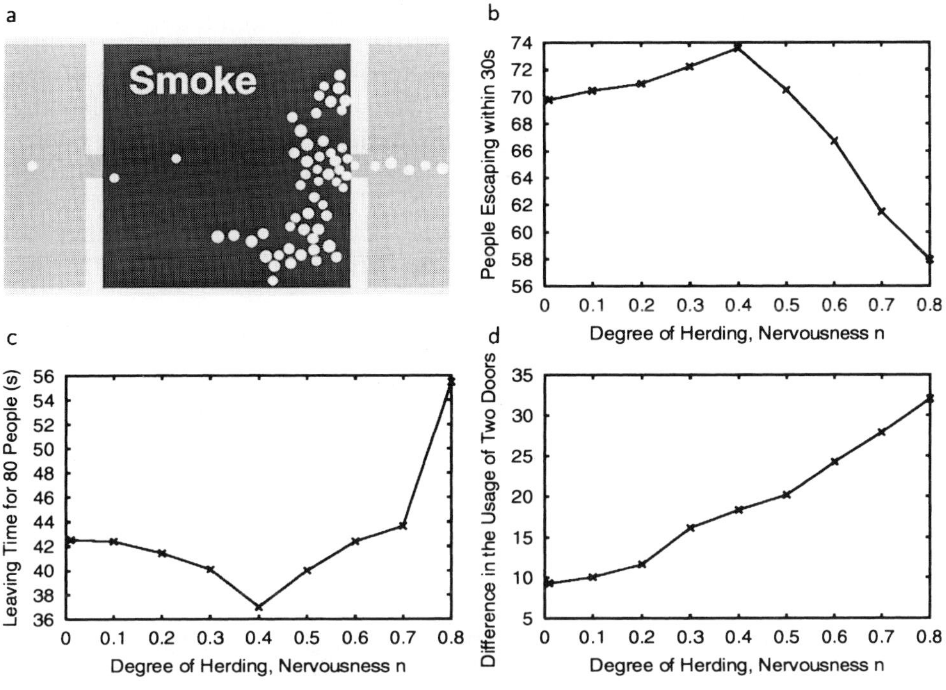

Figure 11. Simulation of $N = 90$ pedestrians trying to escape a smoky room of area $A = 15\text{m} \times 15$ m (black) through two smoke-hidden doors of 1.5 m width, which have to be found with a mixture of individualistic and herding behavior (from [53,95,99]). Java applets are available at `http://angel.elte.hu/~panic/`. a Snapshot of the simulation with $v_i^0 = v_0 = 5$ m/s. Initially, each pedestrian selects his or her desired walking direction randomly. Afterwards, a pedestrian's walking direction is influenced by the average direction of the neighbours within a radius of, for example, $R_i = R = 5$ m. The strength of this herding effect grows with increasing nervousness parameter $n_i = n$ and increasing value of $h = \pi R^2 \rho$, where $\rho = N/A$ denotes the pedestrian density. When reaching a boundary, the direction of a pedestrian is reflected. If one of the exits is closer than 2 m, the room is left. b Number of people who manage to escape within 30 s as a function of the nervousness parameter n. c Illustration of the time required by 80 individuals to leave the smoky room. If the exits are relatively narrow and the degree n of herding is small or large, leaving takes particularly long, so that only some of the people escape before being poisoned by smoke. Our results suggest that the best escape strategy is a certain compromise between following of others and an individualistic searching behavior. This fits well into experimental data on the efficiency of group problem solving [145–147], according to which groups normally perform better than individuals, but masses are inefficient in finding new solutions to complex problems. d Absolute difference $|N_1 - N_2|$ in the numbers N_1 and N_2 of persons leaving through the left exit or the right exit as a function of the degree n of herding. We find that pedestrians tend to jam up at one of the exits instead of equally using all available exits, if the nervousness is large.

5 Optimization of Pedestrian Flows

The emerging pedestrian flows decisively depend on the geometry of the boundaries. They can be simulated on a computer already in the planning phase of pedestrian facilities. Their configuration and shape can be systematically varied, e.g., by means of evolutionary algorithms [121,123,54] (see Fig. 12), and evaluated on the basis of particular mathematical performance measures [18,48]. Apart from the *efficiency* E with $0 \le E \le 1$ defined in formula (9), we can, for example, define the *measure of comfort* $C = (1 - D)$ via the discomfort

$$D = \frac{1}{N} \sum_i \frac{\overline{(\boldsymbol{v}_i - \overline{\boldsymbol{v}_i})^2}}{\overline{(\boldsymbol{v}_i)^2}} = \frac{1}{N} \sum_i \left(1 - \frac{\overline{\boldsymbol{v}_i}^2}{\overline{(\boldsymbol{v}_i)^2}} \right) . \qquad (14)$$

The latter is again between 0 and 1 and reflects the frequency and degree of sudden velocity changes, i.e., the level of discontinuity of walking due to necessary avoidance maneuvers. Hence, the optimal configuration regarding the pedestrian requirements is the one with the highest values of efficiency and comfort.

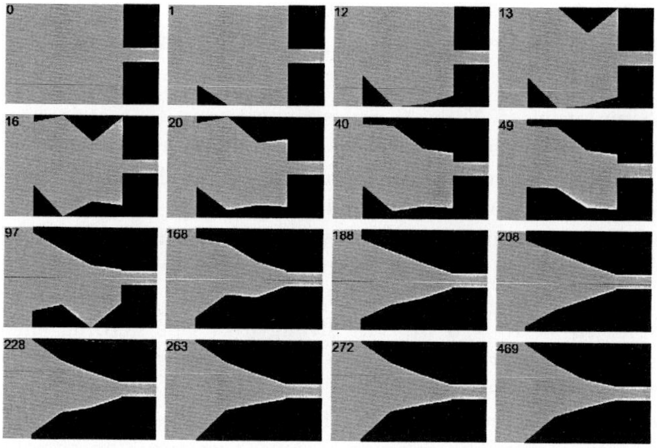

Figure 12. Different phases in the evolutionary optimization of a bottleneck (from [54,19]).

During the optimization procedure, some or all of the following can be varied:

1. the location and form of planned buildings,
2. the arrangement of walkways, entrances, exits, staircases, elevators, escalators, and corridors,

3. the shape of rooms, corridors, entrances, and exits,
4. the function and time schedule of room usage. (Recreation rooms or restaurants are continuously frequented, rooms for conferences or special events are mainly visited and left at peak periods, exhibition rooms or rooms for festivities require additional space for people standing around, and some areas are claimed by queues or through traffic.)

The proposed optimization procedure can not only be applied to the design of new pedestrian facilities but also to a reduction of existing bottlenecks by suitable modifications.

5.1 Normal Situations

Here, we discuss four simple examples of how to improve some standard elements of pedestrian facilities ([18]; see Fig. 13):

1. At high pedestrian densities, the lanes of uniform walking direction tend to disturb each other: Impatient pedestrians try to use any gap for overtaking, which often leads to subsequent obstructions of the opposite walking directions. The lanes can be stabilized by series of trees or columns in the middle of the road (see Fig. 13a) which, in walking direction, looks similar to a wall (see Fig. 14). Also, it takes some detour to reach the other side of the permeable wall, which makes it less attractive to use gaps occuring in the opposite pedestrian stream.
2. The flow at bottlenecks can be improved by a funnel-shaped construction (see Fig. 13b) which, at the same time, allows one to save expensive space. Interestingly, the optimal form resulting from an evolutionary optimization is convex [54] (see Fig. 12).
3. A broader door does not necessarily lead to a proportional increase of pedestrian flow through it. It may rather lead to more frequent changes of the walking direction which are connected with temporary deadlock situations. Therefore, two doors close to the walls are more efficient than one single door with double width. By self-organization, each door is used by one walking direction [18,48–50], which is related to lane formation (see Fig. 15).
4. Oscillatory changes of the walking direction and periods of standstill in between also occur when different flows *cross* each other. The loss of efficiency caused by this can be reduced by psychological guiding measures or railings initializing roundabout traffic (see Fig. 13c). Roundabout traffic can already be induced and stabilized by planting a tree in the middle of a crossing, because it surpresses the phases of "vertical" or "horizontal" motion in the intersection area. In our simulations this increased efficiency up to 13%.

The complex interaction between various flows can lead to completely unexpected results due to the nonlinearity of dynamics. (A very impressive and surprising result of evolutionary form optimization is presented by Klockgether

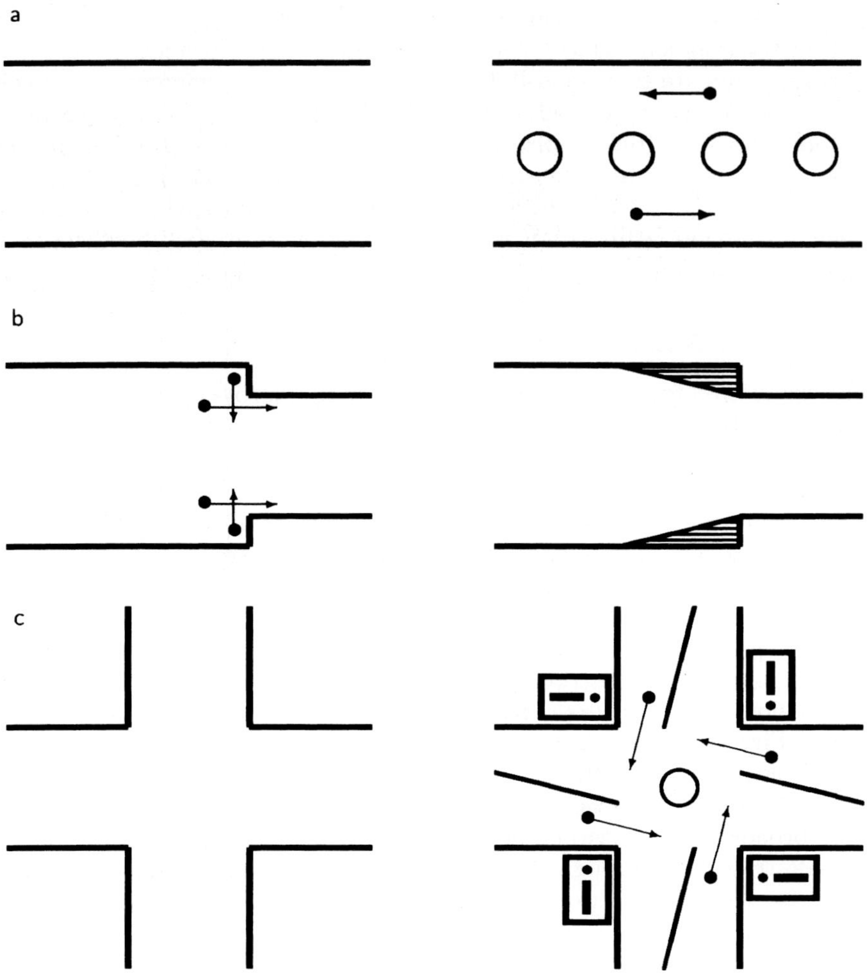

Figure 13. Conventional (left) and improved (right) elements of pedestrians facilities: a ways, b bottlenecks, and c intersections (from [18,19]). The exclamation marks stand for attraction effects (e.g., interesting posters above the street). Empty circles represent columns or trees, while full circles with arrows symbolize pedestrians and their walking directions.

and Schwefel in Ref. [120].) This means, planning of pedestrian facilities with conventional methods does not always guarantee the avoidance of big jams, serious obstructions, and catastrophic blockages (especially in emergency situations). In contrast, a skilful flow optimization not only enhances efficiency

Figure 14. These photographs of a pedestrian tunnel connecting two subways in Budapest at Deák tér illustrates that a series of columns acts similar to a wall and stabilizes lanes by preventing that their width exceeds half of the total width of the walkway.

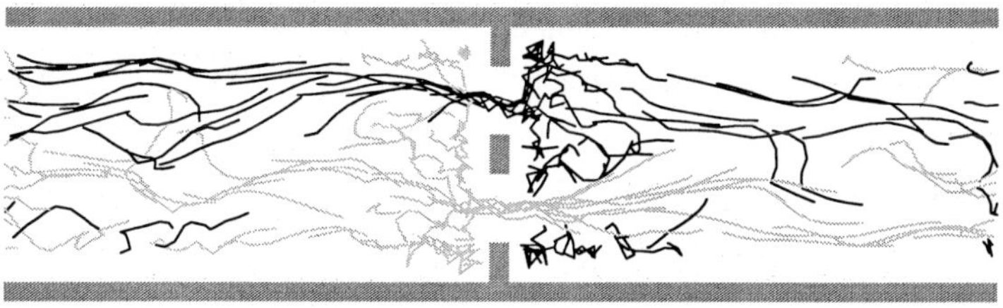

Figure 15. If two alternative passageways are available, pedestrians with opposite walking directions use different doors as a result of self-organization (after [18,19,45,48–50,95]).

but also saves space that can be used for kiosks, benches, or other purposes [18].

5.2 Panic Situations

Similar design strategies can be developed for panic situations, where significantly improved outflows can be reached by columns placed asymmetrically in front of the exits [53,99]. These can prevent the build up of fatal pressures in exit areas and, thereby, also reduce injuries (see Fig. 16 and the Java applets at `http://angel.elte.hu/~panic/`). The asymmetrical placement helps to avoid equilibria of forces (blockages).

Additionally, one can guide people into the directions of usable exits by means of optical and acoustic stimuli, i.e., by suitable arrangements of light and sound sources.

a b

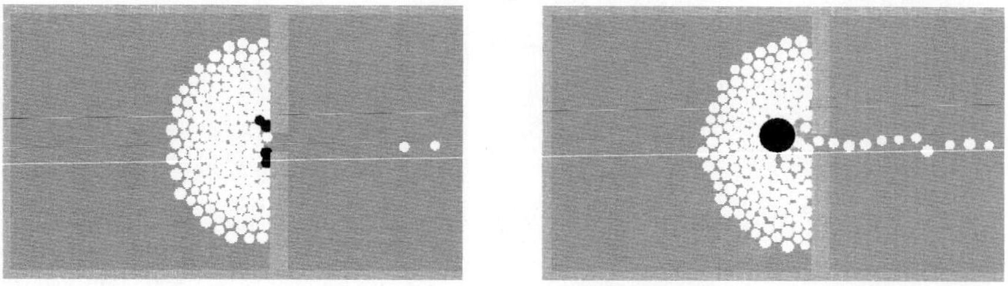

Figure 16. a In panicking crowds, high pressures build up due to physical interactions. This can injure people (black disks), who turn into obstacles for other pedestrians trying to leave (see also the lower curve in Fig. 8c). b A column in front of the exit (large black disk) can avoid injuries by taking up pressure from behind. It can also increase the outflow by 50%. In large exit areas used by many hundret people, several randomly placed columns are needed to subdivide the crowd and the pressure. An asymmetric configuration of the columns is most efficient, as it avoids equilibria of forces which may temporarily stop the outflow. (From [99].)

6 Summary and Outlook

We have developed a continuous pedestrian model based on plausible interactions, which is, due to its simplicity, robust with respect to parameter variations. It was pointed out that pedestrian dynamics shows various *collective phenomena, which every simulation model should reproduce in order to be realistic*. For example, in normal situations one finds

1. lane formation and
2. oscillatory flows through bottlenecks.

These and other empirical findings can be well described by our microscopic simulations of pedestrian streams based on a generalized force model. According to this model, the collective patterns of motion can be interpreted as self-organization phenomena, arising from the non-linear interactions among pedestrians.

We underline that self-organized flow patterns can significantly change the capacities of pedestrian facilities. They often lead to undesireable obstructions, but they can also be utilized to reach more efficient pedestrian flows with less space. Applications to the optimization of pedestrian facilities are, therefore, quite natural.

The proposed force model is also suitable for drawing conclusions about the possible mechanisms beyond escape panic (regarding an increase of the desired velocity, strong friction effects during physical interactions, and herding). After calibration of the model parameters to available data on pedestrian flows, we managed to reproduce many observed phenomena including

1. the breakdown of fluid lanes ("freezing by heating"),
2. the build up of fatal pressures,
3. clogging effects at bottlenecks,
4. jamming at widenings,
5. the "faster-is-slower effect",
6. "phantom panics" triggered by counterflows and impatience, and
7. the ignorance of available exits due to herding.

The underlying behavior could be called "irrational", as all of these effects decrease the chances of survival compared to normal pedestrian behavior. We were also able to simulate situations of dwindling resources and estimate the casualties (see Figs. 8c and 9). Therefore, the model could be used to test buildings for their suitability in emergency situations. It accounts for the considerably different dynamics both in normal and panic situations just by changing a single parameter $n_i = n$. In this way, we have proposed a consistent theoretical approach allowing a continuous switching between seemingly incompatible kinds of human behavior (individualistic, "rational" behavior vs. "irrational" panic behavior). Thereby, however, *we do not want to imply that individuals would always behave irrational in emergency situations.* It has been observed that, even in such situations individuals can behave highly self-controlled, coordinated, rational, and social. Our study just investigates the fundamental collective effects which fluctuations, increased desired velocities, and herding behavior can have, independently of whether all criteria of panics are fulfilled or not.

We believe that the above model can serve as an example linking collective behavior as a phenomenon of mass psychology (from the socio-psychological perspective) to the view of an emergent collective pattern of motion (from the perspective of physics). Our simulations suggest that the optimal behavior in escape situations is a suitable mixture of individualistic and herding behavior. This conclusion is probably transferable to many cases of problem solving in new and complex situations, where standard solutions fail. It may explain why both, individualistic and herding behaviors are common in human societies. For example, herding behavior is also relevant to fashion and stock market dynamics (see Refs. [99,148–153]). Apart from that, the competition of moving particles for limited space is analogous to the situation in various socio-economic systems, where individuals or other entities compete for limited resources as well. Therefore, conclusions from the above findings for self-driven many-particle systems reach far into the realm of the social, economic, and psychological sciences.

Finally, we are calling for quantitative data and, as far as possible, experimental studies of panic situations to make this model even more realistic. For example, one could include direction- and velocity-dependent interpersonal interactions, specify the individual variation of parameters, integrate acoustic information exchange, implement more complex strategies and interactions (also three-dimensional ones), or allow for switching of strategies. One should also complement the program by detailled fire and smoke propagation modules and model hazards, toxicity and behavioral reactions, as evacuation software tools like EXODUS do (see `http://fseg.gre.ac.uk/exodus`).

Acknowledgments. The authors are grateful to the Collegium Budapest—Institute for Advanced Study for the warm hospitality and the excellent scientific working conditions. D.H. thanks the German Research Foundation (DFG) for financial support by the Heisenberg scholarship He 2789/1-1. T.V. and I.F. are grateful for partial support by OTKA and FKFP. Last but not least, Tilo Grigat has helped with formatting this manuscript.

References

1. B. D. Hankin and R. A. Wright, *Passenger flow in subways*, Operational Research Quarterly **9**, 81–88 (1958).
2. D. Oeding, *Verkehrsbelastung und Dimensionierung von Gehwegen und anderen Anlagen des Fußgängerverkehrs*, in *Straßenbau und Straßenverkehrstechnik*, Heft 22 (Bundesministerium für Verkehr, Abt. Straßenbau, Bonn, 1963).
3. L. A. Hoel, *Pedestrian travel rates in central business districts*, Traffic Engineering **38**, 10–13 (1968).
4. S. J. Older, *Movement of pedestrians on footways in shopping streets*, Traffic Engineering & Control **10**, 160–163 (1968).
5. P. D. Navin and R. J. Wheeler, *Pedestrian flow characteristics*, Traffic Engineering **39**, 31–36 (1969).
6. R. L. Carstens and S. L. Ring, *Pedestrian capacities of shelter entrances*, Traffic Engineering **41**, 38–43 (1970).
7. C. A. O'Flaherty and M. H. Parkinson, *Movement on a city centre footway*, Traffic Engineering & Control **13**, 434–438 (1972).
8. U. Weidmann, *Transporttechnik der Fußgänger*, in *Schriftenreihe des Instituts für Verkehrsplanung, Transporttechnik, Straßen- und Eisenbahnbau* (Institut für Verkehrsplanung, Transporttechnik, Straßen- und Eisenbahnbau, ETH Zürich, 1993).
9. M. R. Hill, *Walking, Crossing Streets, and Choosing Pedestrian Routes* (University of Nebraska, Lincoln, 1984).
10. M. Batty, *Predicting where we walk*, Nature **388**, 19–20 (1997).
11. J. J. Fruin, *Designing for pedestrians: A level-of-service concept*, in *Highway Research Record, Number 355: Pedestrians*, pp. 1–15 (Highway Research Board, Washington, D.C., 1971).
12. A. Polus , J. L. Schofer, and A. Ushpiz, *Pedestrian flow and level of service*, Journal of Transportation Engineering **109**, 46–56 (1983).
13. M. Mōri and H. Tsukaguchi, *A new method for evaluation of level of service in pedestrian facilities*, Transportation Research A **21**, 223–234 (1987).
14. H. Schubert, *Planungsmaßnahmen für den Fußgängerverkehr in den Städten*, in *Straßenbau und Straßenverkehrstechnik*, Heft 56 (Bundesministerium für Verkehr, Abt. Straßenbau, Bonn, 1967).
15. D. Boeminghaus, *Fußgängerbereiche + Gestaltungselemente* (Krämer, Stuttgart, 1982).
16. J. Pauls, *The movement of people in buildings and design solutions for means of egress*, Fire Technology **20**, 27–47 (1984).
17. W. H. Whyte, *City. Rediscovering the Center* (Doubleday, New York, 1988).
18. D. Helbing, *Verkehrsdynamik* (Springer, Berlin, 1997).
19. D. Helbing, P. Molnár, I. Farkas, and K. Bolay, *Self-organizing pedestrian movement*, Environment and Planning B **28**, xxx (2001).
20. H. Kirsch, *Leistungsfähigkeit und Dimensionierung von Fußgängerwegen*, in *Straßenbau und Straßenverkehrstechnik*, Heft 33 (Bundesministerium für Verkehr, Abt. Straßenbau, Bonn, 1964).
21. W. M. Predtetschenski and A. I. Milinski, *Personenströme in Gebäuden – Berechnungsmethoden für die Projektierung* (Rudolf Müller, Köln-Braunsfeld, 1971).

22. Transportation Research Board, *Highway Capacity Manual*, Special Report 209 (Transportation Research Board, Washington, D.C., 1985).
23. D. G. Davis and J. P. Braaksma, *Adjusting for luggage-laden pedestrians in airport terminals*, Transportation Research A **22**, 375–388 (1988).
24. W. Brilon, M. Großmann, and H. Blanke, *Verfahren für die Berechnung der Leistungsfähigkeit und Qualität des Verkehrsablaufes auf Straßen*, in *Straßenbau und Straßenverkehrstechnik*, Heft 669 (Bundesministerium für Verkehr, Abt. Straßenbau, Bonn, 1993).
25. S. J. Yuhaski Jr., J. M. Macgregor Smith, *Modelling circulation systems in buildings using state dependent queueing models*, Queueing Systems **4**, 319–338 (1989).
26. J. R. Roy, *Queuing in spatially dispersed public facilities*, presented at the *IV. World Congress of the Regional Science Association International* (Palma de Mallorca, 1992).
27. G. G. Løvås, *Modelling and simulation of pedestrian traffic flow*, Transportation Research B **28**, 429–443 (1994).
28. D. Garbrecht, *Describing pedestrian and car trips by transition matrices*, Traffic Quarterly **27**, 89–109 (1973).
29. A. J. Mayne, *Some further results in the theory of pedestrians and road traffic*, Biometrika **41**, 375–389 (1954).
30. N. Ashford, M. O'Leary, and P. D. McGinity, *Stochastic modelling of passenger and baggage flows through an airport terminal*, Traffic Engineering & Control **17**, 207–210 (1976).
31. A. Borgers and H. Timmermans, *A model of pedestrian route choice and demand for retail facilities within inner-city shopping areas*, Geographical Analysis **18**, 115–128 (1986).
32. A. Borgers and H. Timmermans, *City centre entry points, store location patterns and pedestrian route choice behaviour: A microlevel simulation model*, Socio-Economic Planning Science **20**, 25–31 (1986).
33. H. Timmermans, X. van der Hagen, and A. Borgers, *Transportation systems, retail environments and pedestrian trip chaining behaviour: Modelling issues and applications*, Transportation Research B **26**, 45–59 (1992).
34. D. Helbing, *Stochastische Methoden, nichtlineare Dynamik und quantitative Modelle sozialer Prozesse*, Ph.D. thesis (University of Stuttgart, 1992, published by Shaker, Aachen, 1993).
35. L. F. Henderson, *The statistics of crowd fluids*, Nature **229**, 381–383 (1971).
36. L. F. Henderson, *On the fluid mechanics of human crowd motion*, Transportation Research **8**, 509–515 (1974).
37. L. F. Henderson and D. J. Lyons, *Sexual differences in human crowd motion*, Nature **240**, 353-355 (1972).
38. L. F. Henderson and D. M. Jenkins, *Response of pedestrians to traffic challenge*, Transportation Research **8**, 71–74 (1973).
39. R. L. Hughes, *The flow of large crowds of pedestrians*, Math. Comp. Simul. **53**, 367–370 (2000).
40. R. L. Hughes, *A continuum theory for the flow of pedestrians*, Transportation Research B, in print (2001).
41. D. Helbing, *A fluid-dynamic model for the movement of pedestrians*, Complex Systems **6**, 391–415 (1992).

42. S. P. Hoogendoorn and P. H. L. Bovy, *Gas-kinetic modelling and simulation of pedestrian flows*, Transportation Research Records **1710**, 28–36 (2000).
43. D. Helbing, *A mathematical model for the behavior of pedestrians*, Behavioral Science **36**, 298–310 (1991).
44. D. Helbing, *Traffic modeling by means of physical concepts*, in *Traffic and Granular Flow*, pp. 87–104, D. E. Wolf, M. Schreckenberg, and A. Bachem (Eds.) (World Scientific, Singapore, 1996).
45. D. Helbing, *Pedestrian dynamics and trail formation*, pp. 21–36, in *Traffic and Granular Flow '97*, M. Schreckenberg and D. E. Wolf (Eds.) (Springer, Singapore, 1998).
46. D. Helbing, P. Molnár, and F. Schweitzer, *Computer simulations of pedestrian dynamics and trail formation*, in *Evolution of Natural Structures*, pp. 229–234 (Sonderforschungsbereich 230, Stuttgart, 1994).
47. D. Helbing and P. Molnár, *Social force model for pedestrian dynamics*, Physical Review E **51**, 4282–4286 (1995).
48. D. Helbing and P. Molnár, *Self-organization phenomena in pedestrian crowds*, in *Self-Organization of Complex Structures: From Individual to Collective Dynamics*, pp. 569–577, F. Schweitzer (Ed.) (Gordon and Breach, London, 1997).
49. P. Molnár, *Modellierung und Simulation der Dynamik von Fußgängerströmen*, (Shaker, Aachen, 1996).
50. P. Molnár, *Microsimulation of pedestrian dynamics* in *Social Science Microsimulation*, J. Doran, N. Gilbert, U. Mueller, and K. Troitzsch (Eds.) (Springer, Berlin, 1996).
51. D. Helbing and T. Vicsek, *Optimal self-organization*, New Journal of Physics **1**, 13.1–13.17 (1999).
52. D. Helbing, I. Farkas, and T. Vicsek, *Freezing by heating in a driven mesoscopic system*, Physical Review Letters **84**, 1240–1243 (2000).
53. D. Helbing, I. Farkas, and T. Vicsek, *Simulating dynamical features of escape panic*, Nature **407**, 487–490 (2000).
54. K. Bolay, *Nichtlineare Phänomene in einem fluid-dynamischen Verkehrsmodell* (Master's thesis, University of Stuttgart, 1998).
55. V. J. Blue and J. L. Adler, *Emergent fundamental pedestrian flows from cellular automata microsimulation*, Transportation Research Records **1644**, 29–36 (1998).
56. V. J. Blue and J. L. Adler, *Using cellular automata microsimulation to model pedestrian movements*, in *Proceedings of the 14th International Symposium on Transportation and Traffic Theory*, pp. 235–254, A. Ceder (Ed.) (Pergamon, New York, N. Y., 1999).
57. M. Fukui and Y. Ishibashi, *Self-organized phase transitions in cellular automaton models for pedestrians*, Journal of the Physical Society of Japan **68**, 2861–2863 (1999).
58. M. Fukui and Y. Ishibashi, *Jamming transition in cellular automaton models for pedestrian on passageway*, Journal of the Physical Society of Japan **68**, 3738–3739 (1999).
59. M. Muramatsu, T. Irie, and T. Nagatani, *Jamming transition in pedestrian counter flow*, Physica A **267**, 487–498 (1999).
60. M. Muramatsu and T. Nagatani, *Jamming transition in two-dimensional pedestrian traffic*, Physica A **275**, 281–291 (2000).

61. M. Muramatsu and T. Nagatani, *Jamming transition of pedestrian traffic at a crossing with open boundaries*, Physica A **286**, 377–390 (2000).
62. H. Klüpfel, M. Meyer-König, J. Wahle, and M. Schreckenberg, *Microscopic simulation of evacuation processes on passenger ships*, in *Theory and Practical Issues on Cellular Automata*, S. Bandini and T. Worsch (Eds.) (Springer, London, 2000).
63. C. Burstedde, K. Klauck, A. Schadschneider, and J. Zittartz, *Simulation of pedestrian dynamics using a 2-dimensional cellular automaton*, Physica A, in print (2001).
64. S. Gopal and T. R. Smith, *NAVIGATOR: An AI-based model of human way-finding in an urban environment*, in *Spatial Choices and Processes*, pp. 169–200, M. M. Fischer, P. Nijkamp, and Y. Y. Papageorgiou (Eds.) (North-Holland, Amsterdam, 1990).
65. C. W. Reynolds, *Evolution of corridor following behavior in a noisy world*, in *From Animals to Animats 3: Proceedings of the Third International Conference on Simulation of Adaptive Behavior*, pp. 402–410, D. Cliff, P. Husbands, J.-A. Meyer, and S. Wilson (Eds.) (MIT Press, Cambridge, Massachusetts, 1994).
66. C. W. Reynolds, *Steering behaviors for autonomous characters*, in *Proceedings of the 1999 Game Developer's Conference*, pp. 763–782, A. Yu (Ed.) (Miller Freeman, San Fransisco, 1999).
67. T. Schelhorn, D. O'Sullivan, M. Haklay, and M. Thurstain-Goodwin, *STREETS: An agent-based pedestrian model*, presented at the conference *Computers in Urban Planning and Urban Management* (Venice, 1999).
68. K. H. Drager, G. Løvås, J. Wiklund, H. Soma, D. Duong, A. Violas, and V. Lanèrès, *EVACSIM—A comprehensive evacuation simulation tool*, in the *Proceedings of the 1992 Emergency Management and Engineering Conference*, pp. 101–108 (Society for Computer Simulation, Orlando, Florida, 1992).
69. M. Ebihara, A. Ohtsuki, and H. Iwaki, *A model for simulating human behavior during emergency evacuation based on classificatory reasoning and certainty value handling*, Microcomputers in Civil Engineering **7**, 63–71 (1992).
70. N. Ketchell, S. Cole, D. M. Webber, C. A. Marriott, P. J. Stephens, I. R. Brearley, J. Fraser, J. Doheny, and J. Smart, *The EGRESS Code for human movement and behaviour in emergency evacuations*, in *Engineering for Crowd Safety*, pp. 361–370, R. A. Smith and J. F. Dickie (Eds.) (Elsevier, Amsterdam, 1993).
71. S. Okazaki and S. Matsushita, *A study of simulation model for pedestrian movement with evacuation and queuing*, in *Engineering for Crowd Safety*, pp. 271–280, R. A. Smith and J. F. Dickie (Eds.) (Elsevier, Amsterdam, 1993).
72. G. K. Still, *New computer system can predict human behaviour response to building fires*, Fire **84**, 40–41 (1993).
73. G. K. Still, *Crowd Dynamics* (Ph.D. thesis, University of Warwick, 2000).
74. P. A. Thompson and E. W. Marchant, *Modelling techniques for evacuation*, in *Engineering for Crowd Safety*, pp. 259–269, R. A. Smith and J. F. Dickie (Eds.) (Elsevier, Amsterdam, 1993).
75. G. G. Løvås, *On the importance of building evacuation system components*, IEEE Transactions on Engineering Management **45**, 181–191 (1998).
76. H. W. Hamacher and S. A. Tjandra, *Modelling Emergency Evacuation—An Overview*, to be published (2001).

77. J. P. Keating, *The myth of panic*, Fire Journal, 57–61+147 (May/1982).
78. D. Elliott and D. Smith, *Football stadia disasters in the United Kingdom: Learning from tragedy?*, Industrial & Environmental Crisis Quarterly **7**(3), 205–229 (1993).
79. B. D. Jacobs and P. 't Hart, *Disaster at Hillsborough Stadium: a comparative analysis*, in *Hazard Management and Emergency Planning*, Chap. 10, D. J. Parker and J. W. Handmer (Eds.) (James & James Science, London, 1992).
80. D. Canter (Ed.), *Fires and Human Behaviour*, (David Fulton, London, 1990).
81. A. Mintz, *Non-adaptive group behavior*, The Journal of Abnormal and Normal Social Psychology **46**, 150–159 (1951).
82. D. L. Miller, *Introduction to Collective Behavior*, Fig. 3.3 and Chap. 9 (Wadsworth, Belmont, CA, 1985).
83. J. S. Coleman, *Foundations of Social Theory*, Chaps. 9 and 33 (Belkamp, Cambridge, MA, 1990).
84. N. R. Johnson, *Panic at "The Who Concert Stampede": An empirical assessment*, Social Problems **34**(4), 362–373 (1987).
85. G. LeBon, *The Crowd* (Viking, New York, 1960 [1895]).
86. E. Quarantelli, *The behavior of panic participants*, Sociology and Social Research **41**, 187–194 (1957).
87. N. J. Smelser, *Theory of Collective Behavior*, (The Free Press, New York, 1963).
88. R. Brown, *Social Psychology* (The Free Press, New York, 1965).
89. R. H. Turner and L. M. Killian, *Collective Behavior* (Prentice Hall, Englewood Cliffs, NJ, 3rd ed., 1987).
90. J. L. Bryan, *Convergence clusters*, Fire Journal, 27–30+86–90 (Nov./1985).
91. R. Axelrod and W. D. Hamilton, *The evolution of cooperation*, Science **211**, 1390–1396 (1981).
92. R. Axelrod and D. Dion, *The further evolution of cooperation*, Science **242**, 1385–1390 (1988).
93. N. S. Glance and B. A. Huberman, *The dynamics of social dilemmas*, Scientific American **270**, 76–81 (1994).
94. T. Arns, Video films of pedestrian crowds, (Wannenstr. 22, 70199 Stuttgart, 1993).
95. D. Helbing, *Traffic and related self-driven many-particle systems*, e-print cond-mat/0012229, to appear in Reviews of Modern Physics (2001).
96. J. Ganem, *A behavioral demonstration of Fermat's principle*, The Physics Teacher **36**, 76–78 (1998).
97. D. Helbing, J. Keltsch, and P. Molnár, *Modelling the evolution of human trail systems*, Nature **388**, 47–50 (1997).
98. R. A. Smith and J. F. Dickie (Eds.), *Engineering for Crowd Safety* (Elsevier, Amsterdam, 1993).
99. D. Helbing, I. J. Farkas, and T. Vicsek, *Crowd disasters and simulation of panic situations*, in *Science of Disaster: Climate Disruptions, Heart Attacks and Market Crashes*, A. Bunde, J. Kropp, and H. J. Schellnhuber (Eds.) (Springer, Berlin, 2001).
100. J. S. Coleman, *Introduction to Mathematical Sociology*, pp. 361–375 (The Free Press of Glencoe, New York, 1964).
101. J. S. Coleman and J. James, *The equilibrium size distribution of freely-forming groups*, Sociometry **24**, 36–45 (1961).

102. L. A. Goodman, *Mathematical methods for the study of systems of groups*, American Journal of Sociology **70**, 170–192 (1964).
103. H. H. Kelley, J. C. Condry Jr., A. E. Dahlke, and A. H. Hill, *Collective behavior in a simulated panic situation*, Journal of Experimental Social Psychology **1**, 20–54 (1965).
104. L. P. Kadanoff, *Simulating hydrodynamics: A pedestrian model*, Journal of Statistical Physics **39**, 267–283 (1985).
105. H. E. Stanley and N. Ostrowsky (Eds.), *On Growth and Form* (Martinus Nijhoff, Boston, 1986).
106. H.-H. Stølum, *River meandering as a self-organization process*, Nature **271**, 1710–1713 (1996).
107. I. Rodríguez-Iturbe and A. Rinaldo, *Fractal River Basins: Chance and Self-Organization* (Cambridge University, Cambridge, England, 1997).
108. G. Caldarelli, *Cellular models for river networks*, e-print cond-mat/0011086 (2000).
109. S. B. Santra, S. Schwarzer, and H. Herrmann, *Fluid-induced particle segregation in sheared granular assemblies*, Physical Review E **54**, 5066–5072 (1996).
110. H. A. Makse, S. Havlin, P. R. King, and H. E. Stanley, *Spontaneous stratification in granular mixtures*, Nature **386**, 379–382 (1997).
111. K. Yoshikawa, N. Oyama, M. Shoji, and S. Nakata, *Use of a saline oscillator as a simple nonlinear dynamical system: Rhythms, bifurcation, and entrainment*, American Journal of Physics **59**, 137–141 (1991).
112. X.-L. Wu, K. J. Måløy, A. Hansen, M. Ammi, and D. Bideau, *Why hour glasses tick*, Physical Review Letters **71**, 1363–1366 (1993).
113. T. L. Pennec, K. J.Måløy, A. Hansen, M. Ammi, D. Bideau D, and X.-L. Wu, *Ticking hour glasses: Experimental analysis of intermittent flow*, Physical Review E **53**, 2257–2264 (1996).
114. M. R. Virkler and S. Elayadath, *Pedestrian density characteristics and shockwaves*, in *Proceedings of the Second International Symposium on Highway Capacity*, Vol. 2, pp. 671–684, R. Akçelik (Ed.) (Transportation Research Board, Washington, D.C., 1994).
115. G. H. Ristow and H. J. Herrmann, *Density patterns in two-dimensional hoppers*, Physical Review E **50**, R5–R8 (1994).
116. D. E. Wolf and P. Grassberger (Eds.), *Friction, Arching, Contact Dynamics* (World Scientific, Singapore, 1997).
117. D. Helbing, *Boltzmann-like and Boltzmann-Fokker-Planck equations as a foundation of behavioral models*, Physica A **196**, 546–573 (1993).
118. D. Helbing, *A mathematical model for the behavior of individuals in a social field*, Journal of Mathematical Sociology **19**, 189–219 (1994).
119. D. Helbing, *Quantitative Sociodynamics. Stochastic Methods and Models of Social Interaction Processes* (Kluwer Academic, Dordrecht, 1995).
120. J. Klockgether and H.-P. Schwefel, *Two-phase nozzle and hollow core jet experiments*, in *Proceedings of the Eleventh Symposium on Engineering Aspects of Magnetohydrodynamics*, pp. 141–148, D. G. Elliott (Ed.) (California Institute of Technology, Pasadena, CA, 1970).
121. I. Rechenberg, *Evolutionsstrategie: Optimierung technischer Systeme nach Prinzipien der biologischen Evolution* (Frommann-Holzboog, Stuttgart, 1973).

122. H.-P. Schwefel, *Numerische Optimierung von Computer-Modellen mittels Evolutionsstrategien* (Birkhäuser, Stuttgart, 1977).
123. T. Baeck, *Evolutionary Algorithms in Theory and Practice* (Oxford University Press, New York, 1996).
124. D. Helbing, *Physikalische Modellierung des dynamischen Verhaltens von Fußgängern* (Master's thesis, Georg-August University Göttingen, 1990).
125. D. Helbing, *A mathematical model for behavioral changes by pair interactions*, in *Economic Evolution and Demographic Change. Formal Models in Social Sciences*, pp. 330–348, G. Haag, U. Mueller, and K. G. Troitzsch (Eds.) (Springer, Berlin, 1992).
126. D. Helbing, *Stochastic and Boltzmann-like models for behavioral changes, and their relation to game theory*, Physica A **193**, 241–258 (1993).
127. D. Helbing, *A stochastic behavioral model and a 'microscopic' foundation of evolutionary game theory*, Theory and Decision **40**, 149–179 (1996).
128. N. E. Miller, *Experimental studies of conflict*, in *Personality and the behavior disorders*, Vol. 1, J. McV. Hunt (Ed.) (Ronald, New York, 1944).
129. N. E. Miller, *Liberalization of basic S-R-concepts: Extension to conflict behavior, motivation, and social learning* in *Psychology: A Study of Science*, Vol. 2, S. Koch (Ed.) (McGraw Hill, New York, 1959).
130. W. H. Herkner, *Ein erweitertes Modell des Appetenz-Aversions-Konflikts*, Z. Klin. Psychol. **4**, 50–60 (1975).
131. K. Lewin, *Field Theory in Social Science*, (Harper & Brothers, New York, 1951).
132. D. Helbing and B. Tilch, *Generalized force model of traffic dynamics*, Physical Review E **58**, 133–138 (1998).
133. M. Treiber, A. Hennecke, and D. Helbing, *Congested traffic states in empirical observations and microscopic simulations*, Physical Review E **62**, 1805–1824 (2000).
134. A. K. Dewdney, *Diverse personalities search for social equilibrium at a computer party*, Scientific American **257**(9), 104–107 (1987).
135. D. Helbing and T. Platkowski, *Self-organization in space and induced by fluctuations*, International Journal of Chaos Theory and Applications **5**, 25–39 (2000).
136. D. Helbing, *Die wundervolle Welt aktiver Vielteilchensysteme*, Physikalische Blätter **57**, 27–33 (2001).
137. F. Schweitzer, K. Lao, and F. Family, *Active random walkers simulate trunk trail formation by ants*, BioSystems **41**, 153–166 (1997).
138. D. Helbing, F. Schweitzer, J. Keltsch, and P. Molnár, *Active walker model for the formation of human and animal trail systems*, Physical Review E **56**, 2527–2539 (1997).
139. Y. L. Duparcmeur, H. Herrmann, and J. P. Troadec, *Spontaneous formation of vortex in a system of self motorised particles*, Journal de Physique I France **5**, 1119–1128 (1995).
140. J. Gallas, H. J. Herrmann, and S. Sokołowski, *Convection cells in vibrating granular media*, Physical Review Letters **69**, 1371–1374 (1992).
141. P. B. Umbanhowar, F. Melo, and H. L. Swinney, *Localized excitations in a vertically vibrated granular layer*, Nature **382**, 793–796 (1996).

142. A. Rosato, K. J. Strandburg, F. Prinz, and R. H. Swendsen, *Why the Brazil nuts are on top: Size segregation of particulate matter by shaking*, Physical Review Letters **58**, 1038–1041 (1987).
143. T. Vicsek, A. Czirók, E. Ben-Jacob, I. Cohen, and O. Shochet, *Novel type of phase transition in a system of self-driven particles*, Physical Review Letters **75**, 1226–1229 (1995).
144. A. Czirók, M. Vicsek, and T. Vicsek, *Collective motion of organisms in three dimensions*, Physica A **264**, 299–304 (1999).
145. N. H. Anderson, *Group performance in an anagram task*, Journal of Social Psychology **55**, 67–75 (1961).
146. H. H. Kelley and J. W. Thibaut, *Group problem solving*, in *The Handbook of Social Psychology*, Vol. 4, G. Lindzey and E. Aronson (Eds.) (Addison-Wesley, Reading, MA, 1969).
147. P. R. Laughlin, N. L. Kerr, J. H. Davis, H. M. Halff, and K. A. Marciniak, *Group size, member ability, and social decision schemes on an intellective task*, Journal of Personality and Social Psychology **31**, 522–535 (1975).
148. L. Mann, T. Nagel, and P. Dowling, *A study of economic panic: The "run" on the Hindmarsh building society*, Sociometry **39**(3), 223–235 (1976).
149. M. Youssefmir, B. A. Huberman, and T. Hogg, *Bubbles and market crashes*, Comp. Econ. **12**, 97–114 (1998).
150. J. D. Farmer, *Market force, ecology, and evolution*, e-print adap-org/9812005, submitted to the Journal of Economic Behavior and Organization (1998).
151. T. Lux and M. Marchesi, *Scaling and criticality in a stochastic multi-agent model of a financial market*, Nature **397**, 498–500 (1999).
152. J.-P. Bouchaud, A. Matacz, and M. Potters, *The leverage effect in financial markets: Retarded volatility and market panic*, e-print cond-mat/0101120 (2001).
153. D. Sornette and J. V. Andersen, *Quantifying Herding During Speculative Financial Bubbles*, e-print cond-mat/0104341 (2001).

Modelling Pedestrians in Transfer Stations

W. Daamen[1], P.H.L. Bovy[2], and S.P. Hoogendoorn[2]

[1]Delft University of Technology, Holland Railconsult, P.O. Box 2855, NL 3500 GW Utrecht, The Netherlands

[2]Delft University of Technology, Transportation Section, P.O. Box 5048, NL 2600 GA Delft, The Netherlands

The paper gives an outline of a pedestrian flow model designed for analysis of pedestrian flows in transfer stations. The model is meant as an aid for architects and railway planners in designing the lay-out of such stations and the timetables. The output from the models, such as walking times, densities, flow levels, etc., should be sensitive to variations in the spatial lay-out as well as to variations in the timetable of connecting transport services. The paper briefly compares the features of the new model with existing ones and draws attention to the specific requirements of pedestrian behaviour modelling compared to vehicle modelling. A description is given of the system architecture of the developed model software.

1 Introduction

The traditional way of designing transfer stations is based on rules of thumb. These rules convey many years of experience concerning the behaviour of passengers in transfer stations. However, they only consider static situations. Until now, a scientific foundation for these rules has not yet been provided. Moreover, the dynamic behaviour of pedestrians also needs to be taken into account to predict the impact of a station layout design in practice.

Currently, timetable generation involves the application of approximations for the time people actually need to transfer. Different types of passengers, such as elderly people, or parents with children, need different transfer times. Adopting accurately estimated transfer times will both remove excessive waiting times from a timetable as well as reduce the probability for passengers to miss their connection, thereby increasing the traveller's comfort and the timetable's reliability.

1.1 Purpose of the Research

In association with Holland Railconsult, a simulation tool has been developed to estimate both the mean and variability of the walking times incurred by transferring passengers and to visualise walking patterns inside transfer stations. Simulation studies with this model will reveal the level of comfort for passengers in transfer stations.

1.2 Model Requirements

General traffic characteristics of pedestrians are adjusted to those of passengers in transfer stations and calibrated using Dutch empirical data in order to obtain an appropriate model for the Dutch situation. The model also describes how the transfer services perform. Finally, an object oriented simulation model has been developed and will subsequently be validated and calibrated by performing several case studies.

1.3 Structure of the Paper

This paper outlines the simulation tool which has been developed. Some technical and behavioural modelling details will also be discussed, such as the different approaches of modelling pedestrian movements and the architecture of the total model system.

Firstly, the model systems' purposes are described and an overview is given of the effects which the model will be able to predict. Secondly, the model requirements are discussed, indicating the parameters and situations the model will describe and be sensitive to. Then, an overview of the state-of-the-art is presented. In this part the decision to design and develop a new dedicated pedestrian simulation model is explained. Next, two ways of modelling pedestrians are presented. In doing so, we describe the difference between these approaches, and the application of each of the models. Then, the architecture of the simulation system is described, focusing on the modular structure of the system. The paper concludes with envisaged model applications.

2 Typical Aspects of Pedestrians Flows

In this section we compare the characteristics of pedestrians and pedestrian flows to the characteristics of vehicular traffic flows. Therefore, we consider two different types of differences: the differences of the particles (pedestrians versus cars) and the differences in flow behaviour.

2.1 Differences of the Particles

- *Particle movement*. In order to move, pedestrians have to rely on their own energy supply [14]. For car-based traffic, the motion of the vehicle itself does not directly require the energy of the driver, although the *mental workload* of a driver increases with increasing driving speed [13]. This holds equally for acceleration and deceleration. Therefore, we hypothesise that in comparison to car-drivers, pedestrian's decisions to accelerate or to decelerate are weighed carefully against the advantages of such decisions.
- *Particle acceleration and deceleration*. Pedestrians are able to decelerate at a very high rate, implying that they can stop almost at once. This holds equally

for their acceleration. On the contrary, car drivers are restricted by the possibilities of their vehicles and the discomfort due to high deceleration and acceleration.

- *Spatial freedom.* Cars are restricted to moving in lanes, with occasionally changing from one lane to another. On the contrary, pedestrians are free to choose their way (direction) in two-dimensional space.

- *Sensitivity to traffic and environment.* While cars are bound to lanes (see previous point), they usually have interactions with few vehicles at one moment, most of the time driving in the same direction as the car driver. Only at specific occasions, for instance a lane-change, car-drivers will take traffic other than the vehicles in front of them into account. On the contrary, pedestrians can expect interactions from all directions at all times. Also, the appearance of the direct environment of driving lanes is more restricted than for pedestrian areas.

2.2 Differences of the Flows

- *Legislation.* Due to traffic law enforcement (e.g. drive on the right - overtake on the left, speed limits, overtaking prohibitions, minimal headway enforcement), and the consequent restrictions in driver behaviour, the characteristics of a vehicular flow are more predictable than those of a pedestrian flow.

3 Aspects of Public Buildings

Interesting research objects in pedestrian behaviour are those areas where large crowds occur, for example transfer stations, airports, shopping centres and stadiums. In this article we will focus on transfer stations. Let us indicate some special characteristics of them.

- *Activities besides walking.* Transfer stations are not only used by pedestrians to walk through; most of the time the pedestrians are performing special activities like waiting, buying tickets, or even doing some shopping. These activities can hinder the moving behaviour of other pedestrians, like the occurrence of waiting queues in front of a ticket machine.

- *Time pressure.* Most of the passengers are either hurried while they have to catch their train or extremely relaxed for they have just missed it. These types of pedestrians have totally different moving behaviour: the pressed pedestrians have high speeds and are willing to force their way through a crowd. The relaxed passengers on the contrary are lounging about with no specific destination within the coming period of time.

- *Arrival and destination patterns.* The patterns of arriving and leaving passengers are mostly determined by the public transport timetable. One of the characteristics is the large variance in flows: one moment areas can be empty, while the next moment they are overcrowded, but most of the time this will not last longer than three or four minutes.

4 Purpose of the Model

The pedestrian model presented in this paper is a modelling tool to analyse the impacts of:

- station design and station layout
- timetable design and platform allocation as part of the timetable

on pedestrian flow operations, to assess the performance of the transfer station in terms of bottleneck capacity, experienced level-of-service, and to determine the feasibility of the timetable. This combination of capabilities of the model is innovative.

To this end, the behaviour of pedestrians in unusual or seldom occurring layouts of rooms or larger walking areas needs to be described as well.

4.1 Station Design and Layout

The simulation model can be applied to quantify the level of comfort of pedestrians while they are moving through the station and while they are waiting at the platform or in the hall. These situations can be simulated for existing stations, extensions of existing stations, stations under development and stations under design. The emphasis is at the spatial forms of the station and the amount of space available for the passenger to move. Also, a person can perform other activities at a transfer station, which are or are not related to the transfer function of a station. Examples of these activities are buying a ticket, waiting for the train and shopping. The influences of these activities on the level of comfort of the transferring passengers are also taken into account.

4.2 Timetable Design and Platform Allocation

Timetables or train frequencies have a substantial influence upon the level of comfort of transferring passengers and especially upon their waiting times within the station. High frequent trains in a regular pattern imply a low waiting time and less stress for the passengers to catch their train. Whenever the amount of passengers can not be coped with, not even with this high frequency, the level of comfort will lower quickly, which can lead to dangerous situations. When the supply of passengers does not justify high train frequencies, train services need to be tuned in order to create optimal transfers. These transfers need to be maintained in case of delay or other disturbances, which demands an intelligent train management system and possible non-optimal station platform occupations.

4.3 Dynamics of Pedestrian Flows

As mentioned before, the layout of a station building influences the behaviour of passenger flows. But not only the layout of all the station, also the spatial form of a part of the station or even the layout of one room can influence the behaviour of individual pedestrians. By modelling pedestrians microscopically, whereby the interactions between individuals are taken into account, influences of the spatial environment can be shown. This way, bottlenecks in a certain room, before an escalator or near an entrance or exit can already be found in the design phase and thus saving a lot of expenses.

In short, the developed model is used to evaluate:

- the layout of (parts of) existing transfer stations,
- effects of extensions and/or adaptations of existing transfer stations,
- layout alternatives for newly developed stations,
- alternatives of and changes in different platform allocations and
- alternatives of and changes in different timetables.

5 Model Requirements

Having discussed the model purposes, we will present the specific model requirements. These requirements can be divided into three categories, namely: requirements regarding the situations the model can interpret, requirements regarding the dynamics in the model, and requirements regarding the performance of the model.

5.1 Different Model Situations

The developing process starts with the construction of a model containing only pedestrians, rolling stock and the interactions between pedestrians and rolling stock (i.e. access a train and egress from a train). Afterwards, the concept is used to produce a general architecture, which can be extended with all other modes of transport.

Then, we discuss the different situations the model can interpret. At first, the model needs to be sensible for variations in spatial design, which also includes variations in the spatial form of the different areas. Secondly, variations in the timetable or the train frequencies have effects on the pedestrian flows in the station, which must be visible in the output of the model. Then, variations in the type of trains stopping at the platform result in different times to get on and off the train and thus generate another pattern and headway of incoming passengers. Finally, different types of pedestrians cause different behaviour characteristics leading to different transfer times.

5.2 Model Dynamics

Both pedestrians and public transport facilities are part of dynamic processes. Pedestrians have interactions with each other and the extension in which these interactions are taken into account indicates the detail of the model and the correspondence with the practice. All interactions are modelled in microscopic models, where macroscopic models especially describe the characteristics of pedestrian flows. Both types of models will be included in the tool. Public transport facilities influence the arrival and departure times of the pedestrians and thus influence the resulting pedestrian flows and especially the times pedestrians pass in the railway stations. Also, the characteristics of the different rail vehicles influence the times pedestrians need to enter or exit.

5.3 Model Performance

In order to be able to simulate large transfer stations and large pedestrian flows within an acceptable time, the architecture of the system needs to be designed for this. More specifically, the model should be able to handle large transfer stations of at least 100,000 m^2 area intended for walking purposes. Furthermore, it should be prepared for the presence of about 100,000 persons during the same period of time.

6 State-of-the-Art in Flow Modelling in Transfer Stations

This part of the paper explains the choice for a self developed simulation model. To this end, existing models for prediction pedestrian flow operations in transfer stations were reviewed. Only the model Pedroute, developed by London Underground Limited after the heavy fire at the King's Cross Station, came up. Subsequently, we tried whether a general microscopic traffic simulation model could be used for the envisaged pedestrian flow simulation. We selected two models: an older model *Integration*, and *Aimsun* [6], recently developed by the university of Barcelona. The third option was to develop a dedicated pedestrian simulation model. The advantages and disadvantages of the three options are presented.

6.1 Pedroute

The great advantage of this model is that its functionality largely satisfies the specified requirements. Moreover, the model already exists and licences for using it are available. An extra advantage is the possibility to evaluate the consequences of a fire or a similar emergency situation on the pedestrian flows. Unfortunately, this model does not provide flexible output and dynamic allocation of pedestrians to the network. Also, the input of a timetable for Public Transport is 'difficult', the employment of different types of rolling stock is impossible as is dynamic platform allocation. At last, pedestrians are not modelled microscopically, so specific influences of spatial layouts can not be shown.

6.2 Pedestrianisation of Vehicular Traffic Models

First, we compared the models Integration and Aimsun. Since we are looking for a real microscopic model, Integration does not fit entirely our requirements. Moreover, its user interface is less refined and the input of the model takes some time. So we proceeded our inventory with Aimsun. Aimsun is able to model all types of traffic, like cars, bicycles and pedestrians, but it is especially designed for cars. Pedestrians can only be modelled at crossings and one can not enter specific origins and destinations for pedestrians, as is possible for car traffic. We tried to make cars behave like pedestrians by decreasing their speed, but this did not work out because the minimal speed was 10 km/h which is quite high for a pedestrian.

Further, the interactions between pedestrians and rail vehicles are hard to model and maybe even impossible.

6.3 Developed Pedestrian Simulation Model

With a general simulation language (like Arena or ModSim) all functionalities needed can be included in a self made simulation environment. Besides, the output of the model is flexible and can be used for all different kinds of analyses. The disadvantage is that it still needs to be developed, which takes a lot of time.

Concluding, both Pedroute and the general traffic models can not supply in all required functionalities. For this reason we have developed our own simulation model for pedestrians in transfer stations.

7 Approaches to Modelling Pedestrian Flow Operations

The model will be used to quantify the effects of station layouts on pedestrians in a transfer station. Visualising these effects helps to identify and understand the causes for the presence of bottlenecks, long walking distances and large transfer times. The visualisation is achieved in two ways: *macroscopically* (indicating the level of comfort of a walkway) and *microscopically* (related to the individual pedestrian). Furthermore, the model enables determining the effect disruptions during operation (delays and early arrivals of trains) may have on the transfer times of passengers.

In evaluating the resulting pedestrian flows, the model is considered as a black box, containing a model, which allocates pedestrians to a network (see Figure 1). Inputs for the allocation model are a network with different links and a table of pedestrian origins and destinations. The behaviour of the different types of pedestrians (tourists, workers, etc) is also taken into account.

During simulation, the model performs the following activities repetitively:

- determining origins and destinations of the pedestrians present,
- assigning routes through the station network,
- calculating walking times based on the behaviour models and
- executing situation updates and dynamic evaluations.

The model produces different walking routes used by the pedestrians. Also, walking times are calculated by the model (means and standard deviations). From the number of pedestrians on a link, the mean available space for each pedestrian can be derived which, in turn, is an indication for the performance on the link. This quantity is expressed in a level of comfort. Low levels of comfort indicate the presence of congestion in the network [7].

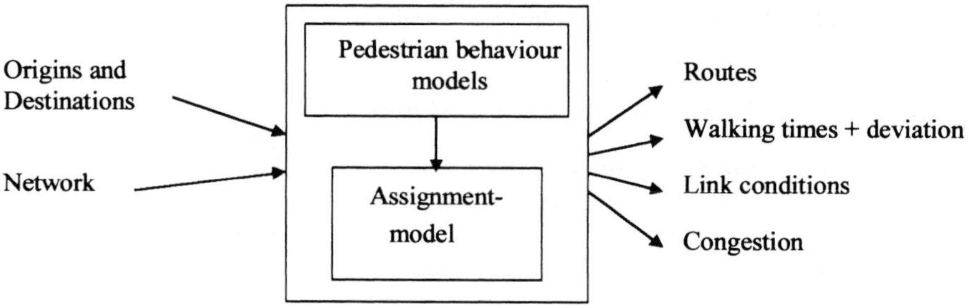

Figure 1: The pedestrian movement evaluation model including inputs and outputs.

Modelling and visualising the dynamic behaviour of pedestrians, especially at railway stations is relatively new. At transfer stations, people behave differently compared to normal circumstances (eg. they may be in a hurry to catch their train or they are milling about because they have missed it). Moreover, the walking conditions in stations differ substantially from those outside (higher densities of people, wider range of walking purposes and speeds). Therefore, the specific behaviour of passengers in transfer stations needs to be studied.

The modelling of walking behaviour of pedestrians can be based on general traffic flow theory [5,11]. In this project, validation of the walking behaviour of pedestrians will be based on observations in Dutch railway stations (for example Utrecht CS).

7.1 Levels of Modelling Pedestrian Flow Operations

Roughly speaking, pedestrian behaviour can be modelled in two ways: *macroscopic* and *microscopic*. Microscopic models treat each pedestrian as a single object having interactions with other pedestrians [1,8,10]. In a macroscopic model, flows of pedestrians moving through the station network are described instead of single pedestrians [8,9].

Macroscopic Model

Using a macroscopic model, we can roughly assess the level of comfort and occurrence of congestion in the network. Subsequently, microscopic simulations are used to analyse these locations in more detail thereby determining the causes for congestion.

In the current version of the model, a (quasi-) macroscopic model[1] is built in, explained by the example in Figure 2.

[1] On the contrary to most (true) macroscopic models, the model described in this article describes the time-space behaviour of individual pedestrians. Although being *microscopic* in that sense, the behaviour of the pedestrians is governed by *macroscopic relations* (speed-density functions), and is therefore classified as a macroscopic model.

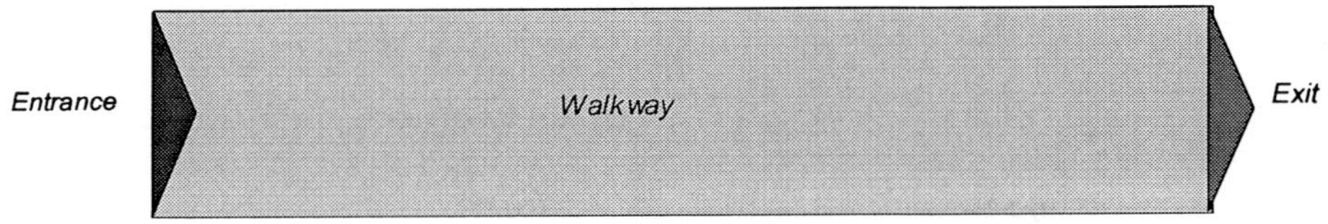

Figure 2: Example facility.

In this example, pedestrians enter the walkway on the left, walk through the walkway and exit the model at the right. This process is visualised in Figure 3.

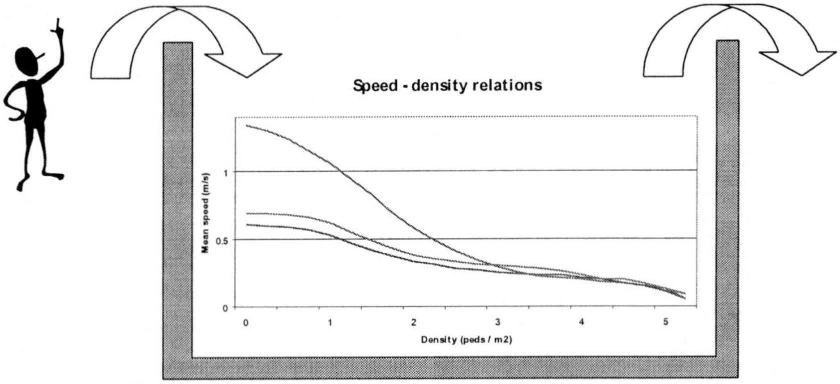

Figure 3: Visualisation of the macroscopic model.

When a pedestrian enters a walkway, the current density is calculated, dividing the number of pedestrians present in the walkway by the net surface of this walkway. Then, the mean speed at this density is derived from the given speed – density relation. The individual speed of the pedestrian is then calculated, by multiplying the mean speed by the ratio of factor free speed of the pedestrian over the mean free speed. The time the pedestrian needs to walk through the area is calculated by dividing the length the pedestrian has to walk to its speed. During this time, the pedestrian is connected to the walkway. After this time the pedestrian will either go to another area or, as is the case here, leave the model.

Microscopic Models

Microscopic or *Walker* models describe the individual time-space behaviour of all pedestrians. Examples of microscopic models are the *social-forces model* [8], and the *Cellular Automata* (CA) models of Blue and Adler [2,3]. In the ensuing of this section, these models are considered in more detail.

The *social forces model* microscopically describes pedestrian behaviour by means of so-called *social force-fields* [8]. This spatially continuous microscopic model inhibits pedestrian's acceleration and direction-changing behaviour. Attracting and repelling forces determine the interaction between the pedestrians and obstacles such as walls. Like the CA-model of Blue and Adler [2,3], the social forces model is able to describe observed features of pedestrian traffic flow, such as self-formation of lanes, due to the interaction of these competitive processes. The CA-model (Blue and Adler [2] is aimed at reproducing macroscopic features of pedestrian flows (such as speed-density curves and dynamic lane formation) using a microscopic model.

Within the context of the research described in this paper, a new microscopic pedestrian flow model MIPSIM has been developed [10]. The model is characterised by pedestrian interaction behaviour, based on stimulus-response mechanisms. Moreover, the model includes local travel choice behaviour, describing how pedestrians choose the bypassing direction based on among other things their destination, and local pedestrian flow conditions. The choice behaviour is based on the concept of *subjective utility maximisation.* Moreover, the model features parallel updating, comparable to recently developed Cellular Automata models for pedestrian flows. However, contrary to these models, pedestrian locations are not bound to discrete cells. The model can therefore easily describe multi-directional pedestrian flows (e.g. uni-directional, bi-directional, crossing flows). Hoogendoorn and Bovy [10] discuss application of the model to a simple hallway. Besides speed-density relations, Hoogendoorn and Bovy [10] study self-organising phenomena in pedestrian flows (dynamic lane formation) following from the microscopic pedestrian flow model.

The precise model specification for both the microscopic model as well as the macroscopic model will depend on among other things the appropriateness of these models compared to the observed Dutch situation, and computation time.

7.2 Application of the Different Model Types

The macroscopic model is used to get an overview of the level of comfort (expressed in the number of pedestrians per square meter) in a large area with a lot of pedestrians. Especially when the arrivals of these pedestrians significantly vary in time (a lot of pedestrians arrive in a short period of time and then for some period of time it is quiet), it is difficult to say whether the level of comfort for the pedestrians is acceptable in each part of the area *without using a dynamic simulation model.* It is even more difficult if the considered area consists of different types of infrastructure (walkways, tunnels, hallways, escalators, stairs, etc.). When the area is a complex network in which different routes are possible, the distribution of the pedestrian flows over these routes is hard to predict using static route-choice algorithms. Moreover, to be able to simulate large areas, it is technically impossible to perform simulations within an acceptable time period when each individual pedestrian is modelled.

Macroscopic models are based on relations between density, mean speed and intensity of pedestrian flows. Some of these relations, especially for public places with 'normal' pedestrian behaviour, are based on observations in practice. However, it is not possible to obtain these observations for *every spatial area form*, for every type of infrastructure under different circumstances. Therefore, we will apply a microscopic simulation model, which is able to predict these macroscopic relations for areas with unusual spatial forms and in different circumstances, like panic situations, in very large crowds or with special types of pedestrians like tourists.

8 System Architecture

In this section the architecture of the developed tool is described. First, an overview is given of the modular structure of the applications which can be modelled. The most important difference is that each module models another type of particles. Also, the different modules can be coupled in order to model a complex system as a transfer system, where we see rail vehicles, buses, taxis and of course pedestrians. Then, the structure of the total simulation environment is clarified and each application being part of this environment is shortly described.

8.1 Modular Structure of the Model

The developed model describes both the behaviour of passengers and different (public) transport modes and the interaction between them. Figure 4 shows an example of an overview of the modular structure of the model.

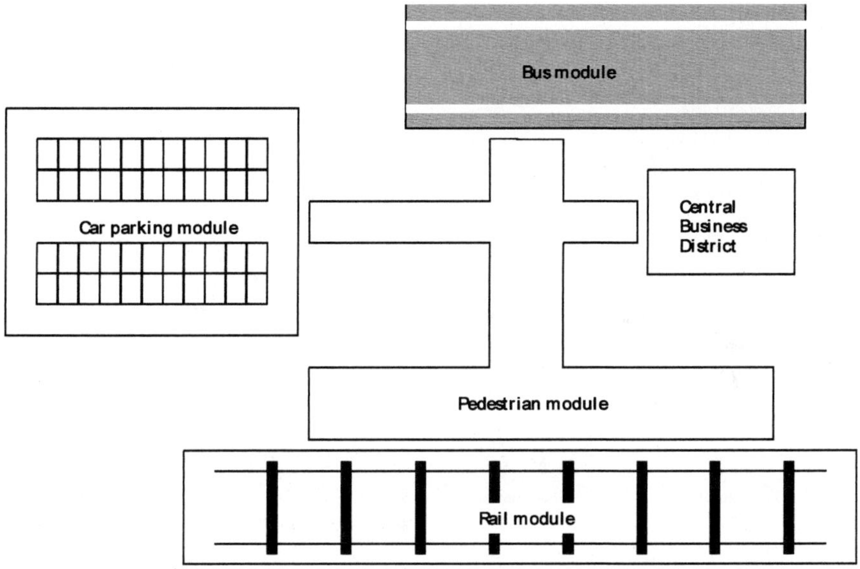

Figure 4: The modular structure of the model (example).

The pedestrian module, called SimPedestrian, includes three theoretical behaviour models: the behaviour of pedestrians on infrastructure available for walking (walkways, stairs, elevators etc.), their behaviour on infrastructure available to perform activities (eg. buying tickets or a newspaper) and their behaviour in relation to other transport modes (eg. getting on and off a train).

Holland Railconsult has developed a model named SimRail, which is currently used to perform railway simulation studies. SimRail will be used as the rail module. In SimRail both the rail infrastructure and the behaviour of vehicles on this infrastructure are modelled. Modules for other transport modes will also be developed, like SimBus, SimCar and especially SimBicycle.

8.2 Modular Structure of the Simulation Environment

As mentioned before, the model will consist of several modules. The first version of the model will only contain the modules for pedestrian behaviour and rolling stock behaviour. The latter module has already been developed and is called SimRail. The pedestrian module is under development. It contains a theoretical behaviour model for pedestrians. Different theoretical models are already available. They are compared to determine which model reflects the walking behaviour at transfer stations most accurately [4].

A software architecture has been designed to combine the different modules of the model. For an initial version of the model, the architecture will be implemented and the pedestrian module has to be completed. An overview of the designed simulation environment is given in Figure 5.

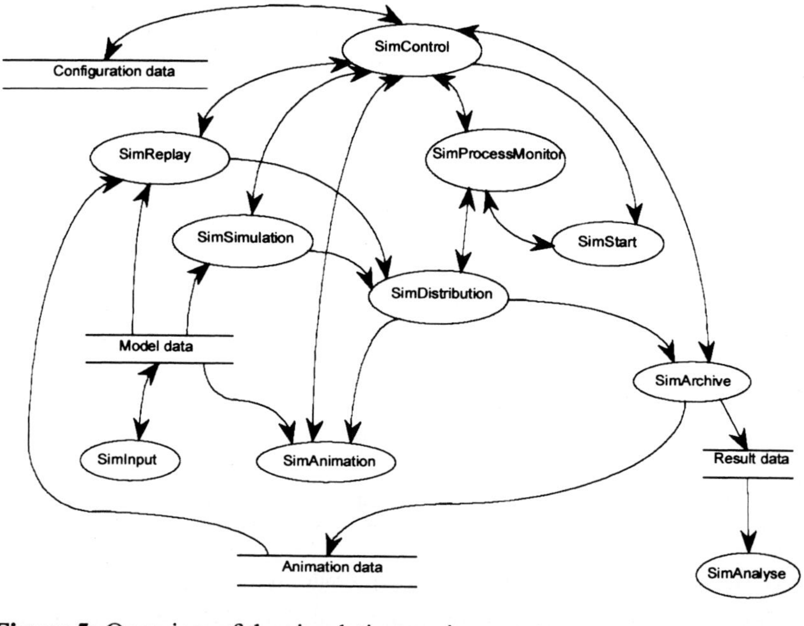

Figure 5: Overview of the simulation environment.

The simulation process can be divided in three phases:

- Initialising: all data is entered.

- Simulating: execution of the actual simulation.
- Analysing: analysis of the results.

Initialising

In this phase, the model is drawn in SimInput. This application converts the drawn data to a database, which is used by other applications. These data consist of the characteristics of the infrastructure (as length, width and spatial form), the characteristics of the pedestrian types (as free speed and familiarity with the environment), characteristics of the public transport vehicles (as length, number and width of doors) and origins and destinations of both pedestrians and public transport vehicles in pedestrian generators and timetables.

Then, information on the simulation to be executed is entered in SimControl. These data are also saved in a database and consist of among other things the models to be simulated, the applications with which the simulation is executed and the length of the simulation.

Simulating

The simulation is started by the application SimControl. Then, the SimSimulation (in this case SimPedestrian) is initialised. A 'ready-to-go' message is sent back to SimControl. SimControl then starts SimAnimation (if required) and SimArchive. Both applications initialise and return a 'ready-to-go' message. SimControl gives a 'Start' signal to SimPedestrian, which starts generating pedestrians. Produced data is sent by SimPedestrian through an application for the distribution of the messages called SimDistribution to SimAnimation and SimArchive. SimAnimation visualises its messages, so the user can view the process. SimArchive registers the data in the messages into a database which is called Result Data.

SimProcessMonitor monitors the total process and watches failures in one of the modules. Also, it checks whether other pc's in the network are available to be included in the simulation environment. However, this application will not be built in the first phase of the model.

Analysing

The data produced by the simulation are stored in the results database. These data are analysed via the application SimAnalysis. This application consists of some forms in MS Access, which calculate for example mean walking times and routes of pedestrians. Also, SimAnalysis produces diagrams and overviews in Excel of for example the progress of the number of pedestrians in some part of the infrastructure over time, thus indicating the level of service for a pedestrian over the time in this part of the infrastructure.

9 Envisaged Model Applications

The system is primarily developed to model pedestrians in transfer stations. But the model is also applicable in other situations with a lot of pedestrians, like airports, central business districts, stadiums and shopping centres. Currently, research effort is focussing on validation of the model. Especially the macroscopic relations will be compared to the behaviour of pedestrian flows in practice. Also, the characteristics of the different microscopic models are to be compared with the behaviour of individual pedestrians (in transfer stations) and their interactions.

During this process, we apply the model to several pilot-projects. In these projects we use the tool to assess existing railway stations and stations under development. Moreover, from these model applications, we aim to establish which output is most useful to quantify the level of service of the pedestrian flows and in what form it is most accessible, not only for researchers, but also for designers and managers.

After these pilot projects, we will apply the tool in other public areas than railway stations. Therefore, specific characteristics of these areas need to be studied and, if necessary, specific pedestrian behaviour is to be added to the model.

Finally, we stress the importance of the developed universal simulation environment, which comprises different models for pedestrian behaviour which can be applied to review all areas where pedestrians occur by quantifying levels of comfort.

10 Acknowledgements

This publication is a result of the research programme Seamless Multimodal Mobility (SMM), carried out within the TRAIL Research School for Transport, Infrastructure and Logistics, and financed by Delft University of Technology in co-operation with Holland Railconsult.

References

1. V.J. Blue and J.L. Adler, *Emergent Fundamental Pedestrian Flows from Cellular Automata Microsimulation*, Transportation Research Record **1644**, 29-36 (1998).

2. V.J. Blue and J.L. Adler, *Bi-directional Emergent Fundamental Pedestrian Flows from Cellular Automata Microsimulation*, in: *Proceedings of the 14th International Symposium on Transportation and Traffic Theory*, (Jerusalem, Israel, 1999).

3. J.B. Blue and J.L. Adler, *Modelling Four-Directional Pedestrian Movements*, in: *Proceedings of the 79th Transportation Research Board 2000,* (2000).

4. W. Daamen, *Onderzoek naar modellen voor het verplaatsen van voetgangers*, Holland Railconsult (in Dutch), (1999).

5. C.F. Daganzo, *Fundamentals of Transportation and Traffic Operations*, (Elsevier Science, 1997).

6. DHV Milieu en Infrastructuur BV, *AIMSUN2, Provisional documentation*, (Universitat Politècnica de Catalunya, Department of Statistics and Operations Research, 1997).

7. J.J. Fruin, *Design for Pedestrians: A Level-of-Service Concept*, Highway Research Record **355**, 1-15 (Highway Research Board, Washington D.C., 1971).

8. D. Helbing, *Verkehrsdynamik – Neue physikalische Modellieringskonzepte*, (Springer-Verlag, 1997).

9. S.P. Hoogendoorn and P.H.L. Bovy, *Gas-Kinetic Modeling and Simulation of Pedestrian Flows*, accepted for presentation and publication in Transportation Research Record **1710**, 28-36 (2000).

10. S.P. Hoogendoorn and P.H.L. Bovy, *Non-local Microscopic Simulation of Pedestrian flows with Local Optimal Choice Behavior*, forthcoming in Transportation Research Record, (2001).

11. A.D. May, *Traffic Flow Fundamentals*, (Prentice Hall, New Jersey, 1990).

12. K. Nagel, *From Particle Hopping Models to Traffic Flow Theory*, Transportation Research Record **1644**, 1-9 (1998).

13. J. van Toorenburg,, *De mens in het autosnelwegverkeer* (in Dutch), Verkeerskunde **34**, nr. 4 en 5, (1983).

14. U. Weidmann, *Transporttechnik der Fussgänger*, ETH, Schriftenreihe IVT-Berichte **90**, Zürich (in German), (1993).

Cellular Automaton Approach to Pedestrian Dynamics - Theory

A. Schadschneider

Institut für Theoretische Physik, Universität zu Köln, 50937 Köln, Germany

We present a 2-dimensional cellular automaton model for the simulation of pedestrian dynamics. The model is extremely efficient and allows simulations of large crowds faster than real time since it includes only nearest-neighbour interactions. Nevertheless it is able to reproduce collective effects and self-organization encountered in pedestrian dynamics. This is achieved by introducing a so-called *floor field* which mediates the long-range interactions between the pedestrians. This field modifies the transition rates to neighbouring cells. It has its own dynamics (diffusion and decay) and can be changed by the motion of the pedestrians. Therefore the model uses an idea similar to chemotaxis, but with pedestrians following a virtual rather than a chemical trace.

1 Introduction

Efficient computer simulations of large crowds consisting of hundreds or thousands of individuals require simple models which nevertheless provide an accurate description of reality. One such class of models, so-called cellular automata (CA), has been studied in statistical physics for a long time [1,2].

In CA space, time and state variables are discrete which makes them ideally suited for high-performance computer simulations. However, CA modelling differs in several respects from continuum models. These are usually based on coupled differential equations which often can not be treated analytically. One has to solve them numerically and therefore the equations have to be discretized. In general, only space and time variables become discrete whereas the state variable is still continuous. One important point is now that CA are discrete from the beginning and that this discreteness is already taken into account in the definition of the model and its dynamics. This allows to obtain the desired behaviour in a much simpler way. On the other hand, the numerical solution of (discretized) differential equations is only accurate in the limit Δx, $\Delta t \to 0$. This is different in the CA where Δx and Δt are finite and accurate results can be obtained since the rules (dynamics) are designed such that the discreteness is an important part of the model.

In order to achieve complex behaviour in a simple fashion one often resorts to a stochastic description. A realistic situation seldomly can be described completely by a deterministic approach. Already minor events can lead to a very different behaviour due to the complexity of the interactions involved. For the problem of pedestrian motion this becomes evident e.g. in the case of a panic where the behaviour of people seems almost unpredictable. But also for "normal" situations a stochastic component in the dynamics can lead to a more accurate description of complex phenomena since it takes into account that we usually do not have full knowledge about the state of the system and its dynamics. Here one has to keep in mind that in general some sort of average over different realizations of the process (e.g. different sequences of random numbers) has to be taken. Even if there are single realizations which yield unrealistic behaviour the average process will be a good description of the real process. Furthermore a stochastic description allows to answer questions like "What is the probability that the evacuation of this building will take longer than 3 minutes ?" in a natural way.

In the following we will present a detailed description of the model and the basic philosophy of our approach. Applications are presented in Part II [3].

2 Other Modelling Approaches

During the last decade considerable research has been done on the topic of highway traffic using methods from physics [4–10]. Cellular automata inspired by the pioneering works [11, 12] compose by now an important class of models. Most studies have been devoted to one-dimensional systems, where several analytic approaches exist to calculate or approximate the stationary state.

On the other hand, pedestrian dynamics has not been studied as extensively as vehicular traffic, especially using a cellular automata approach. One reason is probably its generically two-dimensional nature. In recent years, continuum models have been most successful in modelling pedestrian dynamics. An important example are the *social force models* (see e.g. [7, 10, 13] and references therein). Here pedestrians are treated as particles[1] subject to long-ranged forces induced by the social behaviour of the individuals. This leads to (coupled) equations of motion similar to Newtonian mechanics. There are, however, important differences since, e.g., in general the third law ("actio = reactio") is not fulfilled.

In contrast to the social force models our approach is closer in spirit to the general strategy of modelling (elementary) forces on a microscopic level by the exchange of mediating particles which are bosons. It is therefore similar to *active walker models* [14, 15] used so far mainly to describe trail formation,

[1] In the following we use "pedestrian" and "particle" interchangeably.

chemotaxis (see [16] for a review) etc. Here the walker leaves a trace by modifying the underground on his path. This modification is real in the sense that it could be measured in principle. For trail formation, vegetation is destroyed by the walker and in chemotaxis he leaves a chemical trace. In contrast, in our model the trace is virtual. Its main purpose is to transform effects of long-ranged interactions (e.g. following people walking some distance ahead) into a local interaction (with the "trace"). This allows for a much more efficient simulation on a computer.

Cellular automata for pedestrian dynamics have been proposed in [17–19]. These models can be considered as generalizations of the Biham-Middleton-Levine model for city traffic [12]. Most works have focussed on the occurrence of a jamming transition as the density of pedestrians is increased. All models have only nearest-neighbour interactions, except for the generalization proposed in [19] which is used for analyzing evacuation processes on-board passenger ships. The other models use a kind of "sublattice-dynamics" which distinguishes between different types of pedestrians according to their preferred walking direction. Such an update is not easy to generalize to more complex situations where the walking direction can change. To our knowledge so far most other collective effects encountered empirically [7, 10, 20–23] have not been reproduced using these models. Another discrete model has been proposed earlier by Gipps and Marksjös [24]. This model is somewhat closer in spirit to our model than the cellular automata approaches of [17–19] since the transitions are determined by the occupancies of the neighbouring cells. However, this model can not reproduce all collective effects either. In [25] a discretized version of the social force model has been introduced. The repulsive potentials by the pedestrians are stored in a global potential, with pedestrians reacting to the gradients of this global potential. Although this model is able to reproduce collective effects it is not flexible enough to treat individual reactions to other pedestrians, and collision-avoidance is not always guaranteed for velocities greater than 1.

3 Basic Principles of the Model

First we discuss some general principles we took into account in the development of our model [26]. The implementation of the interactions between the pedestrians uses an idea similar to chemotaxis. The pedestrians leave a virtual trace which then influences the motion of other pedestrians. This allows for a very efficient implementation on a computer since now all interactions are local. The transition probabilities for all pedestrians just depend on the occupation numbers and strength of the virtual trace in his neighbourhood, i.e. we have translated the long-ranged spatial interaction into a local interaction with "memory". The number of interaction terms in other long-ranged models, e.g. the social-force model, grows proportionally to the square of the number of particles whereas in our model it grows only linearly.

The idea of a virtual trace can be generalized to a so-called ***floor field.*** This floor field includes the virtual trace created by the pedestrians as well as a static component which does not change with time. The latter allows to model e.g. preferred areas, walls and other obstacles. The pedestrians then react to both types of floor fields.

To keep the model simple, we want to provide the particles with as little intelligence as possible and to achieve the formation of complex structures and collective effects by means of self-organization. In contrast to older approaches we do not make detailed assumptions about the human behaviour. Nevertheless the model is able to reproduce many of the basic phenomena.

The key feature to substitute individual intelligence is the floor field. Apart from the occupation number each cell carries an additional quantity (field) which can be either discrete or continuous. This field can have its own dynamics given by diffusion and decay coefficients.

Interactions between pedestrians are repulsive for short distances. One likes to keep a minimal distance to others in order to avoid bumping into them. In the simplest version of our model this is taken into account through hard-core repulsion which prevents multiple occupation of the cells. For longer distances the interaction is often attractive. E.g. when walking in a crowded area it is usually advantageous to follow directly behind the predecessor. Large crowds may also be attractive due to curiosity.

With two particle species moving in opposite directions, each with its own floor field, effects can be observed which are so far only achieved by continuous models [13]: lane formation and oscillation of the direction of flow at doors. We consider this model to be another example for the ability of cellular automata to create complex behaviour out of simple rules and the great applicability of this approach to all kinds of traffic flow problems.

In contrast to vehicular traffic the time needed for acceleration and braking is negligible. The velocity distribution of pedestrians is sharply peaked [20]. These facts naturally lead to a model where the pedestrians have a maximal velocity $v_{max} = 1$, i.e. only transitions to neighbour cells are allowed. Furthermore, a greater v_{max}, i.e. pedestrians are allowed to move more than just one cell per timestep, would be harder to implement in two dimensions, especially when combined with parallel dynamics, and reduce the computational efficiency. The number of possible target cells increases quadratically with the interaction range. Furthermore one has to check whether the path is blocked by other pedestrians. This might even be ambigious for diagonal motion and crossing trajectories. Also higher velocity models lead to timescales which are much too small (see Sec. 4).

4 Definition of the Model and its Dynamics

The area available for pedestrians is divided into cells of approximately $40 \times 40\ cm^2$. This is the typical space occupied by a pedestrian in a dense crowd [23]. Each cell can either be empty or occupied by exactly one particle (pedestrian). For special situations it might be desirable to use a finer discretization, e.g. such that each pedestrian occupies four cells instead of one.

The update is done in parallel for all particles. This introduces a timescale into the dynamics which can roughly be identified with the reaction time t_{reac}. In the deterministic limit, corresponding to the maximal possible walking velocity in our model, a single pedestrian (not interacting with others) moves with a velocity of one cell per timestep, i.e. 40 cm per timestep. Empirically the average velocity of a pedestrian is about 1.3 m/s [23]. This gives an estimate for the real time corresponding to one timestep in our model of approximately 0.3 sec which is of the order of the reaction time t_{reac}, and thus consistent with our microscopic rules. It also agrees nicely with the time needed to reach the normal walking speed which is about 0.5 sec. This corresponds to at least v_{max} timesteps if the pedestrian can only accelerate by one unit per timestep. Therefore in models with large v_{max} a timestep would correspond to a real time shorter than the smallest relevant timescale. This makes the model more complicated than necessary and reduces the efficiency of simulations.

4.1 Basic Rules

Each particle is given a preferred walking direction. From this direction, a 3×3 *matrix of preferences* is constructed which contains the probabilities for a move of the particle. The central element describes the probability for the particle not to move at all, the remaining 8 correspond to a move to the neighbouring cells (see Fig. 1). The probabilities can be related to the velocity and the longitudinal and transversal standard deviations (see [26, 27] for details). So the matrix of preferences contains information about the preferred walking direction and speed. In principle, it can differ from cell to cell depending on the geometry and aim of the pedestrians. In the simplest case the pedestrian is allowed to move in one direction only without fluctuations and in the corresponding matrix of preference only one element is one and all others are zero. In the following it is assumed that a matrix of preferences is given at every timestep for each pedestrian. These can e.g. be obtained from some model for route selection which assigns certain routes to each pedestrian.

This ansatz can easily be extended by fixing the direction of preference for each cell separately, e.g. to handle structures inside buildings. Then the particles would use the matrix belonging to the cell they occupy at a given step. However, a similar effect can be obtained much simpler by introducing a second floor field (see Sec. 4.2).

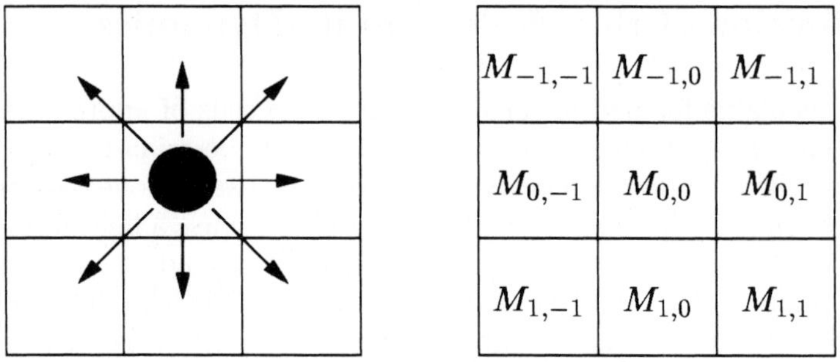

Figure 1. A particle, its possible transitions and the associated matrix of preference $M = (M_{ij})$.

In each update step for each particle a desired move is chosen according to these probabilities. This is done in parallel for all particles. If the target cell is occupied, the particle does not move. If it is not occupied, and no other particle targets the same cell, the move is executed. If more than one particle share the same target cell, one is chosen according to the relative probabilities with which each particle chose their target. This particle moves while its rivals for the same target keep their position (see Fig. 2).

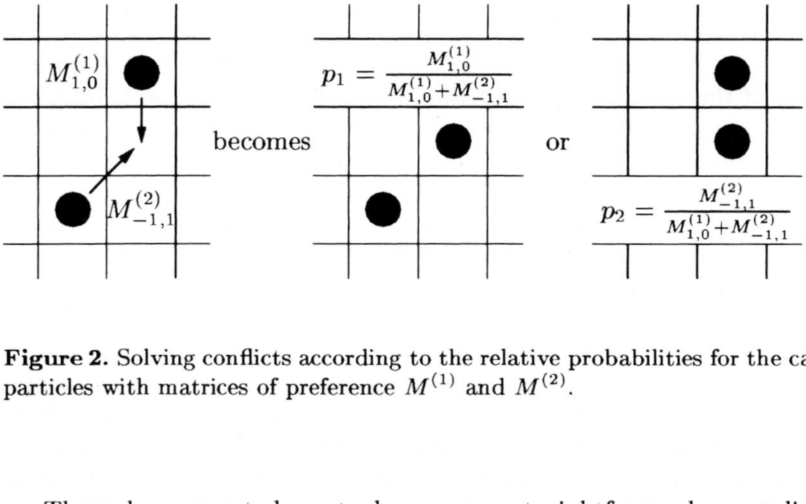

Figure 2. Solving conflicts according to the relative probabilities for the case of two particles with matrices of preference $M^{(1)}$ and $M^{(2)}$.

The rules presented up to here are a straightforward generalization of the CA rules used so far for the description of traffic flow [17, 18]. The main difference is that in principle transitions in all directions are possible and each pedestrian j might have her own preferred direction of motion characterized by a matrix of preferences $M^{(j)}$. The only interaction between particles taken into account so far is hard-core exclusion.

4.2 Floor Field

In order to reproduce certain collective phenomena it is necessary to introduce further longer-ranged interactions. In some continuous models this is done using the idea of a social force [7, 10, 13]. Here we present a different approach. Since we want to keep the model as simple as possible we try to avoid using a long-range interaction explicitly. Instead we introduce the concept of a ***floor field*** which is modified by the pedestrians and which in turn modifies the transition probabilities. This allows to take into account interactions between pedestrians and the geometry of the system (building) in a unified and simple way without loosing the advantages of local transition rules. The floor field modifies the transition probabilities in such a way that a motion into the direction of larger fields is preferred.

The floor field can be thought of as a second grid of cells underlying the grid of cells occupied by the pedestrians. It can be discrete or continuous. As already explained in Sec. 3 we distinguish two types of fields which will be called static and dynamic floor fields, respectively.

The ***dynamic floor field*** D is just the virtual trace left by the pedestrians (see Sec. 3). It is modified by the presence of pedestrians and has its own dynamics, i.e. diffusion and decay. Usually the dynamic floor field is used to model a ("long-ranged") attractive interaction between the particles. Each pedestrian leaves a "trace", i.e. the floor field of occupied cells is increased. Explicit examples where such an interaction is relevant are given in Part II [3]. The dynamic floor field is also subject to diffusion and decay which leads to a dilution and finally the vanishing of the trace after some time.

The ***static floor field*** S does not evolve with time and is not changed by the presence of pedestrians. Such a field can be used to specify regions of space which are more attractive, e.g. an emergency exit (see the example in [3]) or shop windows. This has an effect similar to a position-dependent matrix of preference, but is much easier to realize.

The ***transition probability*** p_{ij} in direction (i, j) (see Fig. 1) now depends on four contributions:

(i) the matrix of preference M_{ij} which contains the information about the aim and average velocity of the pedestrian.

(ii) the value D_{ij} of the dynamic floor field at the target cell. This contribution takes into account the effects of the motion of the other pedestrians. In many applications (see [3]) it is attractive to "follow the crowd", i.e. transitions in directions (i, j) with a large value of the dynamic floor field are preferred.

(iii) the value S_{ij} of the static floor field. It allows to model effects of the geometry. E.g. in a corridor it is usually less attractive to walk close to the walls. Such an effect can be incorporated in a static floor field which decreases near the walls.

(iv) the occupation number n_{ij} of the target cell. A motion in direction (i, j) is only allowed if the target cell is empty ($n_{ij} = 0$) and forbidden if it is already occupied ($n_{ij} = 1$).

One simple possibility to take into account all contributions (i)–(iv) is to define the transition probability in direction (i, j) by

$$p_{ij} = N M_{ij} D_{ij} S_{ij} (1 - n_{ij}). \tag{1}$$

N is a normalization factor to ensure $\sum_{(i,j)} p_{ij} = 1$ where the sum is over the nine possible target cells. There are also slightly more general forms of the transition probabilities which have been studied in [26, 27].

Since the total transition probability is proportional to the dynamic floor field it becomes more attractive to follow in the footsteps of other pedestrians. This effect competes with the preferred walking direction specified by M_{ij} and the effects of the geometry encoded in S_{ij}. The relative influence of the contributions (i)–(iii) is controlled by coupling parameters. These depend on the situation to be studied. Consider for example a situation where people want to leave a large room (see [3]). Normal circumstances, where everybody is able to see the exit, can be modelled by solely using a static floor field which decreases radially with the distance from the door. Since transitions in the direction of larger fields are more likely this will automatically guarantee that everybody is walking in the direction of the door. If, however, the exit can not be seen by everybody, e.g. in a smoke-filled room or in the case of failing lights, people will try to follow others hoping that they know the location of the exit. In this case the coupling to the dynamic floor field is much stronger and the static field has a considerable influence only in the vicinity of the door. This example will be studied in more detail in [3].

4.3 Dynamics of the Floor Field

In contrast to the static floor field S the dynamic floor field D is changed by the motion of pedestrians. Furthermore it is subject to diffusion and decay. Its dynamics consists of three steps:

(a) If a pedestrian leaves a cell (x, y) the dynamic floor field D_{xy} corresponding to this cell is increased by ΔD_{xy}. The increment ΔD_{xy} is a parameter of the model and can either be discrete or continuous.

(b) To model the diffusion, a certain amount of the field is distributed among the neighbouring cells.

(c) To model the decay of the field, the field strength is reduced by a decay constant δ.

In (a) the virtual trace left by the motion of the pedestrians is created. (b) is necessary because pedestrians do not necessarily follow exactly in the footsteps of others. Diffusion leads to broadening and dilution of the trace. (c) implies that the lifetime of the trace is finite and that it will vanish after some time. Diffusion and decay of the dynamic field lead to an effective interaction strength between the pedestrians which decays exponentially with the distance [27].

In [26] we have introduced two variants of the floor field, a discrete and a continuous one. In the discrete case the field strength D_{xy} can be interpreted as the number of bosonic particles ("bosons") at the cell (x, y). In (a) the number of bosons is increased by one. In (b) bosons can move with probability γ to neighbouring cells and in (c) bosons are removed with probability α. In the continuous case the dynamics in (b) and (c) is described by a diffusion-decay equation

$$\frac{\partial D}{\partial t} = d \cdot \Delta D - \delta \cdot D \tag{2}$$

where d is the diffusion constant and δ the decay constant. Details can be found in [26, 27].

4.4 Summary of the Update Rules

The update rules of the full model including the interaction with the floor fields then have the following structure:

1. The dynamic floor field D is modified according to its diffusion and decay rules (see Sec. 4.3).
2. For each pedestrian, the transition probabilities p_{ij} for a move to an unoccupied neighbour cell (i, j) are determined by the matrix of preferences and the local dynamic and static floor fields, e.g. $p_{ij} \propto M_{ij} D_{ij} S_{ij}$ (see Sec. 4.2)..
3. Each pedestrian chooses a target cell based on the probabilities of the transition matrix $P = (p_{ij})$.
4. The conflicts arising by any two or more pedestrians attempting to move to the same target cell are resolved, e.g. using the procedure described in Sec. 4.1.
5. The pedestrians which are allowed to move execute their step.
6. The pedestrians alter the dynamic floor field D_{xy} of the cell (x, y) they occupied before the move (see Sec. 4.3).

These rules have to be applied to all pedestrians at the same time (parallel dynamics). This introduces a timescale into the dynamics which corresponds to approximately 0.3 *sec* of real time. This allows e.g. to translate evacuation times measured in computer simulations into real times.

5 Conclusions

We have introduced a stochastic cellular automaton to simulate pedestrian behaviour. We focused here on the general concept. The effects which can be observed with the basic approach will be presented in [3] together with simple applications.

The key mechanism is the introduction of a floor field which acts as a substitute for pedestrian intelligence and leads to collective phenomena. This floor field makes it possible to translate spatial long-ranged interactions into non-local interactions in time. The latter can be implemented much more efficiently on a computer. Another advantage is an easier treatment of complex geometries. In models with long-range interactions, e.g. the social-force models, one always has to check explicitly whether pedestrians are separated by walls in which case there should be no interaction between them.

The general idea in our model is similar to chemotaxis. However, the pedestrians leave a virtual trace rather than a chemical one. This virtual trace has its own dynamics (diffusion and decay) which e.g. restricts the interaction range (in time). It is realized through a dynamical floor field which allows to give the pedestrians only minimal intelligence and to use local interactions. Together with the static floor field it offers the possibility to take different effects into account in a unified way, e.g. the social forces between the pedestrians or the geometry of the building.

In Part II [3] we will demonstrate that the approach indeed is able to reproduce the known collective effects and self-organization phenomena. Therefore the model is a good starting point for realistic applications [3, 26].

The model can also be applied to more complex geometries and various characteristics of a crowd can be simulated without major changes. So it should be possible to study the effects of panic (see [22] and references therein). In [3] we show results for simple evacuation simulations.

The description of pedestrians using a cellular automaton approach allows for very high simulation speeds. Therefore, we have the possibility to extract the complete statistical properties of our model using Monte Carlo simulations.

Finally it should be emphasized that we have presented only the basic ideas of the approach. For realistic applications modifications might be appropriate, e.g. smaller cell sizes etc. One can also introduce more than just one species of pedestrians (e.g. two groups moving in opposite directions). In this case each species interacts with its own floor field. In the simplest case these fields are independent from each other.

References

1. S. Wolfram, *Theory and Applications of Cellular Automata*, (World Scientific, 1986); *Cellular Automata and Complexity* (Addison-Wesley, 1994).
2. D. Stauffer, J. Phys. A **24**, 909 (1991).
3. C. Burstedde, A. Kirchner, K. Klauck, A. Schadschneider, and J. Zittartz, *Cellular Automaton Approach to Pedestrian Dynamics - Applications*, these proceedings.

4. D.E. Wolf, M. Schreckenberg, and A. Bachem (Eds.), *Traffic and Granular Flow* (World Scientific, Singapore 1996).
5. M. Schreckenberg and D.E. Wolf (Eds.), *Traffic and Granular Flow '97* (Springer, Singapore, 1998).
6. D. Helbing, H. J. Herrmann, M. Schreckenberg, and D. E. Wolf (Eds.), *Traffic and Granular Flow '99, Social, Traffic, and Granular Dynamics* (Springer, Berlin, 2000).
7. D. Helbing, *Verkehrsdynamik, Neue Physikalische Modellierungskonzepte*, (in German) (Springer, Berlin, 1997).
8. D. Chowdhury, L. Santen, and A. Schadschneider, Phys. Rep. **329**, 199 (2000).
9. K. Nagel, J. Esser, and M. Rickert, Annu. Rev. Comp. Phys. **7**, p. 151, ed. D. Stauffer (World Scientific, 2000).
10. D. Helbing: `cond-mat/0012229`.
11. K. Nagel and M. Schreckenberg: J. Physique I, **2**, 2221 (1992).
12. O. Biham, A.A. Middleton, and D. Levine: Phys. Rev. A **46**, R6124 (1992).
13. D. Helbing and P. Molnar: Phys. Rev. E **51**, 4282 (1995).
14. D. Helbing, F. Schweitzer, J. Keltsch, and P. Molnar: Phys. Rev. E **56**, 2527 (1997).
15. D. Helbing, J. Keltsch, and P. Molnar: Nature **388**, 47 (1997).
16. E. Ben-Jacob, Contemp. Phys. **38**, 205 (1997).
17. M. Fukui and Y. Ishibashi, J. Phys. Soc. Jpn. **68**, 2861, 3738 (1999).
18. M. Muramatsu and T. Nagatani, Physica A **275**, 281 (2000); **286**, 377 (2000).
19. H. Klüpfel, T. Meyer-König, J. Wahle, and M. Schreckenberg, in *Theory and Practical Issues on Cellular Automata*, S. Bandini, T. Worsch (Eds.), (Springer, Berlin, 2000).
20. L.F. Henderson, Nature **229**, 381 (1971).
21. L.F. Henderson and D.J. Lyons, Nature **240**, 353 (1972).
22. D. Helbing, I. Farkas and T. Vicsek, Nature **407**, 487 (2000).
23. U. Weidmann, *Transporttechnik der Fussgänger*, Schriftenreihe des IVT **80**, ETH Zürich (1992).
24. P.G. Gipps and B. Marksjös, Math. and Comp. in Simulation **27**, 95 (1985).
25. K. Bolay, Diploma Thesis, Stuttgart University (1998).
26. C. Burstedde, K. Klauck, A. Schadschneider, and J. Zittartz, accepted for publication in Physica A.
27. C. Burstedde, Diploma Thesis, Universität zu Köln (2001); available for download at `http://www.burstedde.de/carsten/diplom.html`.

Cellular Automaton Approach to Pedestrian Dynamics - Applications

C. Burstedde, A. Kirchner, K. Klauck, A. Schadschneider, and J. Zittartz

Institut für Theoretische Physik, Universität zu Köln, 50937 Köln, Germany

We present applications of a cellular automaton approach to pedestrian dynamics introduced in [1, 2]. It is shown that the model is able to reproduce collective effects and self-organization phenomena encountered in pedestrian traffic, e.g. lane formation in counterflow through a large corridor and oscillations at doors. Furthermore we present simple examples where the model is applied to the simulation of evacuation processes.

1 Introduction

In Part I [1] (see also [2]) we have introduced a two-dimensional cellular automaton model for the description of pedestrian dynamics. In the model the space is discretized into small cells which can either be empty or occupied by exactly one pedestrian. These can move to one of the neighbouring cells where the direction of motion is chosen with certain probabilities. These probabilities depend on the

1. occupation of the targets cells: Motion to a cell already occupied is forbidden.
2. matrix of preference: It encodes the direction of motion including the average velocity of the pedestrian.
3. value of the dynamic floor field: This corresponds to a virtual trace left by the other pedestrians.
4. value of the static floor field: This allows to specify preferred regions, e.g. to incorporate effects of the geometry of a building etc.

The introduction of the floor fields allows to translate the long-ranged interactions between pedestrians into local ones. Thus the model is extremely efficient in computer simulations and large crowds can be simulated much faster than real time. Despite this simplicity the cellular automaton allows to reproduce the collective effects and self-organization phenomena encountered in pedestrian traffic. Examples will be given in the following sections.

2 Self-Organization Phenomena

One of the major design principles of our model is to provide the simulated individuals with as little intelligence as possible. There have been approaches[1] where the pedestrians examine their surroundings or look ahead a certain amount of grid cells in order to decide where and how far to move. It has also been proposed to select a target cell and repeatedly choose a different one if it is found occupied.

These approaches imply the need for more complicated algorithms which reduce the simulation speed. Furthermore, it inhibits the unambiguous definition of the update procedure used. We avoid this strategy and rely on the concept of self-organization to model pedestrian behaviour.

We argue that simple, local update rules for the pedestrians and optionally one or more floor fields are sufficient to yield a richness of complex phenomena. Obviously this route is superiour concerning the computational efficiency and even allows for faster-than-real-time simulations of large crowds, e.g. in evacuation processes in public buildings.

2.1 Lane Formation

The most prominent collective phenomenon is the formation of lanes out of an unordered group of pedestrians [5]. This corresponds to a spontaneous breaking of the symmetry of the particle number distribution in space. Our simulations show that an even as well as an odd number of lanes may be formed. The latter corresponds to a spontaneous breaking of the left-right symmetry of the system.

We present simulations of a rectangular corridor which is populated by two species of pedestrians moving in opposite directions (see Fig. 1). Parallel to the direction of motion we assume the existence of walls. Orthogonal to the direction of motion we investigated both periodic and open boundary conditions. The length of the corridor is set to 200 cells. Widths of 15 and 25 cells have been used.

With periodic boundary conditions, the density of pedestrians is fixed for each run. It is ensured that the overall number of pedestrians is evenly divided by the numbers for the different species: with two species, one is moving to the left and the other to the right. For open boundaries, we fix the rate at which pedestrians enter the system at the boundaries. The pedestrians leave the system as soon as they reach the opposite end of the corridor.

Fig. 1 shows the graphical frontend running a simulation of a small periodic system. The lanes can be spotted easily, both in the main window showing

[1] For a more complete discussion of other modelling approaches, see [1] and references therein.

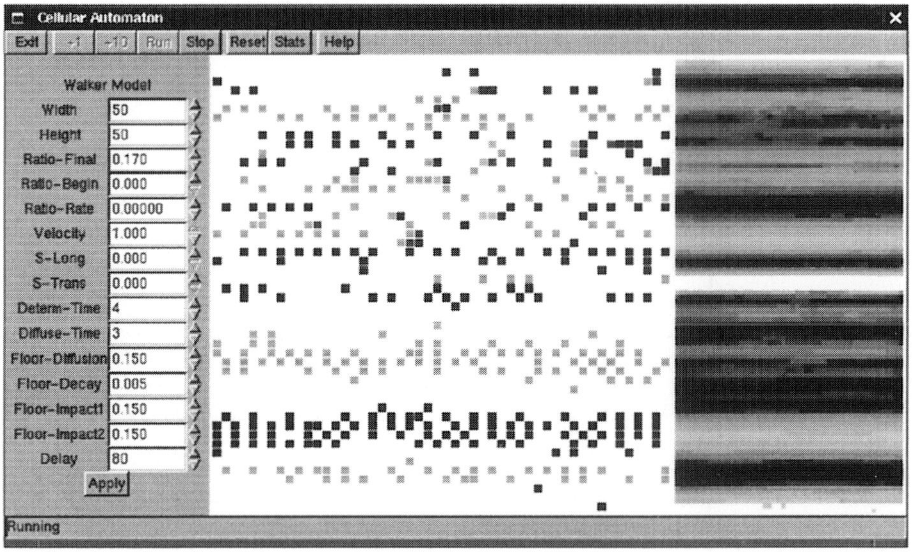

Figure 1. Snapshot of a simulation with $\rho = 0.17$, $w = h = 50$. The left part shows the parameter control. The central window is the corridor and the light and dark squares are right- and left-moving pedestrians, respectively. The right part shows the floor fields for the two species.

the cell contents and the small windows on the right showing the floor field intensity for the two species.

This model clearly provides the option for a complete jam. The jamming probability with periodic boundaries at constant density increases with the length of the system. An open system can be thought of as the limit of an infinitely long periodic system, although density and entry rates do not correspond absolutely.

We performed several runs for different densities and insertion rates, respectively. The focus of our attention is the parameter range where the transition from a stable flow to a complete jam takes place.

To obtain information about the lanes we measured the pedestrian velocities at a cross section perpendicular to the direction of flow. Selected velocity profiles are shown in Figs. 2–4.

It is obvious that the lane formation in the periodic system works far better than in the open system. The floor field leads to an effective attraction of identical pedestrians while different pedestrian species separate. This results in the formation of a stable pattern in the periodic case.

In a certain density regime, these lanes are metastable. Spontaneous fluctuations can disrupt the flow in one lane causing the pedestrians to spread and interfere with other lanes. Eventually the system can run into a jam by

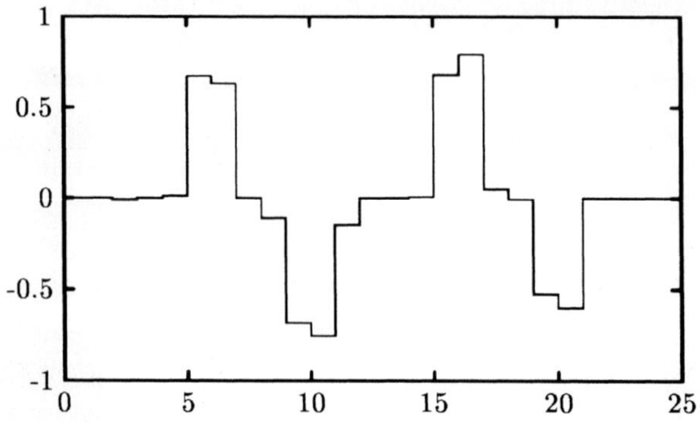

Figure 2. Velocity profile of a periodic system with $\rho = 0.12$.

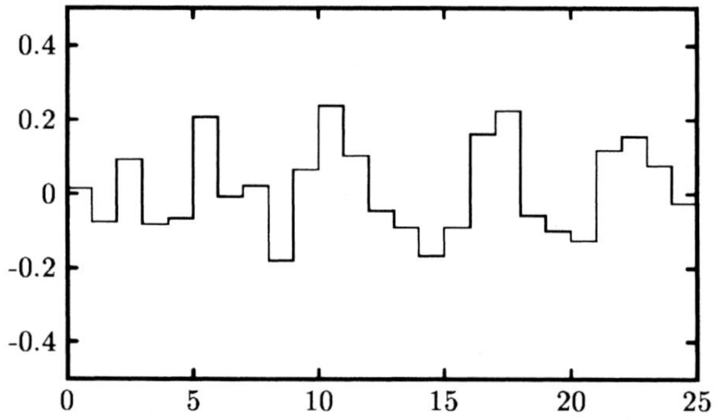

Figure 3. Velocity profile of an open system at $x = \frac{L}{2}$ with $\alpha = 0.04$.

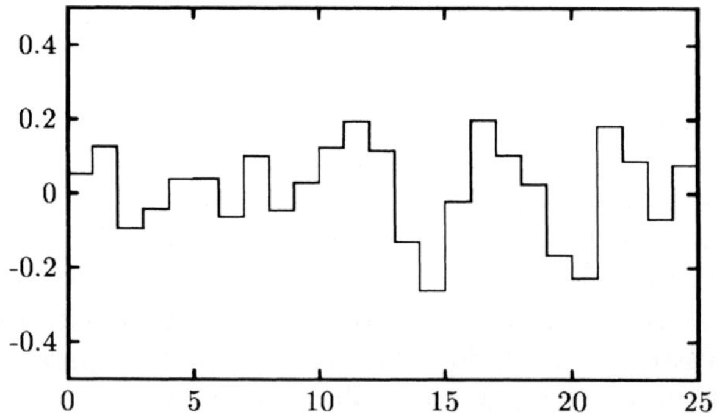

Figure 4. Velocity profile of an open system at $x = \frac{L}{4}$ with $\alpha = 0.04$.

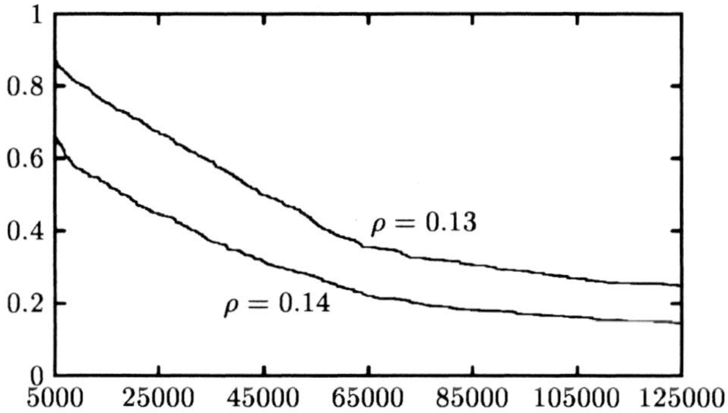

Figure 5. The fraction of systems with nonzero flow versus the simulation time for two densities 0.13 and 0.14.

this mechanism. The average time after which the system is blocked by a jam is an interesting observable which depends on the density of pedestrians. We observe large fluctuations of this quantity which require many samples to find statistically significant information.

To create Fig. 5 we sampled 700 initial conditions for each of two densities and let them evolve for at most 125,000 timesteps. The diagram shows how the percentage of systems with nonzero flow (i.e. jam-free systems) decreases with time. In this regime, the density has a big impact on this quantity.

2.2 Different Geometries

Our model can easily be generalized for the use with different geometrical setups. The transition rules for the particles do not need to be altered, and the computational speed does not suffer.

In Fig. 6 a wall with two doors has been added to the system. It is oriented perpendicular to the direction of flow. With the appropriate choice of the doors' width and the distance between them, they support the separation of the two particle species.

Another feature of pedestrian behaviour are oscillations of the direction of flow when two groups of opposite walking direction are facing each other at a tight spot like a narrow door [5]. If we simulate this setup using our model, we can observe oscillation on two time scales (provided that the parameters are suitably chosen): An exchange between a blocked situation and a flow in both directions is the main result. Inside the blocked period small groups of only one species can break through. This is illustrated in Fig. 7. These breaks can alternate between the two species and can therefore be interpreted as oscillations.

To simulate a crossing, four species of particles are inserted into the system. Their directions of preference differ by 90 degrees. The system is surrounded

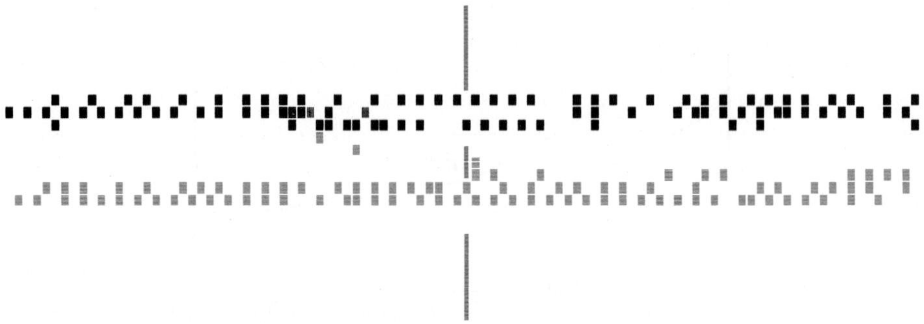

Figure 6. Separation of two species of opposite walking directions into two separate lanes catalyzed by two doors.

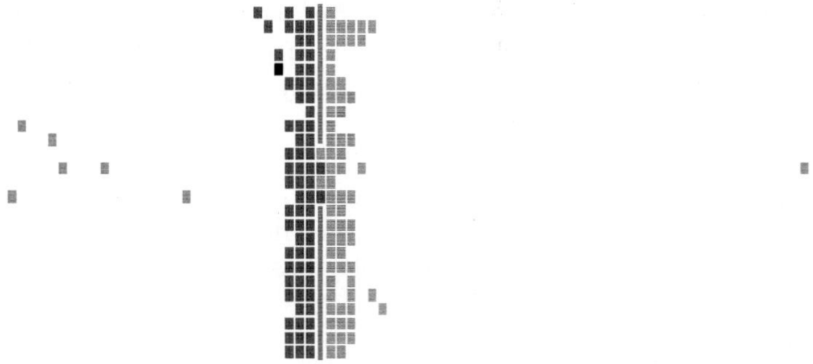

Figure 7. Oscillations of the direction of flow: A group of particles of the same species break through a blockade at a door.

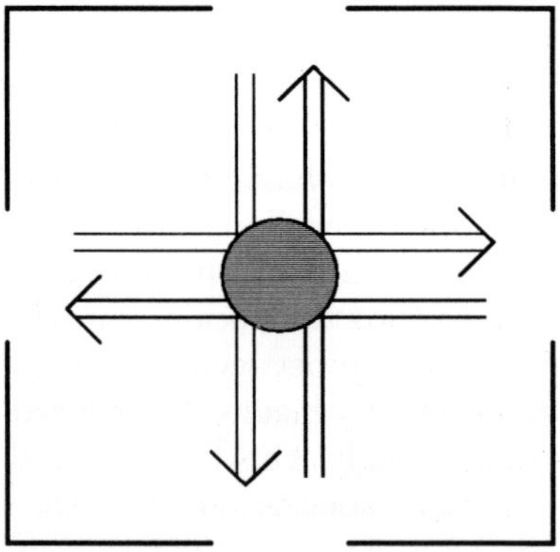

Figure 8. Typical flow pattern which emerges at a crossing.

by walls which have doors in the middle through which the system is closed periodically. Several flow patterns arise from these boundary conditions; the most common one is shown schematically in Fig. 8. In each of the two roads we observe the formation of lanes as discussed in this section, while the orientation is different: In one the particles tend to walk on the right side, in the other on the left. The region of highest disorder lies in the center of the system where the two roads meet. It is marked with a shaded circle.

3 Evacuation Simulations

In the following we describe results of simulations of a typical situation, i.e. the evacuation of a large room (e.g. in the case of a fire). We try to show the relevance of our model for practical applications. It is able to cope with the complex geometrical structures of any given environment and is therefore of use for architectural planing and safety estimations of evacuation times of arbitrary buildings. First we study a large room without any internal structure (e.g. a ballroom) and just one door. As an example for a more complex surrounding we investigate in Sec. 3.2 the stylized geometry of a typical lecture hall. It will be shown that our model captures the main aspects of the dynamical behaviour of large crowds even in such a complex situation.

3.1 Evacuation of a Large Room

Fig. 9 shows a large room, e.g. a ballroom, with one door only. We have simulated the behaviour of more than 100 people, which are initially distributed randomly, leaving this room. It is assumed that they have no knowledge of the exact location of the exit. The only information they get is through the floor fields. The static floor field has been chosen such that its strength decreases radially from a maximum value at the door to zero at the corners opposite to the door. This is already sufficient to achieve a complete evacuation of the room. The measured evacuation times depend on the coupling parameters to the static and dynamic floor fields and the parameters specifying the evolution of the dynamic floor field. Without any fine-tuning of these parameters it is already possible to find reasonable evacuation times.

Fig. 10 shows the situation in the room after several timesteps of the simulation. The left picture shows the people crowding in front of the door in the typical half-circle way. The right picture depicts the underlying dynamical field, i.e. the virtual trace left by the pedestrians. The main traces in the direction of the door are easily spotted.

One interesting result of these simulations concerns the influence of the attractive interaction of the pedestrians. We have found that fluctuations in the measured evacuation times become much more dominant if the coupling to the static field is small [2]. In this case they mainly follow the trace of

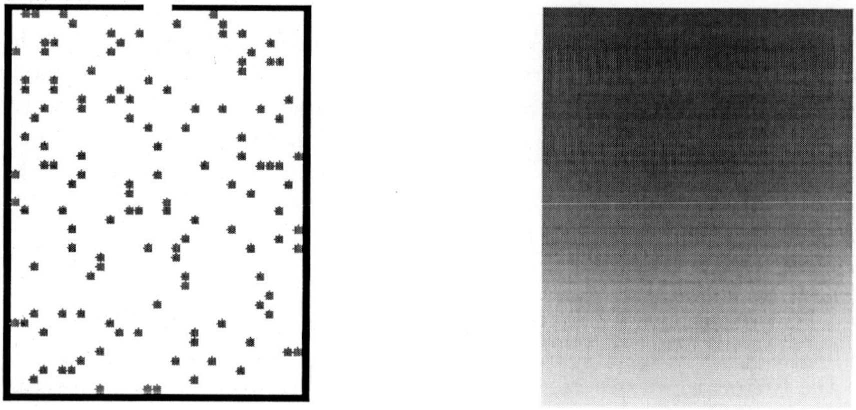

Figure 9. Ballroom at $t = 0$ (left) and static floor field (right). The strength of the field is proportional to its darkness.

other people in the hope that they know the way towards the exit. This corresponds to a situation where the exit can not be seen, e.g. due to failing lights or if the room is smoke-filled. This shows that for the interpretation of evacuation simulations just studying average evacuation times might lead to wrong conclusions.

3.2 Evacuation in Complex Geometries: Lecture Hall

The typical feature of a lecture hall is its compartimentation through large tiers, which act as non-traversable obstacles for the audience. Starting point for the simulations at $t = 0$ is a fully occupied lecture hall, i.e. all avaliable seats are taken (see Fig. 11). The audience then tries to leave the hall through a single door at the front of the room. As in the simpler example of Sec. 3.1 the main guidance in the direction of the door is accomplished by the static floor field described above. The rather complicated geometrical structure of the situation involves some sophisticated problems in the definition of this field. On the one hand, people have to orientate themselves into the direction of the door. On the other hand, they have first to leave the tiers (which is always possible in two directions). Without going into detail one can say that these problems can be solved by the superposition of different suitable static fields. The resulting field is shown in Fig. 11.

The size of the lattice is 33×33 cells with 312 particles on it. Fig. 12 shows typical stages of the dynamics of the model (the left picture displays the hall after 134 and the right picture after 590 timesteps). In this example we have neglected the coupling to the dynamical floor field and therefore the whole evacuation process is driven solely by the static field.

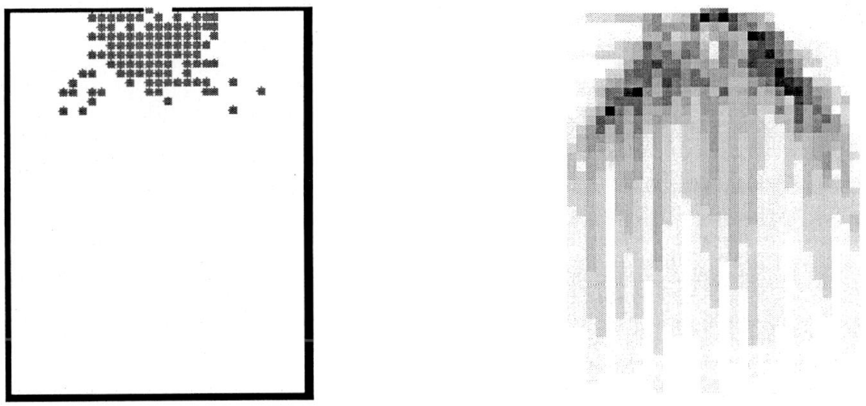

Figure 10. Evacuation of a large room: Typical configuration at an intermediate time (left) and the dynamical floor field created by the people leaving the room (right).

As one can see in these pictures the model is able to describe such a complex evacuation procedure in a very realistic way. The majority tries to leave the hall through the main corridor in the middle of the room. Therefore it is very crowded and the typical jammed motion is easy to spot. However some people try to reach the front door by detouring through the smaller corridors at the sides of the hall, as in reality.

The whole evacuation process takes about 650 timesteps which corresponds to approximately 3 minutes of real time. This seems to be a rather realistic result which has been obtained without any fine-tuning of parameter values.

This simple example already shows that our model is able to describe rather complex behaviour in complicated geometries. Therefore it can play its part in the field of the managemet of evacuation processes. An example for this will be given in the next section.

3.3 Optimization of Evacuation Times

In this subsection we present a simple application of our model in the field of risk analysis for the design of large buildings. Due to the high speed of a single simulation one is able to average over a large number of different evacuation scenarios (which are realised through different sets of random numbers). Therefore it is possible to make predictions about average evacuation times (and its fluctuations!) of different building geometries. As an example for such measurements we choose a slightly different situation then presented in the last subsection. Starting point is again a stylized fully occupied lecture hall, but now with two doors. In the first floor plane the doors are located at the

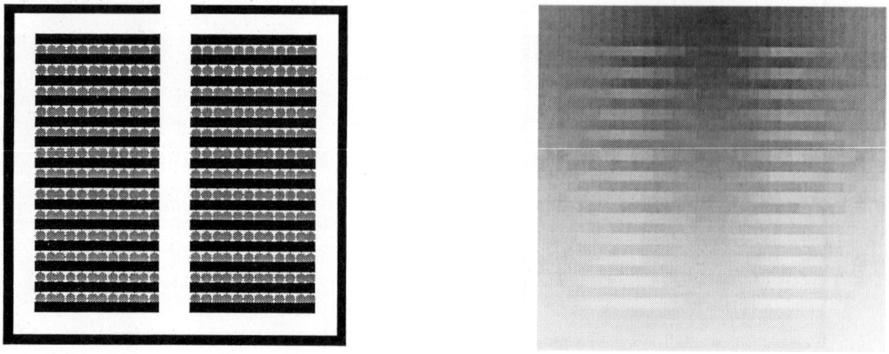

Figure 11. Lecture hall at $t = 0$ (left) and the static floor field used in the simulations (right).

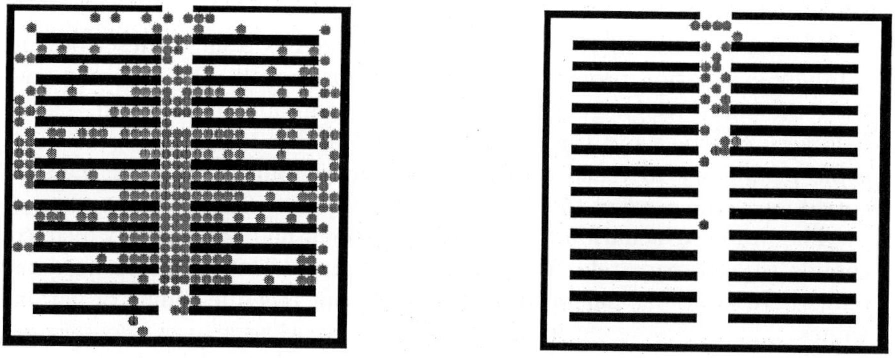

Figure 12. Typical configurations at later times.

front and the bottom of the hall (this geometry will be named "hall A"). In the second floor plane they are placed at the right and left side ("hall B"). These two different geometries are pictured in Fig. 13. We have measured the evacuation times for the 312 people of the fully occupied rooms. The values for the mean evacuation times T and the corresponding variances σ are listed in the table below (in update steps).

	hall A	hall B
T	560	363
σ	85	24

Though this result seems to be rather obvious through common sense, this example shows nevertheless that our model and its rich variations are very well

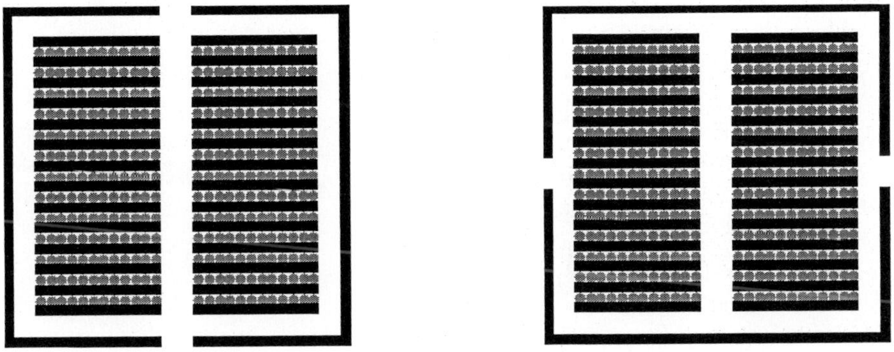

Figure 13. The two floor planes for hall A (left) and hall B (right).

suited to contribute to investigations in the field of risk analysis of evacuation processes.

4 Conclusions

We have presented simple examples for applications of a new two-dimensional cellular automaton model for pedestrian dynamics. In contrast to other CA approaches the model is able to reproduce the collective effects which are characteristic for pedestrian motion.

We can determine the complete statistical properties (including variances) of observables like evacuation times. This knowledge is of major importance if one wants to establish risk management techniques that are nowadays used for the hedging of financial assets all over the world [6].

Due to the high computational speed (simulations can be run 10 to 100 times faster than real time) our model is applicable to time-critical tasks like emergency management.

References

1. A. Schadschneider: *Cellular Automaton Approach to Pedestrian Dynamics - Theory*, these proceedings.
2. C. Burstedde, K. Klauck, A. Schadschneider, and J. Zittartz: accepted for publication in Physica A.
3. C. Burstedde: Diploma Thesis, Universität zu Köln (2001); available for download at `http://www.burstedde.de/carsten/diplom.html`.
4. D. Helbing, I. Farkas and T. Vicsek, Nature **407**, 487 (2000).
5. D. Helbing and P. Molnar, Phys. Rev. E **51**, 4282 (1995).
6. J.-P. Bouchaud and M. Potters, *Theory of Financial Risk* (Cambridge University Press, 2000).

References

1. [illegible]
2. [illegible]
3. [illegible]
4. D. Helbing, [illegible]
5. D. Helbing and [illegible] Rev. E 51, 4282 (1995)
6. [illegible] and M. [illegible] University Press (2000)

Space Syntax Based Agent Simulation

A. Penn and A. Turner

Bartlett School of Graduate Studies, UCL, Gower Street, London, WC1E 6BT, UK

Space syntax derives from a set of analytic measures of configuration that have been shown to correlate well with how people move through and use buildings and urban environments. Space syntax represents the open space of an environment in terms of the intervisibility of points in space. The measures are thus purely configurational, and take no account of attractors, nor do they make any assumptions about origins and destinations or path planning. Space syntax has found that, despite many proposed higher-level cognitive models, there appears to be a fundamental process that informs human and social usage of an environment. In this paper we describe an exosomatic visual architecture, based on space syntax visibility graphs, giving many agents simultaneous access to the same pre-processed information about the configuration of a space layout. Results of experiments in a simulated retail environment show that a surprisingly simple 'random next step' based rule outperforms a more complex 'destination based' rule in reproducing observed human movement behaviour. We conclude that the effects of spatial configuration on movement patterns that space syntax studies have found are consistent with a model of individual decision behaviour based on the spatial affordances offered by the morphology of the local visual field.

1 Introduction

Understanding the way in which pedestrians move around their environment is important for predicting congestion, for evacuation planning, pedestrian traffic and crowd control, as well as assessment of the social and economic function of buildings and urban layouts. In addition, there has been recent interest generated in the computer graphics community due to the maturing of large-scale visualisation using human avatars. Two main approaches have been developed to address the issue of pedestrian movement analysis, modelling and simulation. These might be defined simply as 'configurational analysis' - to cover methods based on representing and quantifying aspects of the spatial configuration or morphology of the environment within which movement takes place, and 'pedestrian simulation' - to cover methods that seek to represent the individual pedestrian or the pedestrian population.

Helbing et al. [1] review various methods of pedestrian simulation in detail. They categorise approaches into two broad groups: large-scale urban simulation and low-level building micro-simulation. For urban modelling, pedestrian movement may be incorporated into Land-Use and Transportation Models (LUTMs), the chief example being TRANSIMS [2,3], where tens of thousands of agent journeys may be simulated concurrently, while small-scale urban and building micro-simulation models are constructed for fire evacuation (for example, Drager et al. [4], Galea et al. [5]) and crowding situations (for example, Helbing et al. [6]). At both levels of resolution, the researcher explicitly chooses the origin-destination pairs for agents, albeit at very different scales. In LUTMs, the movement patterns are calibrated using empirical data, while at the micro-simulation level ideal paths to the destination are generally chosen. Although there are sophisticated methods for determining route choice behaviour (see for example Borgers and Timmermans [7,8] for work in this area), the missing link appears to be an intermediate level of model or simulation where the origin-destination pairs are not pre-programmed, and where there is neither calibration nor a predefined path rationale. There is currently much interest focussed on this aspect of simulation of 'unprogrammed' movement, especially for applications where movement is expected to be exploratory rather than purposive, or where the population might be expected to have a wide variety of purposes and data on these are lacking.

These issues are being approached in different ways by researchers spanning computer graphics and planning. Kerridge [9] has proposed agents with limited environmental perception of their immediate vicinity, by superimposing a grid on space and organising obstructions, so that an agent can employ local behaviour rules. For vehicular micro-simulations these rules include minimum car-to-car separations and speed matching in the one-dimensional queue, for pedestrians similar rules control relative positions in two dimensions. In the field of computer graphics, Thomas and Donikian [10] apply agents that have a ground level knowledge of the environment through a combination of a visibility graph joining sight-lines between key locations and Voronoi diagrams defining routes around the key locations. The agents may then roughly follow the edges of these graphs to provide human-like behaviour. However, the aims of the computer graphics community are somewhat different to those of the simulation and modelling communities in that the requirements for 'believable' behaviour simulations are more qualitative than quantitative.

The main methods of configurational analysis are those based on space syntax analysis defined by Hillier and Hanson [11]. These methods differ substantially in both their aims and their methodology from the simulation and modelling approaches outlined above. Initially, space syntax methods were developed in order to allow architectural space to be represented and its pattern properties quantified so that comparisons could be made between differently designed buildings or urban areas. The aim of the research was to develop an understanding of the way that spatial design and social function were related. It was soon found that a primary effect of spatial configuration on social function resulted from the way that space patterns determined pedestrian movement patterns and so co-presence between people in space [12,13]. These findings show that between 50 and 80% of the variance in pedestrian flows from location to location in an environment can be explained in terms of variations in configurational properties of those locations in the network. Since co-presence is a prerequisite for both communication and transaction in socio-economic life, space syntax research suggests that spatial configuration itself plays a critical role in determining this, but until recently it has been unclear how mechanisms at the level of the individual might give rise to these population level effects. With the exception of some previous research by one of us [14] relatively little effort has been made until now to merge syntax analysis with agent simulation, however, in view of the need to investigate individual level mechanisms that might give rise to the observed population level behaviours we have recently returned to simulation methods [15].

In this paper, we describe recent research in which space syntax is being incorporated into agent micro-simulation models for pedestrian movement. In the following section, we discuss the background of space syntax research to date, and contrast it to the assumptions made by current modelling and simulation approaches. We describe a recent study of a department store by way of example of the application of space syntax methods. Then we turn our attention to how space syntax can be used as the basis for an agent simulation in the form of an exosomatic visual architecture or 'EVA'. An EVA is a computer architecture that contains preprocessed visual information about the environment which agents access via a lookup table. We discuss the implementation of this architecture in our computer simulation 'EVAS' and show how this is being applied to simulate shopper movement patterns. We conclude with a discussion of the way this architecture, on the surface no more than an elegant algorithm to provide agents with elementary vision, takes us back to the principles of space syntax, and possibly offers an insight into the way individual mechanisms give rise to population level effects.

2 Space Syntax Analysis

Modelling and simulation approaches to human movement have tended to make two main assumptions. First that movement is purposive, and second that humans have a good (often perfect) knowledge of the environment that they are traversing. Purposive movement results in models that explicitly represent the origins and destinations, or the attractors and generators of movement. The assumption of perfect knowledge results in rational choice and assignment models in which agents' route choice is either predefined, or can be assigned dynamically in order to optimise some measure of cost such as travel time. The findings of research using space syntax techniques raise doubts about the central role both of these assumptions play in our understanding of pedestrian movement behaviour.

Space syntax methods were first developed to compare the similarities and differences between built environments at both building interior and urban neighbourhood scale. The plan of an environment is represented as a map in which all longest lines of sight are drawn. This map is then translated into a graph in which a line is represented as a node and intersections between lines are shown as links between nodes. Measures of the graph are made that can then be assigned back as variables associated with the location of each line in the original map (Fig. 1).

Figure 1: Line map analysis of London, the colours code from red through blue in the spectrum, for shallow to deep mean depth in the graph.

Research using these techniques has found that both pedestrian and vehicular movement rates are strongly correlated with certain measures of the graph of the line map (Fig. 2). Since the representation captures nothing except the geometry of the configuration of space in the environment, its ability to predict movement rates brings into question the degree to which the location and strength of attractors or generators of movement are central to observed movement behaviour patterns. Hillier et al. have argued that the logical view is that configuration leads to a pattern of movement or a 'passing trade', this then attracts shops to locations where they can take advantage of the passing trade, and the shops then attract additional people [13]. This causal chain involves a feedback loop that Hillier et al. suggest is fundamental to the evolution of urban form and the aggregation of land uses and land values. However it is also clear that the driver for this causal chain is the configuration of space, since countless examples of the failure of shops placed at the heart of modern housing estates in order to 'enliven' them show that the attractor effects of specific land uses fail to overcome spatial isolation.

More recently new methods of representation have been developed which replace the 'line map' with a grid of points within open space, and build a visibility graph in which points are linked if they are visible to each other [16]. Measures of the global properties of the graph proceed as before. Visibility Graph Analysis (VGA) of this sort gives a much higher resolution analysis of the configurational properties of open space, and in certain cases has been found to significantly better predict movement patterns than the previous line based analysis.

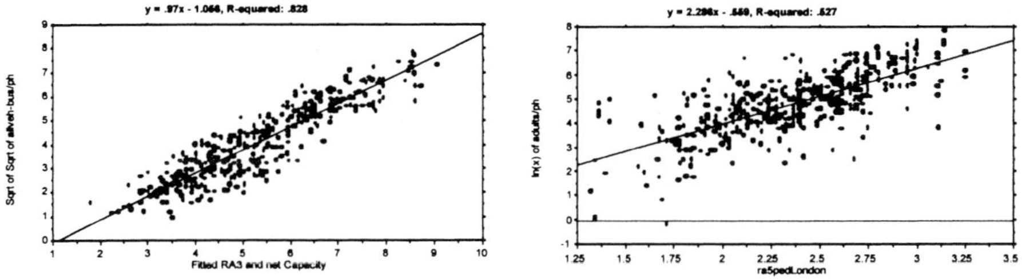

Figure 2: (Left) Correlation between vehicular flow rates and a fitted variable consisting of radius 3 mean depth in the graph, and net road width in metres (r^2=.83 n=405, p<.0001); (Right) Correlation between radius 5 mean depth and adult pedestrian movement per hour (r^2=.53, n=466, p<.0001), both scatters are for all road segments in five study areas in central London.

A specific example of this is given by a recent study of shopper behaviour in a large London department store (Fig. 3).

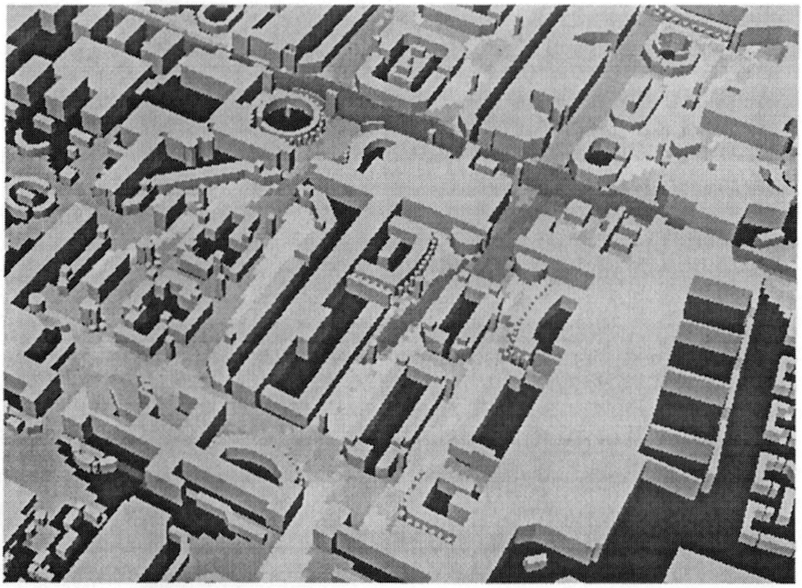

Figure 3: VGA analysis of a London department store. Colours signify mean depth in the graph, red is shallow, blue is deep.

In this study we were asked to investigate the effects of congestion on sales and till receipts. The problem faced by the store owners was simple. The number of shoppers entering the store on a Saturday was double that on a weekday, however the till receipts only increased by 60%. It was thought that the problem was due to a reduction in 'passing trade' within the store caused by congestion reducing movement. In our study we carried out detailed observations of shopper movement and stopping patterns, stalking individual customers, and observing average movement rates in all parts of the store. We also observed congestion patterns and browsing and shopping behaviour. All of these data were then put together with information on sales and receipts from the Electronic Point of Sale (EPOS) system and with the VGA data on the configuration of the building plan, and analysed statistically.

Our findings were very clear. The measure of mean depth in the VGA is strongly correlated with observed movement flows (Fig. 4 left), however the number of items sold at a till is not directly correlated with movement past the till, but is correlated with static occupancy in the vicinity of the till (Fig. 4 right). Further investigation allowed us to develop a better understanding of what contributed to stopping behaviour, which was the key to selling goods. Static occupancy was found to be associated with two factors, levels of movement and the measure of the linear metres of goods on display in the immediate area.

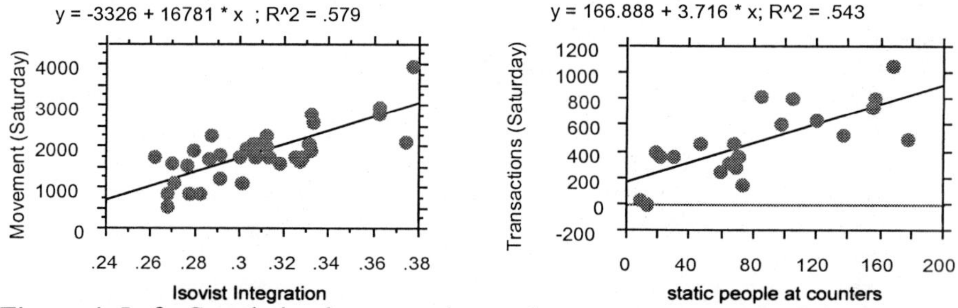

Figure 4: Left: Correlation between observed movement mean depth in the VGA (r^2=.58, p<.0001); Right: Correlation between number of transactions and numbers of static people within a buffer of the counter (r^2=.54, p<.0001).

This was an important finding since there are two camps amongst retailers. The 'stack it high and sell it cheap' camp is founded on the notion that the goods and price competition attract shoppers, whilst the 'passing trade' camp is founded on the notion that movement, and therefore circulation space, is fundamental. What we found suggests that both camps are in fact correct. A shop needs a constant supply of moving people past its goods, and the spatial layout of the store is a critical component in achieving this, however movement alone does not make a sale. In order to sell one needs to stop people so that they can browse and select what they want to buy. Stopping people depends on the supply of people in the first place (and so on movement) and on the quantity of goods on display.

These findings allowed us to clarify the nature of the problem faced by the store management. First, they were correct that congestion was having an effect, however, they were wrong about the mechanism involved. It was not that congestion stopped movement and so reduced the passing trade. It was something like the opposite: congestion appeared to be stopping people from stopping – the pressure of congestion kept people moving and so interfered with the conversion of passers-by into shoppers. Thus, determining where crowding will occur is of prime importance in analysis of shopping environments.

It is important to note that these findings are analytic rather than simulation based, and that the analysis of spatial configuration using the VGA method provides a good predictive model for movement. This account of movement incorporates only information on the configuration of space, and has no input on the location of attractors or generators of movement, no representation of individuals and their intentions, or of origins and destinations. However, as soon as one wants to take account of the way that stopping behaviour occurs, or to go a step further and look at the effects of congestion on stopping and shopping behaviour, one needs to simulate individual behaviour and interactions. For this reason we have now gone on to develop micro-simulations of individual movement behaviour specifically aimed at situations, such as shopping, where people browse or have multiple destinations, rather than move on routes with well characterised origins and destinations.

3 Implementation

Our simulation approach starts from what we have found in our space syntax based analytic studies. This is that movement patterns are strongly affected by spatial configuration. We have therefore developed an agent simulation architecture based on the visibility graph analysis, in which agents have access to pre-computed information about what is visible from any given location in the map. In addition information can be attached to the nodes of the graph describing attributes of the visible nodes. Amongst these attributes are space syntax measures of the configuration properties of the graph at each node, but we can also attribute information regarding goods on display or other static aspects of the environment. We call this architecture an 'exosomatic visual architecture' since it in effect provides agents with a form of exosomatic (outside the body) memory common to all agents in an environment [17].

The idea of pre-computing a look-up table to represent static parts of the environment has been developed by others (see for example Tu [18]). However, the system developed here differs in that it uses the visibility graph as the basis for the look-up table, and for computation of global spatial relations in the environment. The key factor here is that by using the visibility graph and by computing metrics of that graph (including those used in the space syntax approach), the look-up table encodes not only object locations, but also information about the accessibility structure of the environment. This means that in effect the agents can infer the affordances of the environment, or at least information on the global spatial relations of different locations visible from their current position in the environment.

This allows rules governing agent movement not only to read local information from the look-up table, but also to use that to fulfil global intentions. It is this property that converts the look-up table architecture from a method for representing agent perception to one for representing a form of agent memory. Perception is entirely local and a look-up table (such as [18] cited above) that gives information only about what can be seen directly, essentially only replaces perception. However, in this system the visibility graph allows the look-up table to do three additional things. Firstly, it can store extended local, telling the agent about spaces within their field of view with high potential for further movement. Secondly, it can store global information - for example, the global mean depth of all locations visible from the agent's point of view. Thirdly, it allows the entire graph to be traversed, and so for the computation of rational routes to remote locations.

By giving agents access to the VGA graph and associated attribute data we are effectively giving them a form of vision, but without the computational overhead generally associated with even rudimentary forms of agent perception [19]. However, since the graph captures global information we are also giving them a form of cognition or memory. Our implementation of this architecture uses a 'bin' representation of the environment. The visibility graph is used to allocate point values to locations, and also provides a technique to retrieve data quickly for the visual cone of an agent. For example, for each location (each node in the graph) one stores a set of 'bins', containing the nodes visible in different directions (see Fig. 5). Each bin simply contains a set of reference numbers, indicating other nodes. Our current implementation has 32 bins giving the agents a visual acuity of about 11°. By selecting the number of bins an agent has access to one can give the agent a variable cone of vision.

As well as a set of bins at each location, a set of attributes is stored for each location, and these attributes may be retrieved with the bin information. One attribute of the visibility graph that appears to be potentially useful is the 'clustering coefficient'. The clustering coefficient is a simple measure of a graph, first proposed by Watts and Strogatz [20] to measure 'small world' social networks. If one defines the 'neighbourhood' of a node in the graph to be the set of nodes immediately connected by a link to the current node, then the clustering coefficient is the number of links between all members of the neighbourhood divided by the total number of links that could possibly exist given that number of nodes. The clustering coefficient in a visibility graph gives an idea of how much visual information is lost when one moves away from a specific location to any neighbouring location. We have previously suggested therefore that it gives a measure of 'junctioness' of a location [16]. We propose that humans are probably well able to recognise junctions (from years of 'training' of what a junction looks like). We also suspect that humans typically search for promising junctions when seeking out new parts of an environment. Other research in this field by Conroy has found that human subjects in immersive virtual environments pause and scan the environment at junctions, but move more directly on segments of the environment between junctions [21]. Hence, we propose that an agent that searches out junctions is likely to resemble a human moving around a layout in exploratory mode.

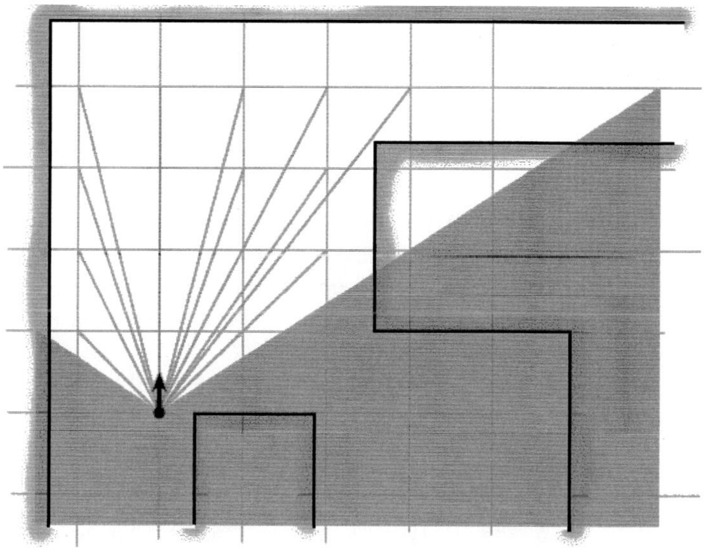

Figure 5: Using the visibility graph to select possible locations for a next target destination.

Using the clustering coefficient to guide an agent eliminates the need for the agent to have a highly developed cognitive map of an environment - all the agent has to do is be able to recognise junctions and to have a rough map of say, previously visited areas, for it to behave in a sensible exploratory fashion. In fact this seems to provide a system that will achieve globally sensible and rational looking movement patterns while employing minimal rule sets.

Of course, the clustering coefficient is just one measure of a visibility graph, and many other metrics may be used to determine junctioness, or any other feature of the space that might be considered important for agent wayfinding. We make no assertion here that the way that real humans actually wayfind is associated with either clustering coefficients or with the visibility graph. However, first we must investigate whether a rule as simple as that described above produces agent behaviour that is in any way similar to real observed human movement patterns.

4 Results

Figure 6 illustrates the cumulative movement traces for real observed shoppers in the department store described above. These observations, together with sampled shopper flow counts at 49 'observation gate' locations concentrated within the 'food halls' area of the store (illustrated in Fig. 6) are used to evaluate agent performance in our simulations. For each 'gate' location the flow of shoppers and of agents is compared statistically. Fig. 7 shows the correlation between agent movement and observed shopper movement rates, using agents that headed for areas of low clustering coefficient as described above. Although there is a positive trend, the correlation is disappointing at $r^2 = .13$.

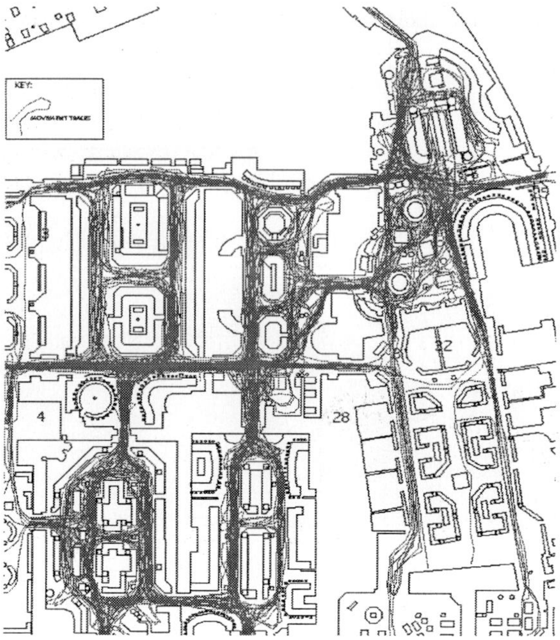

Figure 6: Observed movement trails for shoppers on a Saturday in the 'food halls' area of the store.

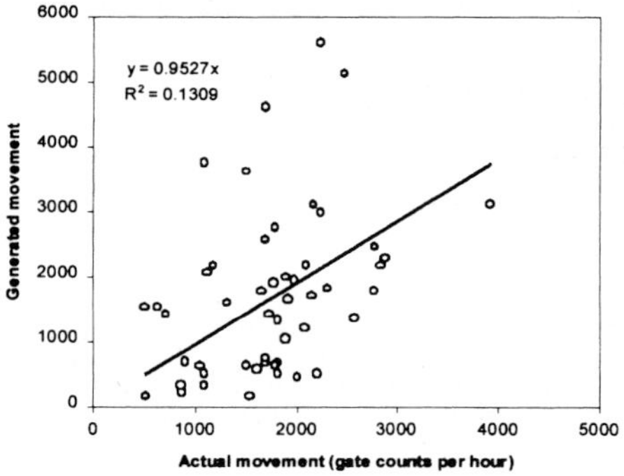

Figure 7: The correlation between agent (generated) movement using agents with the clustering coefficient rule, and observed shopper movement in the department store (r^2=.13, n=49, p<.05).

In view of this poor result we decided to try a simpler method similar to that devised by Mottram [15] for animating virtual environments with moving people. In our implementation a node is selected at random from all those within the agent's 170° view cone. Their next destination is selected at random from all visible nodes within that view cone. A new destination node is selected every few steps giving an opportunity to change direction. Fig. 8 shows the rule set in diagrammatic form.

Agents were loaded into the system at the main entrances to the building, using observed flows of people to weight the loading at the different entrances. One agent enters every other iteration of the system. Each agent is removed after 2000 generations. The system was run for 10000 iterations, but the first 2000 iterations were discarded to allow the system to reach a steady state. The results of this experiment were far more interesting. Fig. 9 is a scattergram showing the correlation between agents and observed shoppers at $r^2 = .56$. This is a surprisingly powerful result given that the rule is essentially random, and no account is taken in the simulation of the attractor properties of goods in the store.

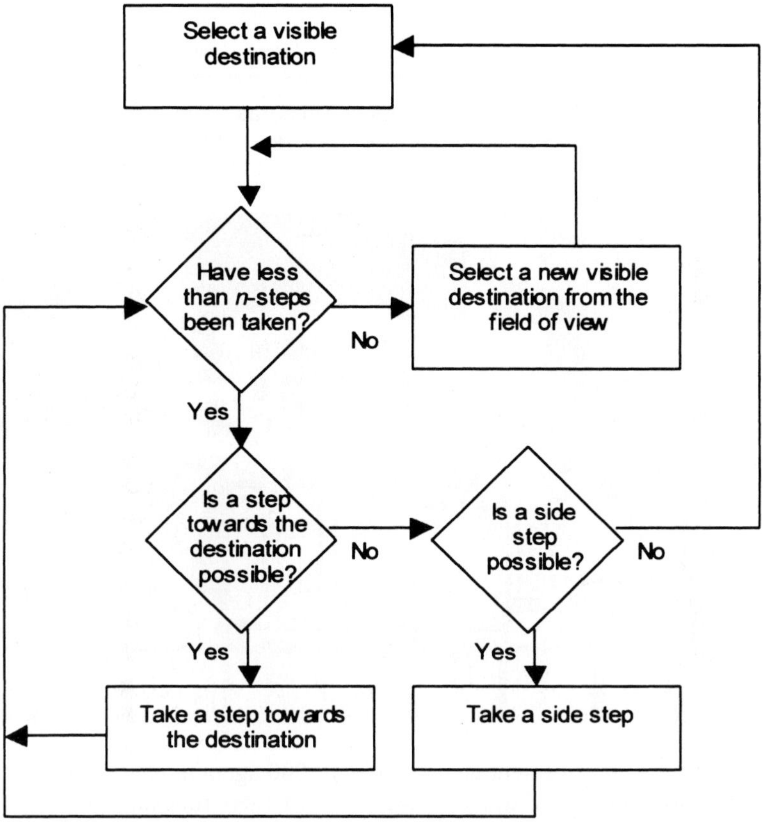

Figure 8: Flow-chart of the agent decision process for the random next step rule.

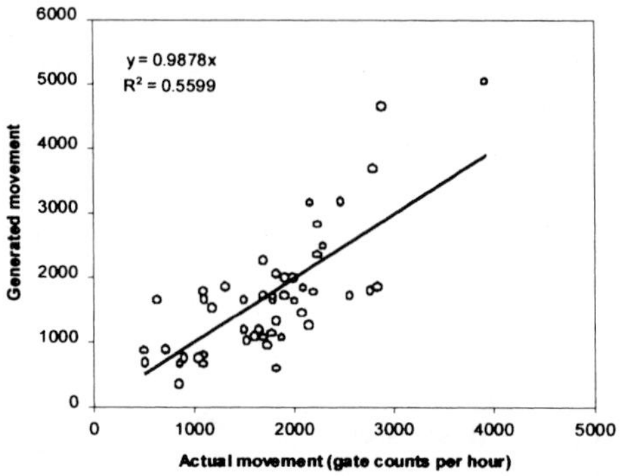

Figure 9: The correlation between agent (generated) movement and observed shopper movement in the department store (r^2=.56, n=49, p<.001).

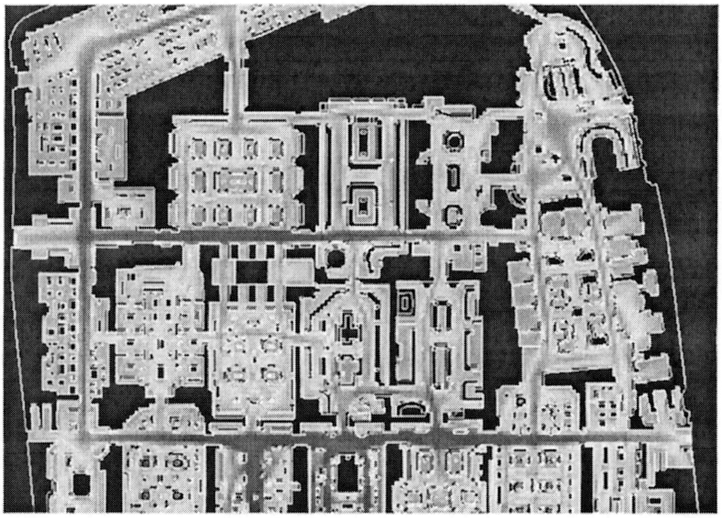

Figure 10: Agent movement density after 10 000 agent moves using the random next step rule. Each grid square is incremented every time an agent steps on it.

Figure 10 shows the cumulative agent traces for the EVAS agents using the random next-step method, and a comparison with Fig. 6 shows considerable similarity to the observed real movement patterns. In this illustration the red end of the spectrum denotes high and the blue end low agent flows. At a qualitative level the average flow patterns are highly convincing, and even isolate various local characteristics visible in the observed human behaviour in Fig. 6.

5 Discussion

These results were simpler than anticipated. Although we expected agents without specific goals to correlate positively with observed human behaviour in an exploratory shopping environment, we were expecting that some form of relatively local forward planning would be taking place. For this reason we chose to test the 'clustering coefficient' method as a proxy for detecting junctions, and to use these to provide the immediate 'next step' goal for agents. However, the finding that a much simpler 'random' next step rule significantly outperformed the 'clustering' rule, and that it provides a relatively good correlation with observed human movement patterns, is surprising. Initial results of experiments applying agents using the same rules to two different building morphologies and types, an art gallery and a supermarket, have shown similar degrees of correlation to observed human movement flows, and it seems that approximately 50% of the variance in flows can be accounted for by the dynamics of this agent simulation.

It is clear that the way the random next step rule works is to lead agents to continue moving forwards along linear spaces such as corridors or aisles when the majority of visible nodes within their view cone lie ahead. If an agent comes up against a wall the majority of visible nodes will lie to the right and to the left, and so it is likely that the next step chosen at random will involve a turn one way or the other. If the agent comes to a cross-road junction, the majority of nodes will lie in three predominant directions, left, right and ahead. Since the view cone is 170 the probability involved in the random selection of visible nodes depends almost entirely on the extent of the view in each of those three directions. Of course, since the selection of the next heading is random, there is a great deal of variation in individual agent behaviour at any step, however, since the selection of the heading is repeated every three steps, the laws of probability smooth out these variations, and what emerges is a pattern of movement very closely related to the linear arrangement of space in the environment, including highest flows on corridors, and lower flows in more broken-up areas.

These findings allow us to draw three tentative conclusions. 1. The affordance offered by an environment for human movement is location and orientation dependent. As an agent moves through an environment their local perception of choices available for movement is directly related to their current location and heading and thus changes at each step. In other words, a fixed environment is objectively different from different points of view; 2. The perceptual range for humans is large scale. By this we mean that, unlike the motion of gas molecules which are more or less locally determined by interaction with their immediate neighbours, humans have long distance vision and make next step decisions based on the longer distance affordances of the environment; this leads to 3. The spatial morphology of the environment acts on human movement mainly through its linear arrangement. Linearly arranged spaces offer more possible 'next step' destinations along their length than across their width, and thus they tend to gather and organise linear movement patterns.

This last conclusion brings us back to one of the most surprising findings of space syntax research: the empirical success of the 'line map' representation in explaining human movement patterns. One of us has argued elsewhere [22] that human cognition is likely to employ a representation of linear spatial elements since these elements are the least subject to changes during forward movement and so a compressed cognitive representation is likely to store linear spatial elements as single entities. The findings presented here however, suggest a still simpler explanation for the observed linear logic of human movement behaviour. It is possible that the effective space in which agents with long distance vision make immediate 'next step' decisions that guide movement is 'collapsed' over distance itself: we choose our immediate next destination from the full range of locations that we can currently see, without particular consideration for how far away they are. We then continually review these decisions as we move. When we are moving along a linear corridor space most opportunities are likely to lie ahead, however when we reach a junction the opportunities are defined by the morphology of the local visual field. Conroy found that it was exactly at these junction locations that people paused and looked around before moving on [21], suggesting that it was at these locations that the cognitive load of decision making required a slower pace of movement and gathering of additional information. Recent work by the authors, currently in preparation, suggests a mean of three steps before the direction review may be optimal in the EVAS agent simulation.

In conclusion, our discussion suggests that the next steps in the development of the agent simulation could also be simpler than originally anticipated. These next steps address two factors, first the inclusion of stopping and shopping behaviour, and second the inclusion of crowding and congestion effects. We anticipate that, taken together, the feedback effects of stopping and congestion on agent movement flows will further improve the correlation with observed shopper movement patterns.

The authors should like to acknowledge funding from the UK's Office of Science and Technology Foresight Challenge Award VR Centre for the Built Environment and Platform funding from the UK's Engineering and Physical Sciences Research Council.

References

1. D. Helbing, P. Molnár, I.J. Karkas, and K. Bolay, *Self-organizing pedestrian movement*, Environment and Planning B **28**, forthcoming, (2001).

2. R.J. Beckman et al. *The TRansportation ANalysis SIMulation system (TRANSIMS): The Dallas-Ft. Worth case study,* Los Alamos Unclassified Report LA–UR–97–4502, Los Alamos National Laboratory, (1997).

3. J.L. Casti, *Would Be Worlds: How Simulation is Changing the Frontiers of Science,* (John Wiley & Sons Inc., New York, 1997).

4. K.H. Drager, G. Løvås, J. Wiklund, H. Soma, D. Duong, A. Violas, and V. Lanères, *EVACSIM — a comprehensive evacuation simulation tool*, in: *Proc. 1992 Emergency Mangement and Engineering Conference*, pp. 101–108 (1992).

5. E.R. Galea, M. Owen, and P.J. Lawrence, *Computer modelling of human behaviour in aircraft fire accidents*, Toxicology **115**, 63–78 (1996).

6. D. Helbing and P. Molnár, *Self-organization phenomena in pedestrian crowds*, in: *Self-Organisation of Complex Structures: From Individual to Collective Dynamics*, F. Schweitzer, (Ed.), (Gordon & Beach, London, UK, 1997)

7. A. Borgers and H. Timmermans, *A model of pedestiran route choice and demand for retail facilities within inner-city shopping areas*, Geographical Analysis **18**, 115–128 (1986).

8. A. Borgers and H. Timmermans, *City centre entry points, store location patterns, and pedestrian route choice behaviour, A microlevel simulation model*, Socio-Economic Planning Science **20**, 25–31 (1986).

9. J. Kerridge and N. McNair, *PEDFLOW — a system for modelling pedestrian movement in occam*, in: *Architectures, Languages and Techniques,* B.M. Cook, (Ed.), (IOS Press, Amsterdam, 1999).

10. G. Thomas and S. Donikian, *Modelling virtual cities dedicated to behavioural animation*, Computer Graphics Forum **19**, C71–C80 (2000).

11. B. Hillier and J. Hanson, *The Social Logic of Space,* (CUP, Cambridge, 1984).

12. B. Hillier, R. Burdett, J. Peponis, and A. Penn, *Creating life, or, does architecture determine anything?* Architecture and Behaviour/Architecture et Comportement **3** (3), 233–250 (1987).

13. B. Hillier, A. Penn, J. Hanson, T. Grajewski, and J. Xu, *Natural movement; or, configuration and attraction in urban space use*, Environment and Planning B: Planning and Design **20**, 29-66 (1993).

14. A. Penn and N. Dalton, *The Architecture of Society: stochastic simulation of urban movement*, in: *Simulating Society: The computer simulation of social phe-*

nomena, N. Gilbert and J. Doran, (Eds.), Ch 5, pp. 85-126 (UCL Press, London, 1994).

15. C. Mottram, A. Penn, R. Conroy, and A. Turner,1999, *Virtual beings: emergence of population level movement behaviours from individual rulesets*, in: *The Proceedings of the Second International Space Syntax Symposium*, Universidad de Brasil, Brazil **1**, 10 (1999).

16. A. Turner, M. Doxa, D. O'Sullivan, and A. Penn, *From isovists to visibility graphs: a methodology for the analysis of architectural space*, Environment and Planning B **28** (1), 103–121 (2001).

17. A. Penn and A. Turner, *System for the intelligent modelling of public spaces*, UK Patent Office, application #0020850.4, 28/8/2000, (2000).

18. X. Tu, *Artificial Animals for Computer Animation: Biomechanics, Locomotion, Perception, and Behaviour,* (LNCS **1635**, Springer-Verlag, Berlin, 1999).

19. I.A. Bachelder and A.M. Waxman, *Mobile robot visual mapping and localisation: a view-based neurocomputational architecture that emulates hippocampal place learning*, Neural Networks **7** (6/7), 1083–1099 (1994).

20. D.J. Watts and S.H. Strogatz, *Collective dynamics of 'small-world' networks*, Nature **393**, 440–442 (1998).

21. R.A. Conroy, *Spatial Navigation in Immersive Virtual Environments*, (PhD Thesis, University of London, 2001).

22. A. Penn, *Space Syntax and Cognition: or why the axial line?,* Proceedings of the Third Space Syntax Symposium, Georgia Tech., Atlanta GA, May 2001, forthcoming (2001).

Flow Capacities from Cellular Automata Modeling of Proportional Splits of Pedestrians by Direction

V.J. Blue[1] and J.L. Adler[2]

[1]New York State Department of Transportation, Poughkeepsie, NY 12603, USA

[2]Consultant, Silver Spring, MD 20902, USA

This paper describes the design and application of an Artificial Life cellular automata-based (CA) microsimulation that yields speed-flow-density characteristics, or fundamental properties, for various proportional splits of pedestrians by direction. It is shown that a CA model using a relatively small rule set is capable of effectively capturing collective behaviors of pedestrians who are autonomous at the micro-level. The model captures emergent pedestrian stream characteristics for one, two, cross-, and four directional flows and reveals how much stream formation and directional splits are important factors in determining walkway capacities. For all flows studied capacities occur at about 30 percent occupancy, a guide for channeling maximum flows. The results agree with and extend well-established fundamental properties of pedestrian flows according to the directional splits.

1 Introduction

Cellular Automata (CA), can be characterized as an Artificial Life approach to simulation modeling [1]. The CA local rules prescribe the behavior of each automaton over a cellular lattice creating an approximation of actual individual behavior. Group organization and complex dynamics evolve as entities with varying behaviors act and interact with other entities over time and space.

This paper describes a CA model of pedestrian flow. Behavior of individual pedestrians is represented by a small set of rules. Emergent collective behavior is an outgrowth of the interaction of the autonomous pedestrians, without any centralizing coordination. By "designing" the CA-based pedestrian from the bottom-up at the interfaces with one another, higher-level functions, like route selection and trip behavior, can be added later without fundamentally changing the inter-pedestrian dynamics.

Using CA models to study pedestrian flow is attractive for a number of reasons. The CA interactions of the pedestrians are based on intuitively understandable behavioral rules. They are easily implemented on digital computers and run exceedingly fast. CA models function as discrete idealizations of the partial differential equations that describe fluid flows and allow simulation of flows and interactions that were previously intractable [2].

Over the past several years, researchers have demonstrated the applicability of cellular automata (CA) microsimulation to vehicular flows. These CA models have included traffic with two-lane flow with passing [3]. CA taffic models have been shown to provide a good approximation of complex traffic flow patterns including the formation of shock waves in traffic jams [4].

Researchers have found the task of modeling pedestrian flows more daunting. In several ways, pedestrian movements are more complex than vehicle flows:

- Pedestrian concourses and corridors may support movement in several directions
- Pedestrians are capable of changing speed quickly, in one second accelerating to full speed from a standstill or braking to a stop from full speed.
- Unlike vehicular flow, as density increases on pedestrian walkways, there can be substantial activity as lane changing, exchanges, and speed variations occur frequently, including the dynamic formation of streams of various widths.

There are many emerging pedestrian models which is evident from this conference. In CA-like models, early on, in 1985 Gipps and Marksjo [5] developed a model that focused on moving pedestrians over a grid of hexagonal cells.

Blue and Adler have previously demonstrated that unidirectional, bi-directional, cross-directional, and 4-directional pedestrian flows emerged from CA simulation experiments that correspond to the fundamental parameters [6-11]. Established corresponding fundamental pedestrian flow curves are published in a chapter dedicated to pedestrian characteristics for crosswalk and facility design in the Highway Capacity Manual [12]. The CA approach is also well-suited to capture the behavior of pedestrian streams that are observable within corridors of high pedestrian concentration. Behavior of pedestrians in crowds has not been extensively treated in the literature except on the macroscopic scale [12].

Formation of directional streams is shown in Fig. 1(a) where people can been seen forming four distinct directional streams. Without any external "policeman" organizing them into streams, people dynamically self-organized into this emergent pattern. From many observations, it would appear that bi-directional flow in only two separated lanes is a special case of a more general dynamical model. Multiple streams are not an anomaly and occur in many environments including corridors, as shown in Fig. 1(a) in Grand Central Station, and in multidirectional environments. Fig. 1(b) shows a CA simulation where multiple streams form with bi-directional flows.

Figure 1: (a) emerging dynamic multiple streams at Grand Central Station in NYC and (b) dynamic multiple lanes (DML) in a bi-directional 50-50 split simulation.

2 Rule Set

The CA model for pedestrian flows is based on vehicular CA models developed by Rickert et al. [3]. The model centers on a two-stage parallel update process whereby lane assignment and forward motions change the positions of all pedestrians in two parallel update stages in each time step. Parallel updates avoid succession interdependencies encountered in sequential updates by determining all the new positions before anyone moves. All the entities are then repositioned together. Only the pedestrians in the immediate neighborhood affect the movement of a pedestrian, which, though myopic, is relatively realistic. Because of limited space, the rule set is not given here. Readers are referred to other works by the authors [6-11] that delineate the rules in detail. When two opposing pedestrians are deadlocked, the rate of exchanging places is set at 50 percent in each time step in this set of experiments.

The model is designed to account for variations in walking speeds observed in the real world. Each time step is one second in duration. Walking speed varies among pedestrians, so a distribution of walking speeds is used:

(a) *Fast Walkers* -- maximum speed of 4 cells per time step (1.8 m per time step)

(b) *Standard Walkers* – maximum of 3 cells per time step (1.3 m per time step)

(c) *Slow Walkers* – maximum rate of 2 cells/time step (0.85 m per time step).

A 5% fast, 90% standard; 5% slow (5:90:5) distribution of walkers is used to represent the pedestrian population. This distribution had the best realization of the fundamental diagram compared with other distributions in the single-direction case [6] and is consistent with means and standard deviations of speed used by others [12].

3 Pedestrian Simulation Experiments

The pedestrian walkway is modeled as a circular lattice (i.e., a rectangular grid that wraps around at the ends). The circular lattice enables a set of pedestrians to interact at constant density while maintaining strict conservation of pedestrians after they are initially randomly assigned to the lattice. The cell size, $0.21\mathrm{m}^2$, discretizes the lattice at the individual pedestrian level and is scaled according to the minimal requirements for personal space as described in the Highway Capacity Manual [12]. The scale is also used to generate the speed-flow-density relationships that emerge. A 50 x 50 lattice of cells is used in these simulation experiments.

One second is the duration of each time step. Each simulation is 1,000 time steps with the first 100 time steps discarded to initiate the simulation and the latter 900 (15 minutes) used to generate performance statistics. Each set of experiments is run at 19 densities ranging from 0.05 to 0.95 percent occupancy in increments of 0.05. For statistical accuracy, ten replications at each density are run and the parameters are computed as the average over these replications. The emergent profile by directional split is a map of the corresponding flows over the range of densities.

4 Results

Although it seems relevant, the type of pedestrian flow is not generally credited as an important factor in determination of capacity, where capacity of flow is the peak flow possible in a given environment. The literature is devoid of in-depth treatment of this important topic. For instance, in presenting capacity of pedestrian flow, the HCM mentions bi-directional splits but has no discussion of 4-directional pedestrian interactions and splits. In this context, the CA model sheds significant insight into the effects of directional split on capacity of a facility to allow for pedestrian flows.

In Fig. 2 a range of directional splits are analyzed. Best is unidirectional flow, equal to two separated bi-directional flows. Next best are bi-directional flows where pedestrians tend to follow someone going in the same direction, forming streams of various widths as shown in Fig. 1and termed dynamic multi-lane (DML) flows here. Least capacity is interspersed flows (ISP) where no lane formation is encouraged at all, generally a transient case. Fig. 2 shows the flow rates of DML and ISP flows for 90-10 and 50-50 splits. Stream formation (DML) improves flow especially in the 50-50 split where stream formation is more easily accomplished over a 90-10 split. The 90-10 DML curve shows a 15 percent decrease from unidirectional flow, which agrees with the HCM that mentions this particular case. Interestingly, at high density the ISP 50-50 split has unrealistically high flows due to the large number of place exchanges occurring.

In Fig. 3 cross-directional and 4-directional flows are added to uni- and bi-directional DML flows. The DML and cross-directional flows have splits of 50-50 and 90-10, while the 4-directional flow is split 25 percent in each direction. As would be expected the bi-directional DML flows outperform the cross-directional flows which in turn outperform the 4-directional flows. The capacity in each case is reduced as flows become less capable of forming streams. Stream formation in cross- and 4-directional flows can occur in crowds. The model shows some capability of finding capacity values given the assumptions for exchange rates and directional splits used in the experiments. Further work is needed to explore 4-directional stream formation.

5 Discussion and Conclusions

While observing pedestrian flows in the real world, it is evident that the emergence and movement of pedestrian streams has a profound effect on volume, capacity, and speeds. The results from the CA model provide the ability to analyze these phenomena in greater detail. Results from the model illustrate it is more efficient to form bi-directional streams (DML) than not to (ISP). When multiple streams form, they are not as efficient as two separate streams, which are equivalent to two uni-directional flows. Cross- and 4-directional flows also have lower capacities.

Peak flows or capacities occur at a density of 30 percent ocupancy of cells in each case. While it is not easily understood why capacity occurs at 30 percent occupancy, it is a useful finding when trying to manage crowds no matter what the percent split by direction. By managing density at bottlenecks to near a 30 percent occupancy, flows will be maximized, no matter what the directional split. Technologically, it is difficult to count people today, even in video clips, with the accuracy of the model. Though complete calibration remains elusive, the model allows us to understand and to estimate the loss of capacity as directionality changes. The results also point out data investigation needs.

Though the standard design guide, the Highway Capacity Manual [12], shows a linear speed-density curve, and corresponding parabolic curve for flow-density, it is generally acknowledged that the unidirectional and separated flow curves would more closely follow the Bell-curve flow-density relationship realized by the model (see [4] and [9] for discussion). This model would further imply that the HCM could indicate that the various types of flows, unidirectional, and bi-directional (a) separated, (b) interspersed, and (c) stream- producing dynamic multiple lanes, cross-directional, and 4-directional flows have differing speed-flow-density curves and capacities.

Safe design practices for increased walking densities from a growing worldwide population would give this model practical purpose. We are expanding the model to network flows using bi-directional sidewalks [7-9, 11] and four-directional flows [10] at street corners. Illustrative simulations can be viewed at www.ulster.net/~vjblue.

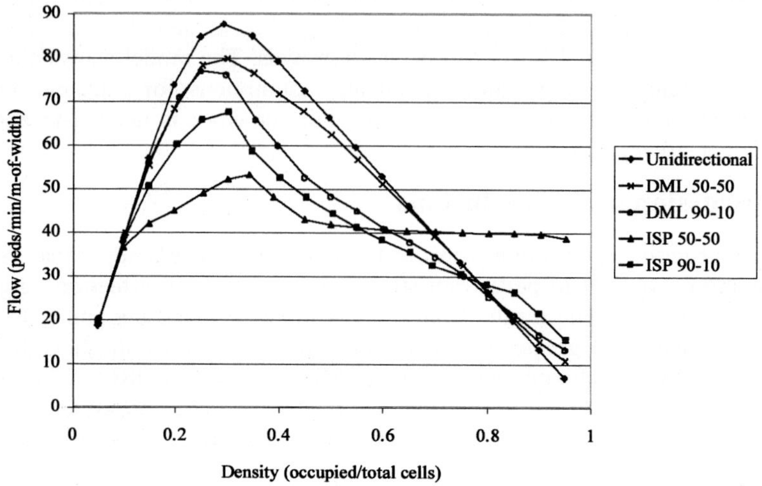

Figure 2: Capacities of uni- and bi-directional flows (DMLand ISP).

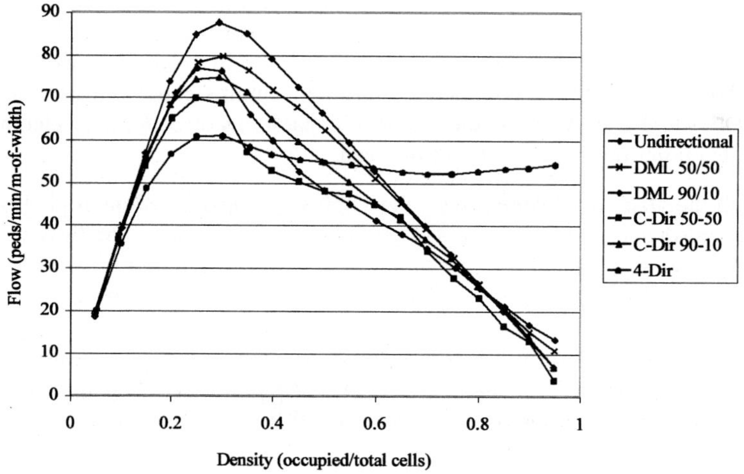

Figure 3: Capacities of uni-, bi- (DML), cross -, and 4-directional flows.

References

1. C. Adami, *Introduction to Artificial Life*, (Springer-Verlag, New York, 1998).

2. S. Wolfram, *Cellular Automata and Complexity,* (Addison-Wesley Publishing Company, 1994).

3. M. Rickert, K. Nagel, M. Schreckenberg, and A. Latour, *Two-Lane Traffic Simulations Using Cellular Automata*, Physica A **231**, 534 (1996).

4. K. Nagel, *Particle Hopping Models And Traffic Flow Theory*, Phys. Rev. E **53**, 4655 (1996).

5. P.G. Gipps and B. Marksjo, *A Micro-Simulation Model For Pedestrian Flows*, Mathematics and Computers in Simulation **27**, 95 (1985).

6. V.J. Blue and J.L. Adler, *Emergent Fundamental Pedestrian Flows From Cellular Automata Microsimulation*, Transportation Research Record, Journal of the Transportation Research Board **1644**, 29 (1998).

7. V.J. Blue and J.L. Adler, *Bi-Directional Emergent Fundamental Pedestrian Flows From Cellular Automata Microsimulation,* in: *Transportation and Traffic Theory: Proceedings of the 14th International Symposium on Transportation and Traffic Theory*, A. Ceder, (Ed.), p. 235 (Pergamon, Amsterdam, 1999).

8. V.J. Blue and J.L. Adler, *Cellular Automata Microsimulation Of Bi-Directional Pedestrian Flows*, Transportation Research Record, Journal of the Transportation Research Board **1678**, 135, (1999).

9. V.J. Blue and J.L. Adler, *Cellular Automata Microsimulation For Modeling Bi-directional Pedestrian Walkways*, Transportation Research B **35/3**, 293 (2001).

10. V.J. Blue and J.L. Adler, *Modeling Four-Directional Pedestrian Flows,* Transportation Research Record, Journal of the Transportation Research Board **1710**, 20 (2000).

11. V.J. Blue and J.L. Adler, *Cellular Automata Model of Emergent Collective Bi-Directional Pedestrian Dynamics*, in: *Artificial Life VII*, M. Bedau, J. McCaskill, N. Packard, and S. Rasmussen, (Eds.), p. 437 (MIT Press, 2000).

12. *Special Report 209: Highway Capacity Manual*, Transportation Research Board, National Research Council, Washington, D.C., Chapter 13, (1994).

Microscopic Pedestrian Wayfinding and Dynamics Modelling

S.P. Hoogendoorn, P.H.L. Bovy, and W. Daamen

Delft University of Technology, Faculty of Civil Engineering and Geosciences, Transportation and Traffic Engineering section, Delft, The Netherlands

Gaining insights into pedestrian flow operations and assessment tools for pedestrian walking speeds and comfort are important in for instance planning and geometric design of infrastructural facilities. Additionally, management of pedestrian flows requires knowledge of pedestrian flow behavior. This paper puts forward a microscopic pedestrian flow model. In the model, pedestrians of different types (age, gender, travel purpose, experience) are randomly generated at origin areas. Pedestrians will walk from their current location to the destination area, while minimizing a cost function, expressing for instance the travel time and cost incurred by walking too close to obstacles or walls. On the contrary to discrete choice analysis, the number of choice options the pedestrian has is nearly infinite. To correctly describe pedestrian wayfinding, the choice problem is formulated as an optimal control problem, aimed at finding the path which optimises the trip utility. On the tactical and operational pedestrian behaviour levels, the model is characterized by non-local stimulus-response mechanisms describing pedestrian anticipation and interaction behavior. Moreover, local route choice behavior is included, describing how pedestrians choose their direction based on among other things their destination, and local pedestrian flow conditions. This local choice behavior is based on the concept of subjective utility maximization. Moreover, the model features parallel updating. The model can describe multi-directional pedestrian flows (e.g. uni-directional, bi-directional, crossing flows). This paper discusses application of the model to a number of test-case scenarios. We discuss model properties such as the dependence of the speed-density relation on pedestrian composition, and self-formation of lanes in case of bi-directional pedestrian flows. Moreover, differences in pedestrian wayfinding for inexperienced and experienced pedestrians, and the consequent differences in flow operations are discussed. It turns out that the location of congestion depends on the experience of pedestrians.

1 Introduction

Understanding pedestrian behavior is of dominant importance in planning and design of airports, public transport stations, and shopping malls. When designing time tables, insights into walking times experienced by pedestrians when accessing or transferring may play an important role as well. Modeling tools, such as the pedestrian simulation model described in this paper, can support both infrastructure designers as well as public transport planners in their tasks thereby optimizing their design [4]. Finally, also the management of pedestrian flows demands understanding of both the collective pedestrian flows as well as the individual pedestrian movements in the flow.

Pedestrian walking behavior concerns microscopic aspects at the level of the individual pedestrians (route choice behavior, walking speeds, passing behavior), as well as macroscopic properties of the pedestrian flow, such as the formation of dynamic lanes and clusters, capacities of doors, escalators, etc. Current car-based flow theories and models are not able to fulfil the needs of pedestrian flow modeling.

2 Pedestrian Flow Modelling Approaches

Several modeling approaches have been put forward to describe pedestrian flow at different levels (e.g. route-choice behavior, macroscopic flow operations, microscopic pedestrian behavior). Among these approaches are regression models, queuing models, route-choice models, gas-kinetic and macroscopic models, and microscopic models.

Regression models use statistically established relations between flow variables (e.g. flow per meter width and speed) and use these to predict pedestrian flow operations under specific circumstances. The form of these statistical relations is generally different for different kinds of infrastructure (hallways, stairs) and flow compositions, and needs to be established explicitly for the facilities of interest. *Queuing models* use Markov-chain models to describe how pedestrians move from one node of the network (mostly a room) to another. Random waiting times are incurred on the network links (generally doors), due to queues building up when pedestrian demand is larger than the door capacity. These models have been used mostly to describe evacuation behavior from buildings. *Route-choice models* describe pedestrian wayfinding through the walking facility based on the concept of utility maximization. In this case, pedestrians choose their (intermediate) destinations in order to maximize the utility (travel time, comfort, etc.) of their trip. *Macroscopic models* (see [11]) and *gas-kinetic models* (see [9]) use the analogy with fluid or gas dynamics to describe how density $r(t, \boldsymbol{x})$ and velocity $v(t, \boldsymbol{x})$ change over time using partial differential equations (Navier-Stokes or Boltzmann-like equations).

The emphasis of the research described in this paper will be on *microscopic models* describing the time-space behavior of individual pedestrians. Examples of these so-called walker models are the ***social force model*** (see [6], and [7]), and the *Cellular Automata (CA) models* (see [1], and [2]). In the ensuing of this section, these models are considered in more detail.

The social force model microscopically describes pedestrian behavior by means of so-called ***social force-fields***. This spatially continuous microscopic model inhibits pedestrian's acceleration and direction-changing behavior. Attracting and repelling forces determine the interaction between the pedestrians and obstacles such as walls. Like the CA-model, the social force model is able to describe observed features of pedestrian traffic flow, such as ***self-formation of lanes***, due to the interaction of these competitive processes. Nevertheless, we argue that some of the underlying model assumptions oversimplify the process of pedestrian wayfinding through the traffic flow. For instance, the attracting or repelling forces are invariant with respect to the pedestrian characteristics, particularly to their direction. In real-life, pedestrians are likely to be attracted more by pedestrians moving into their own direction than by pedestrians walking into the opposite direction.

The CA-model is aimed at reproducing macroscopic features of pedestrian flows (such as ***speed-density curves*** and ***dynamic lane formation***) using a microscopic model. On the contrary, the macroscopic model of [11] uses these speed-density relations in a macroscopic flow model, comparable to flows in gases, provided that the speed-density relations are comformable mappable (independent of the directions of the pedestrians). That is, the model of [11] does not explicitly consider pedestrian interactions nor their consequent walking behavior.

Although the CA-model is based on behavioral rules prescribing pedestrian walking behavior, adapting the CA-model to suit the description of multi-directional pedestrian flows is complicated by a number of issues, among which is the grid-like partition of the walking area and the limited possibilities to describe different kinds of pedestrians (commuters, shopping people, people with luggage). With respect to the latter issue, we point out that CA-models can distinguish pedestrians with different free speeds and destinations.

We argue that in describing pedestrian flows accurately and realistically, all relevant aspects in the pedestrian behavior, and the consequent differences between pedestrians, need to be considered. Depending on the envisaged model application, either a macroscopic, or a microscopic approach is appropriate.

3 Pedestrian Travel Decision Hierarchy

Let us consider an infrastructural facility (e.g. a public transit station, an urban walkway, a shopping mall). Pedestrians will enter and leave these facilities at certain designated areas. In this paper, the areas where pedestrians enter

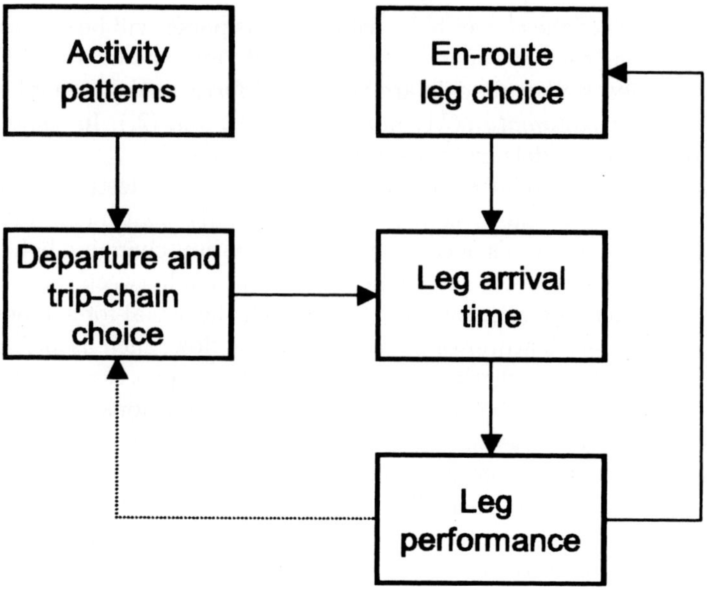

Figure 1. Dependence between activity patterns, dynamic trip-chain travel demand, and dynamic leg demand.

the facility are referred to by the term *origin areas*. In the ensuing, we will distinguish between *intermediate* and *terminal destination areas*. Pedestrians leave the facility at their terminal destination area. At the *trip-generation level* pedestrian trips are generated. In a multi-modal transportation network, the generated trips (or rather, origin-destination relations) depend on among other things:

- *Activity choices*
- *Trip-chain* and *departure time choices*
- *Leg-performance* and consequent *en-route leg choice behavior*

Figure 1 shows the schematics of general travel processes. It shows how activity patterns determine the need for transportation, which in turn incurs dynamic mode-specific travel demand. In the following, we will discuss the different decision levels in more detail.

3.1 Activity Choice Behavior

Travelers mostly embark upon a trip to perform an activity (e.g. to work, learn, eat, or to relax). In other words, activity planning determines collective travel demand patterns. Depending on the type of considered infrastructural facility, different activities cause pedestrian traffic. For instance, in case of a railway station during the morning peak hour, most of the pedestrians are likely to be commuters traveling to work. On the contrary, pedestrian traffic in a shopping mall is mostly generated by leisure-related activities. In this respect, note that walking can either be the induced transportation required to perform an activity (commuting), or an activity itself (e.g. shopping).

3.2 Trip-chain and Departure Choice Behavior

Once the activities are planned, and the origin and destination of the trips needed to perform these activities are known, travelers will have to decide which *chain of transport modes* (referred to as a *trip chain*) is most proficient to travel from origin to destination.

Trip-chains consist of *legs*, connected by nodes. A leg can reflect *any kind of transport movement*, for instance by car, public transport, bicycle, or by walking. We do not explicitly distinguish access, egress, and transferring. This implies for instance that a transfer is not a node, but just another leg in the trip-chain. At the nodes, travelers switch from the current leg to the next leg.

In general, different trip-chain alternatives will be available to the traveler. For instance, a commuter can walk to his car, drive to the Park and Ride facility, where he can get on the train to the city center, from which he can walk towards the office. Alternatively, the commuter can use the subway to travel to the railway station. In the latter case, the commuter will walk from the subway to the train platform. The choice for a specific multi-modal travel chain, and the preferred arrival time (depending on the activity type), determine the time the traveler starts the trip.

3.3 Leg-performance and En-route Leg Choice Behavior

Once the pre-trip travel choices (i.e. trip-chain and departure time) are made, and the traveler embarks on his trip, the performance of the constituent legs of the trip-chain determines the travel process. For instance, when non-recurrent congestion is experienced on the motorway, the trip to the Park and Ride facility may be delayed to the extent that the traveler is too late to catch the train. He may then decide against using the Park and Ride service altogether, and continue his trip by car. Similar events can occur for public transport legs.

Whether or not the trip-chain is modified en-route, it is clear that the performance of the current leg (mainly in terms of travel time) determines the arrival time at the next leg. In other words, the leg-specific dynamic travel demand is determined by the performance of the preceding legs in the trip-chain. This holds equally for travel demand at the destination, generally, the location where the activity is to be performed.

Summarizing, it is clear that pedestrian travel demand at the origin areas of the pedestrian infrastructure is determined by behavioral processes at different levels. This holds for both a leg traversed by walking (e.g. access, transfer, or egress), as well as an activity involving walking (e.g. shopping). At this point, note that in case of a leg, the following leg determines the destination. For instance, when the following leg is a bus-trip, the traveler will walk towards the bus-station.

4 Pedestrian Choice Behavior Framework

In the remainder of this paper, we assume that dynamic pedestrians travel demands, as well as the destinations of the pedestrians are known. We hypothesize that observed walking behavior is the result of pedestrian decision making at the strategic, tactical, and operational level. Figure 2 shows the relations between these levels.

The *strategic level* includes the (pre-trip) route choice behavior of the pedestrian: it describes how pedestrians choose their (subjective) optimal routes from the collection of available route alternatives between origin and destination. The behavioral processes that are relevant can be described by route choice modeling mechanisms, similar to the models used to describe route choice behavior in traffic networks. Note that at the strategic level, pedestrians have incomplete information concerning the conditions on the different route alternatives. Consequently, they have to rely mostly on local observations and experience.

At the *tactical level*, pedestrians make short-term decisions, based on the prevailing conditions. In illustration, a pedestrian will choose whether an obstacle is passed on the right-hand-side, or on the left-hand-side, or what the optimal location of a pedestrian is (Where are few pedestrians? Where do pedestrians walk in the same direction? Will I not drift away from my destination?). Travel choices at this level are mostly based on macroscopic features of the pedestrian flow (velocities, densities, flows). Similar to the strategic level, pedestrians have alternatives at the tactical level. Contrary to the strategic level, pedestrians mostly have sufficient information to make a user-optimal decision. Therefore, it is plausible to assume that the pedestrian will choose an alternative based on the principle of subjective utility maximization.

The *operational level* describes among other things acceleration behavior of pedestrians, and reaction or anticipation on slow pedestrians or pedestrians walking in the opposite direction, including how pedestrians will adapt their direction to achieve the goals set at the tactical and operational levels. At this level, interaction with other pedestrians plays an important role.

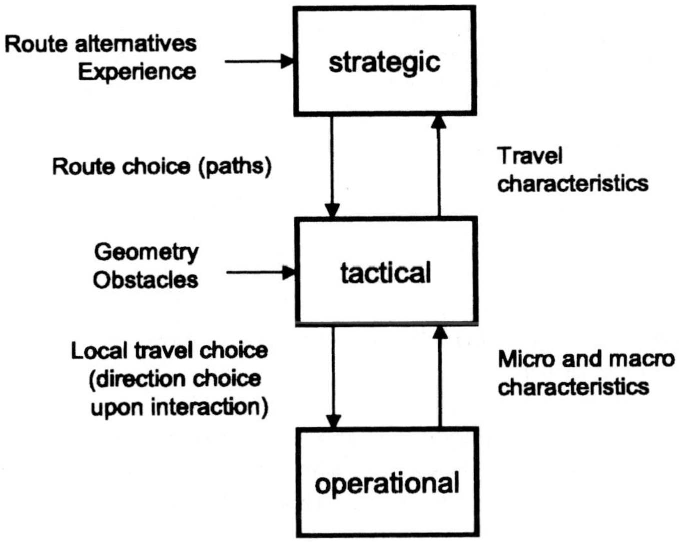

Figure 2. Strategic, tactical, and operational levels in pedestrian walking behavior.

5 Trip Generation and Wayfinding

Similar to wayfinding in transportation networks, we can assume that pedestrians *aim to maximize the utility of their trip* in order to get from their origin to their destination. In this section, we present an approach to describe pedestrian wayfinding behavior based on this notion. The wayfinding approaches discussed in this section pertains mainly to route choice within the walking area itself. We assume that the origin destination demand patterns are known a priori. Moreover, we assume that on pedestrians walk along predetermined paths, which are described by an ordered set of subsequent destination areas.

5.1 Pedestrian Trip Generation and Services

We assume that pedestrians are generated at so-called origin areas D_0. The time of generation is determined by some random process, for instance a Poisson process. The location of the pedestrian within the origin area is randomly chosen as well. When generated, pedestrian attributes (age, gender, trip purpose, route familiarity), which in turn yield parameters describing pedestrian behavior (free-speed, spatial use, acceleration time, reaction time, etc.) are determined. Moreover, the OD-path of the pedestrian is set. This path is an *ordered set of the origin area* D_0, of intermediate destinations D_n, with $n = 1, ..., N-1$, and a final destination D_N. The intermediate destinations can reflect *activities which a pedestrian embarks upon during his trip* (e.g. buying a ticket, looking a the time-table, etc.). To these intermediate destination areas, *service times* can be addressed, reflecting the average time needed for the pedestrian to be served or to complete the activity. In other words, once pedestrians are located in the destination area D_n, and the next destination area D_{n+1} is considered, pedestrians incur a stochastic destination dependent waiting time to include activities such as waiting to be served at the ticket office, or looking at a direction sign. That is, when a pedestrian has arrived at the destination D_n, a waiting time T_D is drawn from the waiting time distribution. The pedestrian will remain at his location until his waiting time has passed. Upon continuing his trip, the pedestrian will start walking towards the next destination.

In principle, some intermediate destination areas can be interchanged in practice. For instance, a pedestrian aiming to buy a ticket may be inclined to first get a newspaper at the kiosk, when the line at the ticket office is too long. We will however not consider this type of behavior in this paper. In the remainder, we will consider pedestrian wayfinding given *fixed OD-path.*

In the remainder, we discuss the following research problem: *given the current location $\boldsymbol{x}_\alpha(t)$ of pedestrian α, which trajectory $\boldsymbol{x}_\alpha(s)$ for $s > t$ is chosen given expected traffic conditions, the preferences and attitude of pedestrian α.*

5.2 Optimal Wayfinding Modelling Approach

We hypothesize that a *subjectively optimal travel choice* is made. In utility theory, perceived utility is assumed to be the weighted sum of trip attributes, such as expected travel time, travel time variance, distance travelled, safety, comfort, etc. In this section, we describe an approach to described pedestrian wayfinding by using mathematical optimal control. In doing so, we will define the *state* $\boldsymbol{X}_\alpha(t)$ of pedestrian α at time instant t, which for instance can comprise the location, and the velocity, i.e. $\boldsymbol{X}_\alpha(t) = [\boldsymbol{x}_\alpha(t),\ \boldsymbol{v}_\alpha(t)]$. For our purpose of path optimization, it suffices to use the current location $\boldsymbol{X}_\alpha(t) = \boldsymbol{x}_\alpha(t)$. In the remainder, we will use the lowercase notation $\boldsymbol{x}_\alpha(t)$ to denote the state.

Besides the state, we define the *control vector* $\boldsymbol{u}_\alpha(t)$. This vector describes how pedestrian α can influence its state $\boldsymbol{x}_\alpha(t)$. We assume that the pedestrian can choose the speed and the direction, i.e. $\boldsymbol{u}_\alpha(t) = [v_\alpha(t), \omega_\alpha(t)]$, with $0 \leq v_\alpha(t) \leq V_\alpha$, where V_α denotes the maximum speed possible, given prevailing walking conditions (pedestrian densities, grade, type of infrastructure, etc.). The *state-dynamics* can now be established very easily, and are defined by the following system of ordinary differential equations

$$\frac{d}{dt}\boldsymbol{x}_\alpha = \boldsymbol{f}_\alpha(\boldsymbol{x}_\alpha, \boldsymbol{u}_\alpha) = \begin{pmatrix} v_\alpha \cos\omega_\alpha \\ v_\alpha \sin\omega_\alpha \end{pmatrix} \tag{1}$$

In the remainder of this section, we will drop the index α for notational convenience.

5.3 Generalized Costs

Let $D \subset I\!R^2$ denote the *destination area* of pedestrian α. Let T denote the arrival time of pedestrian α at the destination area. Moreover, let $\boldsymbol{x}_{[t,T)}$ denote a *state path* from $\boldsymbol{x}(t)$ to $\boldsymbol{x}(T)$, and let $\boldsymbol{u}_{[t,T)}$ denote the *control path*. When traversing this path, α will incur *cost*. These costs represent travel time, walking comfort, energy use, etc. In its most generic form, the cost for the state path $\boldsymbol{x}_{[t,T)}$ are defined by the following expression

$$J = \int_t^T L[s, \boldsymbol{x}(s), \boldsymbol{u}(s)]ds + \phi[T, \boldsymbol{x}(T)] \tag{2}$$

where L and ϕ denote the *running cost* and the *terminal cost* respectively. The running cost L reflects among other things [3]:

1. distance and travel time between origin and destination
2. proximity of obstacles or other physical obstructions; closeness to walls
3. number of sharp turns and rapid directional changes
4. stimulation of environment

Clearly these factors are not mutually consistent, and their importance in wayfinding will probably vary substantially between the different groups of pedestrians. For instance, commuters that are in a hurry to catch their next connection will prefer the shortest path connecting the origin and the destination. On the contrary, shoppers will be more likely to choose routes that provide some form of stimulation (shop windows). In the remainder, we will only consider travel time and the proximity of other obstacles. Nevertheless, the approach discussed here can be easily extended to include the other factors that have either been mentioned or not.

To include the travel time in the running cost, we simply include the value 1. In including the proximity of obstacles, let us denote the minimal distance from the location $\boldsymbol{x}$ to the obstacle or wall i defined by the area O_i by $\delta(\boldsymbol{x}, O_i)$. We assume that the cost for walking close to an obstacle is related to this distance, for instance according to the function $a \exp[-b\delta(\boldsymbol{x}, O_i)]$, where $a > 0$ and $b > 0$ are constants. The running cost then becomes

$$L(t, \boldsymbol{x}, \boldsymbol{u}) = L(\boldsymbol{x}) = 1 + \sum_i a \exp[-b\delta(\boldsymbol{x}, O_i)] \tag{3}$$

The terminal cost ϕ reflects the cost incurred when the pedestrian *is not at his destination* at terminal time T. We will use the following specification of these costs

$$\phi(T, \boldsymbol{x}) = \eta^{-1}\delta(\boldsymbol{x}, D) \tag{4}$$

where $\eta > 0$ is some small constant. Clearly, when the pedestrian is at its destination, $\delta(\boldsymbol{x}, D) = 0$ and $\phi(T, \boldsymbol{x}) = 0$. However, when the pedestrian is not at the destination, the terminal cost can become very large.

Since eqn. (1) defines an explicit relation between the state path $\boldsymbol{x}_{[t,T)}$ and the control path $\boldsymbol{u}_{[t,T)}$, the cost can also be described in terms of the control path only, i.e. $J = J(\boldsymbol{u}_{[t,T)})$. We hypothesize that pedestrians will choose the path $\boldsymbol{x}^*_{[t,T)}$ by choosing the control path $\boldsymbol{u}^*_{[t,T)}$ that minimizes the cost $J(\boldsymbol{u}^*_{[t,T)})$, i.e.

$$\boldsymbol{u}^*_{[t,T)} = (v^*_{[t,T)}, \omega^*_{[t,T)}) = \arg\min\{J(\boldsymbol{u}_{[t,T)})\} \tag{5}$$

5.4 Optimal Control

Two equivalent approaches exist to determine the optimal control laws, Pontryagins maximum principle, and the dynamic programming approach. In this paper, the latter is used since this approach readily provides the optimal speed and direction for all t and $\boldsymbol{x}$. To apply dynamic programming, we define the so-called *optimal cost* or *value function* $W(t, \boldsymbol{x})$ by the minimum cost incurred when applying the optimal control functional $\boldsymbol{u}^*_{[t,T)}$ to get from the current location $\boldsymbol{x} = \boldsymbol{x}(t)$ at time t to the destination D

$$W(t, \boldsymbol{x}) = J(\boldsymbol{u}^*_{[t,T)}) \tag{6}$$

Without going into details, we can show that the value function follows the following partial differential equation, referred to as the *Hamilton-Jacobi-Bellmann* equation

$$-\frac{\partial W}{\partial t} + H^*(t, \boldsymbol{x}, \boldsymbol{\nabla}_{\boldsymbol{x}} W) = 0 \tag{7}$$

where $H^*(t, \boldsymbol{x}, \boldsymbol{\nabla} W)$ denotes the Hamilton function, defined by

$$H^*(t, \boldsymbol{x}, \boldsymbol{\nabla} W) = \min_{\boldsymbol{u} \in U} \left[L(t, \boldsymbol{x}, \boldsymbol{u}) + \boldsymbol{\nabla}_{\boldsymbol{x}} W \cdot \boldsymbol{f}(t, \boldsymbol{x}, \boldsymbol{u})\right] \tag{8}$$

where U denotes the set of *admissible control values*.

Let us first consider how a pedestrian optimally chooses its direction. We can easily show that when $\boldsymbol{u}^* = (v^*, \omega^*) = \arg\min\left(L + \boldsymbol{\nabla}_{\boldsymbol{x}} W \cdot \boldsymbol{f}\right)$, that ω^* satisfies

$$\mathbf{e}^*(t, \boldsymbol{x}) = \begin{pmatrix} \cos\omega^*(t, \boldsymbol{x}) \\ \sin\omega^*(t, \boldsymbol{x}) \end{pmatrix} = -\frac{\boldsymbol{\nabla}_x W(t, \boldsymbol{x})}{||\boldsymbol{\nabla}_x W(t, \boldsymbol{x})||} \tag{9}$$

In other words, the pedestrian chooses the direction ω^* that optimizes the rate at which the generalized cost J is decreased.

We can show that with respect to the speed, the optimal control is of the *bang-bang* type, implying that v^* is either zero or V. That is, when the *marginal costs* of taking a single step into the current direction are positive, the pedestrian chooses the smallest speed possible, namely 0. On the contrary, when the marginal costs are positive, the pedestrian chooses the maximum speed possible, i.e. V. More precisely, we have

$$v^*(t,\boldsymbol{x}) = \begin{cases} 0 \ \langle \nabla_{\boldsymbol{x}} W(t,\boldsymbol{x}), \boldsymbol{e}\rangle \geq 0 \\ V \ \langle \nabla W_{\boldsymbol{x}}(t,\boldsymbol{x}), \boldsymbol{e}\rangle < 0 \end{cases} \tag{10}$$

where $\boldsymbol{e}$ denotes the *current direction* of the pedestrian. However, since the state-dynamics assume that the direction adaptation is instantaneous, we have $\boldsymbol{e} = \boldsymbol{e}^*$ and thus $\langle \nabla_{\boldsymbol{x}} W(t,\boldsymbol{x}), \boldsymbol{e}\rangle = \langle \nabla_{\boldsymbol{x}} W(t,\boldsymbol{x}), \boldsymbol{e}^*\rangle < 0$, yielding $v^*(t,\boldsymbol{x}) = V$.

When applying the optimal direction (9) and speed (10), we find that the Hamilton function equals

$$H^*(t,\boldsymbol{x},\nabla_{\boldsymbol{x}} W) = L(t,\boldsymbol{x},\boldsymbol{u}^*) + \nabla_{\boldsymbol{x}} W \cdot \boldsymbol{f}(t,\boldsymbol{x},\boldsymbol{u}^*) = 1 - V \left\|\nabla_{\boldsymbol{x}} W\right\| \tag{11}$$

Subsequently, we substitute eqn. (11) into eqn. (7), and numerically solve the resulting partial differential equation. This is however not straightforward, since the HJB-equation has *no continuous differentiable solution*, but only generalized solutions. While the latter are not unique, we therefore need to determined one special generalized solution of the HJB eqn. (7), namely the so-called ***viscosity solution***. It turns out that one approach to determine this viscosity solution is to determine the ***vanishing viscosity solution*** W_ε of the equation

$$\varepsilon \Delta W_\varepsilon - \frac{\partial W_\varepsilon}{\partial t} + H^*(t,\boldsymbol{x},\nabla_{\boldsymbol{x}} W_\varepsilon) = 0 \tag{12}$$

and let $W = \lim_{\varepsilon \downarrow 0} W_\varepsilon$.

5.5 Time-independent Optimal Wayfinding

Assuming that pedestrians will stop walking only if they reach their destination, it can be shown, that when both the running cost L, the terminal cost ϕ, and the state-dynamics $\boldsymbol{f}$ are time-independent, the HJB equation simplifies to

$$H^*(\boldsymbol{x},\nabla_{\boldsymbol{x}} W) = 1 - V \left\|\nabla_{\boldsymbol{x}} W\right\| = 0 \tag{13}$$

with boundary conditions

$$W(\boldsymbol{x}) = 0 \quad \text{for} \quad \boldsymbol{x} \in D \tag{14}$$

Example 1 Consider the infrastructure depicted in figure 3, where pedestrians are generated at the left hand side, and walk towards the destination at $x_1 = 40m$ and $17.5m \leq x_2 \leq 22.5m$. Besides the obstacles that pedestrians find on their path, the figure shows the iso-optimal cost curves (indicating the travel time plus additional cost incurred due to the presence of obstacles), where it was assumed that the pedestrian speed is constant and equal to $1.5m/s$. According to the optimal control law, pedestrians walk in the direction perpendicular to the iso-cost curves.

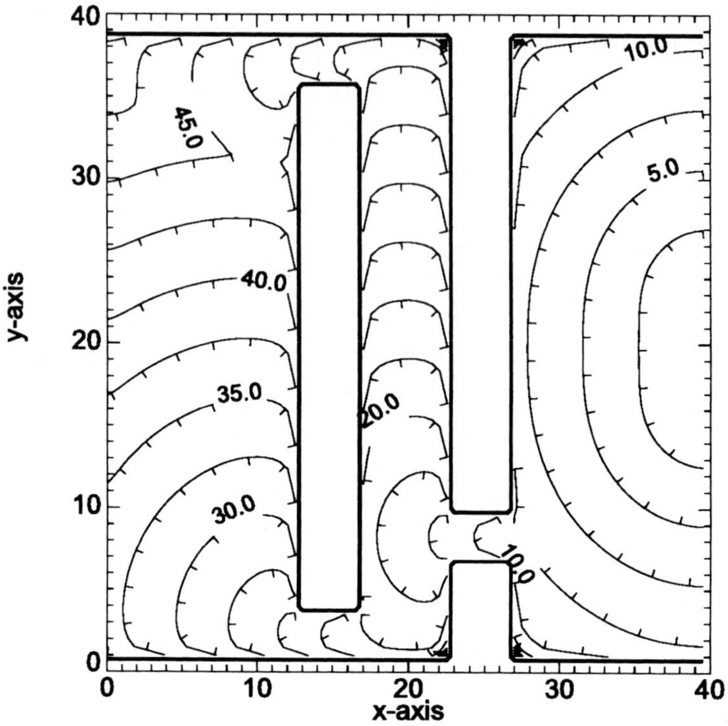

Figure 3. Example value function $W(t, \boldsymbol{x})$ assuming that $V = 1.5$ m/s and destination at $x_1 = 40$ and x_2 somewhere between 17.5 m and 22.5 m. The numbers indicate the minimal travel time (in seconds) to the destination. The lines perpendicular to the iso-cost curves define the optimal directions of the pedestrians.

5.6 Pedestrian Wayfinding for Constrained Operations

The previous section showed how we could determine the optimal walking paths for pedestrians under the conditions that pedestrian flow conditions are unconstrained. However, in practice pedestrian traffic conditions will deteriorate due to other pedestrians walking in the same area. In illustration, when pedestrian demands are large, the passage at the bottom of the first obstacle (from left to right) depicted in figure 3 may be oversaturated, yielding congestion. In this case, it is very likely that pedestrians will reconsider the former route choice by including the additional delays due to the prevailing congestion into their path choice.

To include the prevailing and future conditions in the pedestrian wayfinding behavior, two options are considered in the ensuing:

1. Pedestrian include only instantaneous information regarding prevailing conditions into their path choice behavior.
2. Pedestrians will anticipate on future conditions, by incorporating not only the prevailing conditions into their choice behavior, but also the anticipated behavior of the other pedestrians and the resulting anticipated traffic conditions.

In the remainder of this section, we will consider the case where pedestrians react on instantaneous information. The case where *pedestrians anticipate on future flow conditions* involves using predictions of the pedestrian flow conditions. This can be achieved by *simultaneously solving the time-dependent HJB equation (7) and a dynamic system forecasting pedestrian flow operations*, such as the conservation of pedestrians equation

$$\frac{\partial r}{\partial t} + \nabla_{\boldsymbol{x}} \cdot \boldsymbol{q}^* = 0 \quad \text{with} \quad r(t_0, \boldsymbol{x}) = r_0(\boldsymbol{x}) \tag{15}$$

with the pedestrian density flux $\boldsymbol{q}^*$ defined by

$$\boldsymbol{q}^* = \boldsymbol{e}^* r V_e(r) \tag{16}$$

where $V_e(r)$ defines the average speed as a function of the pedestrian density [14], and assuming that the average speeds equal $V(t, \boldsymbol{x}) = V_e[r(t, \boldsymbol{x})]$. For details, we refer to [9].

To include the *reaction to prevailing traffic conditions*, during the microscopic simulation, average speed fields $V(t, \boldsymbol{x})$ are determined based on the average speeds of the pedestrians. This field is substituted in the state-equation (1), and the optimal control problem is solved. Ideally, the pedestrian route-choice is updated each simulation step. That is, we solve the time-independent HJB-equation (13) for each simulation step at time $t_j = j\Delta t$. Note that the computations can be accelerated by using the value function $W(t_{j-1}, \boldsymbol{x})$ determined at the previous time-period as the initial conditions for iterative procedure used to calculate the value function. Alternatively, it

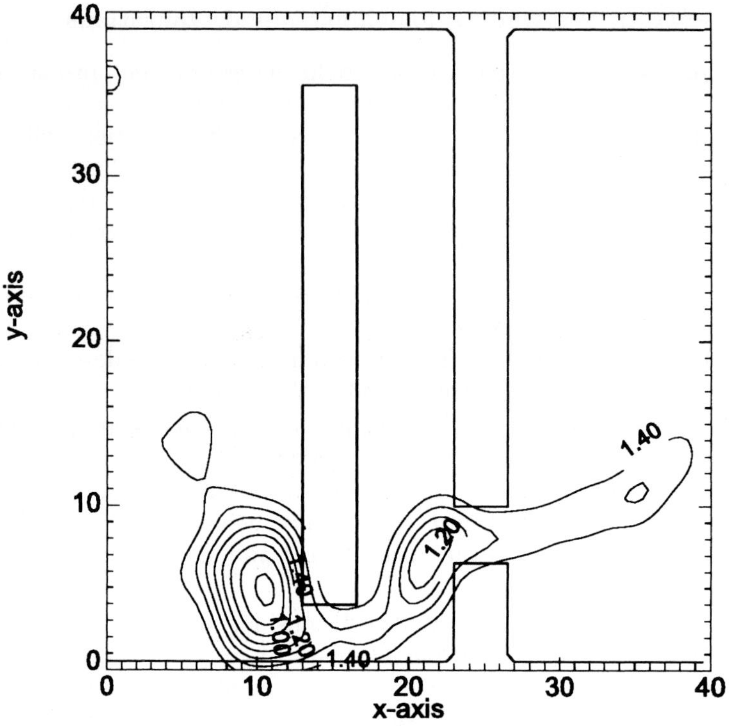

Figure 4. Example of average speed function $V(t, \boldsymbol{x})$, based on the results of microscopic simulation described in the ensuing of the report. The figure clearly shows that congestion has occurred in front of the bottleneck caused by the left-hand-side obstacle.

can be hypothesized that pedestrians will not continuously reconsider their route-choice, but only at discrete time intervals, say at each 10 s.

Example 2 Figure 4 shows the average speed function $V(t, \boldsymbol{x})$ determined from the results of microscopic simulation. When substituting this average speed field in the state-dynamics and subsequently solving the time-independent HJB equation, we find the optimal cost function $W(t, \boldsymbol{x})$ depicted in figure 5.

Without going into detail, we can show that by using instantaneous information, issues also observed in road networks, such as oscillatory choice behavior, are observed as well.

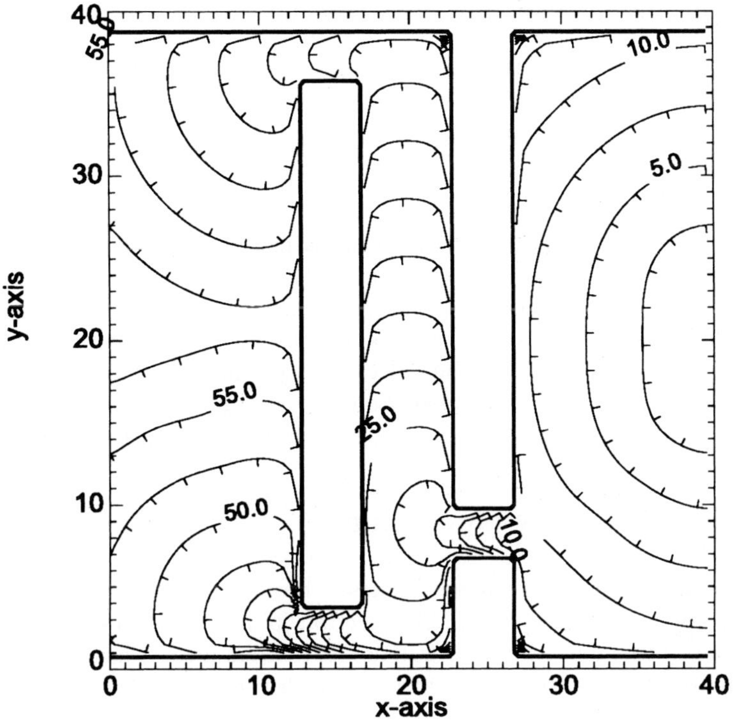

Figure 5. Value function $W(t, \boldsymbol{x})$ determined by considering average speeds resulting from microscopic simulation. Compared to the no-information case, pedestrians will be inclined pass the left-hand-side obstacle at the top-side, given the iso-cost curves.

5.7 Additions to Wayfinding Approach

Hoogendoorn and Bovy [10] discuss several additions and alternative modelling approaches to pedestrian wayfinding. Among these are

1. Inclusion of poor visibility and incomplete information
2. Inclusion of inexperienced pedestrians that are unfamiliar with the considered infrastructure

In the first case, the state-dynamics are adjusted such that the pedestrians only expect travel time delays in these areas that are visible from their current location. Besides this incomplete information, in the second case it is also assume that the obstacles that cannot be observed are also not taken into account by the pedestrians in their wayfinding behavior. It turns out that in the latter case, path enumeration approaches (see [5], and [10]) are often more convenient than the optimal control approach discussed in this paper.

6 Pedestrian Flow Operations

In the previous section, we have discussed how pedestrians choose the path that is optimal (strategic level). This section jointly describes pedestrian behavior at both the operational as well as the tactical level. That is, it describes how we model pedestrians accelerating towards their free speed, or pedestrian changing their direction to get to their destination location, and how pedestrian react or anticipate on slower pedestrians, given the conditions in their direct environment.

To this end, the dynamics of a single pedestrian α are described. More precisely, the location, speed, and direction of pedestrian α at time $t + dt$ are described as a function of the characteristics of pedestrian α and other pedestrians β at time t. This is done for all pedestrians in the flow. Since it is assumed that the pedestrian characteristics at instant $t + dt$ only depend on the state at instant t, updating can be done for all pedestrians at once (parallel updating). That is, the model does not presume an order at which the pedestrians are handled by the simulation, on the contrary to other microsimulation models.

6.1 Processes Determining Pedestrian Dynamics

At the combined operational / tactical level, the dynamics of pedestrian α are determined by the following processes:

1. Pedestrians moving according to their velocity $\boldsymbol{v}_\alpha(t)$.
2. Longitudinal acceleration towards the pedestrian free speed $v_\alpha^*(t)$.
3. Changing the current direction $\omega_\alpha(t)$ to the desired direction $\omega_\alpha^*(t)$, in order to get to the nearest point in the destination region D_α.

4. Changing velocity $\boldsymbol{v}_\alpha(t)$, while either *reacting* (local) *to* or *anticipating* (non-local) *on* other pedestrians.
5. Pedestrians that come too close to a wall or another obstacle O_i aim to increase this distance as quickly as possible.

In the remaining subsections, processes 1-5 are discussed in detail, in terms of ordinary differential equations.

6.2 Updating Pedestrian Locations

In the preceding chapters, we have already discussed how pedestrians update their location given the current velocity. Consider pedestrian α located at $\boldsymbol{x}_\alpha(t)$ at instant t. The location $\boldsymbol{x}_\alpha(t+dt)$ is prescribed by the velocity $\boldsymbol{v}_\alpha(t)$ via the following differential relation

$$\frac{d}{dt}\boldsymbol{x}_\alpha = \boldsymbol{v}_\alpha \tag{17}$$

6.3 Exponential Acceleration Laws in Cartesian Co-ordinates

Hoogendoorn and Bovy [10] assume *exponential acceleration laws* (see [7]) for both *radial* and *angular* (direction changing) acceleration. In doing so, they establish the following relation, that describes how pedestrians change their speed and their direction smoothly to the free speed v_α^* and desired direction ω_α^*

$$\frac{d}{dt}\boldsymbol{v}_\alpha = \boldsymbol{e}_\alpha\left(\frac{v_\alpha^* - v_\alpha}{\tau_\alpha}\right) + \boldsymbol{n}_\alpha v_\alpha\left(\frac{\omega_\alpha^* - \omega_\alpha}{\gamma_\alpha}\right) \tag{18}$$

where

$$\boldsymbol{e}_\alpha = (\cos\omega_\alpha, \sin\omega_\alpha) \quad \text{and} \quad \boldsymbol{n}_\alpha = (-\sin\omega_\alpha, \cos\omega_\alpha) \tag{19}$$

respectively denote the unit *direction vector*, the *unit normal vector*, perpendicular to the pedestrian's direction; τ_α and γ_α respectively denote the radial and angular acceleration times. Note that the second term of the right-hand-side of eqn. (18) reflects the inclusion of smooth angular adaptation (compare to [7], where it is assumed that pedestrians can instantaneously adapt their direction).

6.4 Local and Non-local Interactions on a Microscopic Level

Pedestrians are *anisotropic particles*, implying that they (mainly) react or anticipate to stimuli in front of them. Having described how pedestrians move, and accelerate, in this section, let us consider how pedestrians react and anticipate on other pedestrians in the flow downstream of them. In this respect, we can distinguish between *local reaction* and *non-local anticipation*. We assume that pedestrians perform (remedial) actions only if they are stimulated to do so. That is, both reaction and anticipation are described by a stimulus-response mechanism, the main advantage of which is the improved computational efficiency: only in case of an interaction (stimulus), the pedestrian will consider taking evasive maneuvers, thereby considering locations and speeds of other pedestrians in the flow.

Anticipation Path Let us define the so-called *anticipation path* $\mathcal{L}_\alpha$ of pedestrian α by a parameterized curve in the two-dimensional space, spawning from the current location $\boldsymbol{x}_\alpha(t)$ to the anticipation point $\boldsymbol{x}'_\alpha(t)$, defined by

$$\boldsymbol{x}'_\alpha(t) = \boldsymbol{x}_\alpha(t) + [l^0_\alpha + v_\alpha(t)T_\alpha]\boldsymbol{e}_\alpha(t) \tag{20}$$

where the anticipation distance $l_\alpha = l^0_\alpha + v_\alpha(t)T_\alpha$ is defined by a fixed anticipation distance l^0_α, and a speed dependent distance $v_\alpha(t)T_\alpha$, depending on the reaction time T_α of α. The anticipation point describes the maximum *look ahead region* which α will consider. In other terms, the pedestrian will not anticipate on traffic conditions that are beyond this point. Note that the extent in which the pedestrian will look ahead depends on both the reaction time and the pedestrian speed. In other words, when the pedestrian is walking at a high speed, he will look ahead further, implying that, despite the higher speed, the pedestrian will be able to react in time to downstream conditions to avoid colliding with another pedestrian.

If at instant t, one or more other pedestrians are on the anticipation path, pedestrian α will consider an evasive manoeuvre. Whether a manoeuvre is actually performed depends on among other things the speed and direction of the impeding pedestrians, and if there is indeed an opportunity to do so, i.e. whether there is a gap available. Which manoeuvre is performed, depends on the conditions in the direct environment of the pedestrian (e.g. density, average speeds and directions of other pedestrians).

Interaction Events To establish whether α interacts with other pedestrians β in the flow, we determined whether the distance between the anticipation path $\mathcal{L}_\alpha$ and the locations $\boldsymbol{x}_\beta$ is smaller than the sum of radius R_α and radius R_β of α and β. This is done by considering the *projection* of the location of β on $\mathcal{L}_\alpha$; if the distance between the location and the projection is smaller than $R_\alpha + R_\beta$, an interaction has occurred. Hoogendoorn and Bovy [9] distinguish between different types of interactions, depending on the direction and speed of the impeding pedestrian β compared to the direction of α. They argue that *passive interactions* (β is walking faster in the direction of α than α) are to be excluded. We will therefore only consider *active interactions* (β is moving slower in the direction of α than α).

Remedial Actions upon Interacting When α interacts with a slower pedestrian β, α either needs to:

1. *Deceleration.* Adapt its current speed $v_\alpha(t)$ to assure that α does not collide with β; this may require α to come to a complete standstill, if β walks in the opposite direction
2. *Bypassing.* Change the current direction $\omega_\alpha(t)$ to bypass β.
3. *Combination of α and β.*

Whether α changes direction and / or adapts his velocity depends on both the *possibility to pass* the impeding pedestrian β without needing to decelerate, and the *gains which can be expected from passing.* Similar to car-based traffic, passing possibilities depend on the available gaps in the pedestrian flow. When no gaps are available, we assume that α either adopts its velocity to match the velocity component of the impeding pedestrian's velocity with respect to its direction (in case the impeding pedestrian walks in the same direction), or comes to a complete standstill (in case the impeding pedestrian walks in the opposite direction). In the first case, an additional assumption is that the velocity of the pedestrian is diminished further when the distance towards the impeding pedestrian is too small. That is, we assume that the pedestrian increases his distance to the impeding pedestrian to match the preferred distance $l_\alpha(v_\alpha)$ by further reducing his speed. In either case, the pedestrian will still adapt his direction. As a consequence, even a pedestrian who is standing still changes his direction when interacting or adapting the direction towards the desired direction.

Choosing the Passing Direction We assume that when pedestrian α is impeded by β, he will first choose his *target direction* ξ_α. On the one hand, α will choose the direction in which the *local conditions are optimal* (e.g. few pedestrians, pedestrians walking into the same direction), while on the other hand, α will not diverge too much from his desired direction $\omega^*_\alpha(t)$, or make a directional changes which is too sharp. For the sake of simplicity, we assume that a chooses amongst a finite number of discrete directions i, which are characterized by the angle ξ_i for $i = 1, ..., I$. For each of these directions, *anticipation zones* Ω_i, representing the traffic conditions in the direction ξ_i are established. The shapes of these regions are not fixed, and can be either partially overlapping or disjoint.

For each of these zones, the utility U_i can be established by considering the conditions in these zones Ω_i. These conditions reflect attributes of both the candidate direction ξ_i as well as the zone Ω_i, such as the density, directions and velocities of the pedestrians, and the drift from the desired direction of α. These attributes are determined by considering the pedestrians in the zone, yielding the utilities that determine the probability that a chooses direction ξ_i.

We describe the probability that α chooses direction i by a *discrete choice model.* This implies that the utility U_i is determined by a *systematic component* Y_i, determined by the attributes reflecting the conditions in the zones

Ω_i, and a random component E_i. The latter describes the ***intra-personal*** and ***inter-personal*** differences of pedestrians, as well as observation errors of the pedestrians, and is described by a random number. Assuming that E_i is a ***Gumbel distributed random variable***, the resulting model is a ***logit-type model***, in which case the probability π_i that direction i can be determined by the following expression

$$\pi_i = \frac{\exp(-Y_i)}{\sum_j \exp(-Y_j)} \tag{21}$$

given the systematic utility components Y_i.

Which direction the pedestrian chooses is determined by chance. We hypothesize that pedestrian α takes the following attributes into account, when determining the systematic utility Y_i of direction ξ_i:

1. *Density* Y_{i1}. It is hypothesized that pedestrian α will choose to pass pedestrian β on the side with the smallest pedestrian density. The densities are determined by considering the number of pedestrians in the anticipation zones Ω_i.
2. *Likelihood of interaction* Y_{i2}. Pedestrian a prefers the direction in which he is less likely to interact with another pedestrian. The likelihood of interaction is determined by first determining the component $c(\boldsymbol{v}_\alpha, \boldsymbol{v}_\beta)$ of the velocities of each pedestrian β in the zone along the velocity vector $\boldsymbol{v}_\alpha$ of α. Subsequently, for each pedestrian β satisfying $c(\boldsymbol{v}_\alpha, \boldsymbol{v}_\beta) < 1$, the likelihood of interaction is determined by calculating $|\boldsymbol{v}_\alpha - \boldsymbol{v}_\beta|$, and subsequently adding the results.
3. *Difference with desired direction* Y_{i3}. When passing pedestrians, pedestrian α will keep in mind his destination. In other words, the pedestrian aims to prevent drifting away from his destination too much. Therefore, it can be argued that the pedestrian has some preference to pass on the side in which the desired direction is pointing.
4. *Preference for a specific passage side* Y_{i4}. Depending on the customs in the considered situation, pedestrians may have a preference to pass each other on either the right-hand-side or the left-hand-side.
5. *Presence of obstacles* Y_{i5}. When an obstacle or wall is present in the region, it is unlikely that the pedestrian will consider this region Ω_i to change its direction to.
6. *Rapid directional changes* Y_{i6}. Pedestrians dislike making frequent directional changes. That is, $|d\omega_\alpha/dt|$ is preferably as small as possible.

We assume an aggregate linear utility function for the entire group of pedestrians, which is assumed homogeneous. That is, having determined the attributes of zone Ω_i, the systematic utility Y_i is determined by

$$Y_i = \sum_j a_j Y_{ij} \tag{22}$$

where a_j are weighing factors, describing the relative importance of the different attributes. The model structure implies that optimality of the alternative is determined by trade-off mechanisms.

Passing Probability Once the pedestrian has chosen the preferred passing direction, the possibility of passing in that direction needs to be determined. Clearly, this possibility depends on the number of pedestrians (density) and their spatial distribution in the region Ω_i^* where the pedestrian aims to walk too. For the sake of simplicity, we have assumed that the passing event can be described by a random variable, where the probability of passing p is a function of the pedestrian density $r_i(t)$ in the chosen anticipation zone Ω_i^*.

Time Discretization The models described by the expressions given in the preceding sections are continuous in time and in space. Although this yields distinct conveniences, time-continuous models cannot be implemented on a digital computer: time-discretization is needed. The proposed approach is very simple, and is based on explicit forward differences in time. This implies that the simulation period is divided into a finite set of intervals $[t_j, t_{j+1})$, usually with a fixed duration Δt. Assuming that the state of the system (i.e. the locations, and velocities of the pedestrians) at instant t_j is known, the state at instant t_{j+1} can be approximated using the continuous time equations described in the preceding sections. In case the dynamics consist of (ordinary) differential equations, forward differences are used.

6.5 Summary of the Model

Summarizing, pedestrian dynamics are governed by four processes, namely location updating, accelerating towards the free speed, changing direction to traverse the optimal route to the destination, and interactions and consequent remedial actions with slower pedestrians or pedestrians in the opposing direction. Contrary to the CA-model [1], [2], the model equations are continuous in time and space, the main advantage of which is that in any time-step, movements of any angular value can be made. Secondly, the model features local optimal behavior of pedestrians, implying that pedestrian will optimize their direction by considering conditions in their direct environment, based on the concept of subjective utility optimization.

7 Simulation Results

In this section, we will discuss simulation results obtained using the microscopic simulation model. For the first case, we will study the relation between the speed-density function and pedestrian composition, and pedestrian lane formation, by considering bi-directional pedestrian flow through a hallway of $40m \times 8m$ for different pedestrian compositions. The second case aims at illustrating pedestrian wayfinding modelling approaches.

Having experimented with different time-steps, it turned out that a time step of 0.1 s presented a good trade-off between numerically accuracy and computation speed.

7.1 Model Calibration

One of the main problems with any (vehicular or pedestrian) microscopic simulation model is the calibration of the parameters. While some of the parameters are directly observable from the flow, others can only be determined indirectly. In microscopic modeling, it appears to be common practice to change parameters of the models until some macroscopic features of the flow (such as speed-density curves, or flow-density curves) are approximated with sufficient accuracy. For the pedestrian model described here, model calibration was equally difficult, and mostly done using trial and error. This section discusses the resulting model parameters:

- *Space use.* We assume that each pedestrian occupies the same amount of space A. Using the empirical results from [14], we have used $A = 0.15m^2$ per pedestrian.
- *Free speeds.* The free speeds of the pedestrians are randomly drawn from a Gaussian distribution. Using the empirical results of [14], a Gaussian distribution with mean $1.34m/s$ and standard deviation of $0.26m/s$ is appropriate. In the exceptional case that non-positive free speeds are drawn, these are corrected.
- *Longitudinal and lateral acceleration times.* It is assumed that pedestrians need approximately $1s$ to accelerate towards their free speed, i.e. the radial acceleration time equals $\tau_\alpha = 1s$. For the angular acceleration time, the same value is chosen $\gamma_\alpha = 1s$.
- *Anticipation parameters.* With respect to the parameters describing pedestrian anticipation, we have used $T_\alpha = 0.5s$, $l_\alpha^0 = 0.5m$, and $R_\alpha = 2.5m$. Two anticipation zones have been used, with $\xi_{1,2} = \pm 0.1\pi$.
- *Passing probability.* The passing probability $\pi(r_i)$ is assumed to be a linear function of the zone density r_i, i.e. $\pi(r_i) = 1 - cr_i$. Clearly, when all space is occupied by pedestrians $(r - r_{jam})$, we have $cr_{jam} = 1$ and $\pi(r_{jam}) = 0$ (i.e. there are no passing possibilities).

7.2 Results from Microscopic Simulation

In the remainder of this section, we will illustrate the results from the simulation experiments by considering the resulting speed-density and flow-density curves, and the surface plots. While the former show the ability of the model

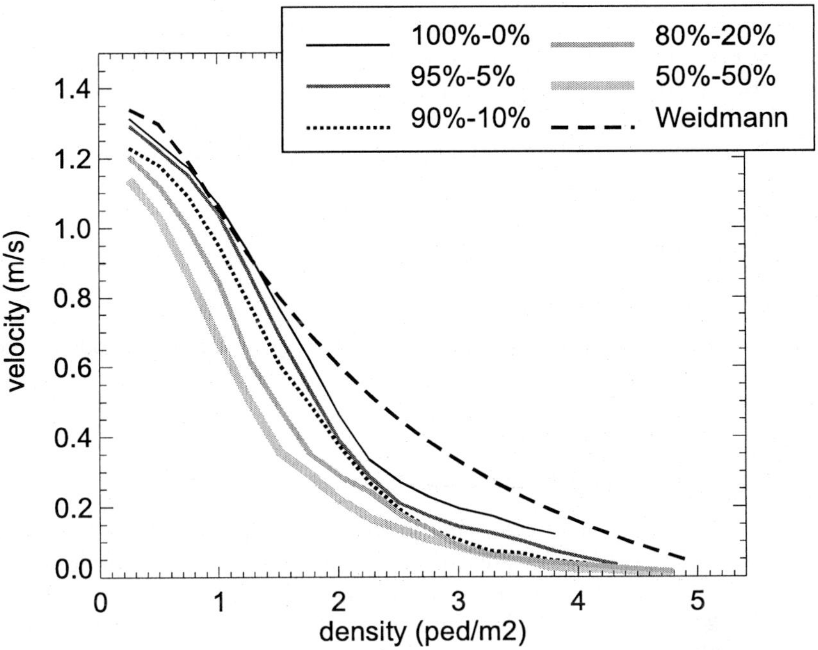

Fig. 6. Speed-density function for different compositions in bi-directional simulated pedestrian flows (averaged for both directions).

to reproduce speed-density curves that have been established empirically, the latter illustrate another interesting feature of pedestrian flows, namely self-organization and chaos. Pedestrian wayfinding is illustrated by showing snapshots from the simulation model.

Speed Density Curves Let us first consider the speed density relations that can be obtained from the model results. Both the speeds and the densities are obtained by dividing the simulated area into squares of $2m \times 2m$. For each of these cells, the density can be determined by counting the number of pedestrians in these cells. Similarly, the average velocity can be determined for each cell. Having determined the density and velocity for each cell, the time-average for $2s$ is considered (20 simulation steps of $0.1s$ each). This is done for the entire simulation period. To ensure coverage for an entire range of densities, several simulation runs have been used with varying pedestrian traffic demands. The simulated area represents a bottleneck situation.

Figure 6 shows the results of this approach. The figure shows speed-density relations for uni-directional, and bi-directional flows with varying compositions. Also, the empirical speed-density relation of [14] (for uni-directional flows) is depicted. Clearly, both relations are similar. We argue that the small differences can be reduced by improved model calibration. Similarly, we can determine the flow-density curves from the simulation experiments. This provides insights into the capacity of the pedestrian flow as a function of the composition. It also appears that the capacity determined from simulation is a function of the composition. Starting from uni-directional flows, the capacity reduces as the fraction of pedestrians in the opposite direction increases. This reduction pertains until a critical composition in reached (approximate 65%-35%), after which the capacity increases slightly (until a composition of 50%-50%). Let us finally note that the simulated capacity $C = 1.2 ped/m/s$ is equal to the capacity value reported by [14]. However, the capacity reduction caused by traffic in opposing directions is smaller than our simulation experiments suggest (see among others [12]).

7.3 Self-organization of Dynamic Lanes

In real-life traffic flow, self-organizing phenomena have been observed [7]. More specifically, from empirical results it has been shown that dynamic lanes are self-formed in bi-directional pedestrian flows. These lanes have a considerable width, and consist of pedestrians walking into the same direction. These features have been observed in a number of Japanese studies (e.g. [13]). Similarly, in crossing flows, clusters are self-formed. These clusters are relatively large regions in which pedestrians with (approximately) the same direction are walking.

The reason of self-formation is two-fold. Firstly, pedestrians walking in the same direction are less likely to interact than pedestrians walking in opposite directions. As a consequence, pedestrians are more likely to be able to walk undisturbed in a region where pedestrians are walking into the same direction. Until arriving at such a region (lane), pedestrians will interact with and other pedestrians walking in opposite directions, and pass them. In other words, even when passing direction choice behavior is random (no preferences for passage sides), it is likely that dynamic lanes are formed (see [7], [1], and [2]). However, the lane-formation is also dependent on the local choice behavior of pedestrians, assuming that they prefer walking towards regions where the other pedestrians are walking in the same direction, rather than walking towards a region where pedestrians are walking in the opposite direction, or perpendicular to his direction.

In this section, we will show that the presented microscopic model exhibits self-formation, specifically for bi-directional pedestrian flows (see figure 7). Moreover, we will show that these self-organized phenomena have a chaotic-like nature. To this end, we consider the case in which 50% of the pedestrians are walking from left to right, while 50% of the pedestrians are walking from right to left. In the sequel, flows 1 and 2 respectively refer to pedestrians walking from left to right and from right to left.

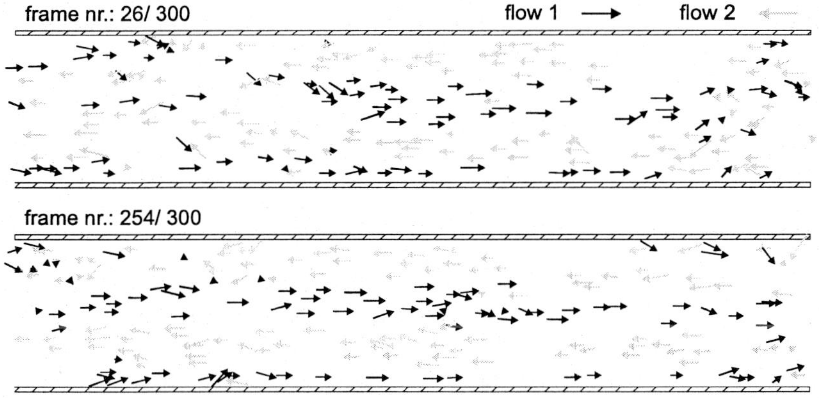

Figure 7. Animation frames for different time steps ($t = 26s$ and $t = 254s$) showing the formation of dynamic lanes in a $40m \times 8m$ hallway. The lanes consist of pedestrians having the same destination. When the lanes are formed, their characteristics (location, width) are near constant during the remainder of the simulation.

Figure 8 (example I) shows the formation of the previously discussed dynamic lanes. In this case, it appears that flow 1 self-organizes into two unidirectional lanes. In this case, one of the lanes is not straight, but moves from the down left corner to the upper right corner of the hallway. Flow 2 constitutes three smaller lanes, two of which move along the walls. Figure 8-c,d (example II) shows that flows 1 and 2 form three and two lanes respectively. The emerging patterns are quite similar in both examples.

Figure 8-e,f shows very different results. In this case, seven lanes are formed, namely three for flow 1 and four for flow 2 (although the width of one lane is very small). Clearly, this pattern differs considerably from the patterns observed from the previous examples I and II.

The formation of dynamic lanes results in the so-called conformable mapability of the flow-density curve which has been observed in empirical pedestrian traffic flow operations [11]. This property reflects the fact that the flow-density relation is mostly independent of the flow composition. That is, although the flow is composed of pedestrians walking in different directions, the speed density curve is approximately the same (although some differences exist).

Moreover, from the simulation experiments performed during our research, we conclude that, although in all cases dynamic lanes are formed, no unique characteristics of the stable lanes formed (e.g. width, number, direction in which lanes move) could be determined. It is therefore plausible to hypothesize that the model behaves like a chaotic system, where the state of the system converges to one of the possible final (and sometimes stable) states (so-called attractors). Small differences with respect among other things to the initial conditions (e.g. spatial distribution of pedestrians at the start of the simulation), and microscopic pedestrian behavior, determine the final system state. However, it is impossible to predict to which state the system will converge beforehand.

7.4 Pedestrian Wayfinding

This final model application shows the differences in wayfinding behavior of inexperienced and experienced pedestrians. To this end, we assume that *inexperienced pedestrians* have some vague idea on the location of their destination, but rely on their vision only to determine which route to take. In other words, when an obstacle or wall blocks the view, and the destination cannot be seen from the current destination, pedestrians choose the route around the visibile obstacle which has the shortest distance to the destination, not considering obstacles or walls that cannot be observed. In illustration, figure 9 shows how inexperienced pedestrians that enter the area at $x_1 = 0m$ choose to pass the left-hand-side obstacle on either the top or the bottom route, irrespective of the fact that in a lot of cases, the latter is shorter, due to the right-hand-side obstacle. More details on inexperienced pedestrian wayfinding are presented in [10].

On the contrary, the approach to pedestrian wayfinding discussed in this paper incorporates obstacles that pedestrians cannot see from their current location. Figure 10 shows pedestrians that make an *optimal route choice*. Compared to the results depicted in figure 9, figure 10 shows that more experienced pedestrians choose the bottom route compared to the inexperienced pedestrians.

It turns out that the route choice behavior has a considerable effect on the pedestrian traffic flow conditions. Considering inexperienced pedestrians, congestion occurs at the right-hand-side bottleneck. For experienced pedestrians, congestion occurs at the bottleneck on the bottom of the left-hand-side bottleneck. When pedestrian react on the prevailing traffic conditions, this congestion in turn yields a shift in pedestrian wayfinding behavior, implying that more pedestrians are included to pass the left-hand-side obstacle on the top (see figure 11).

8 Summary and Outlook

In this paper we have put forward a microscopic simulation model for pedestrian flows. Besides pedestrian flow operations of individual pedestrians, the model includes approaches to pedestrian wayfinding based on the optimal control paradigm. The approach involves specifying the cost due to travel time losses, walking too close to obstacles, rapid directional changes, etc., incurred

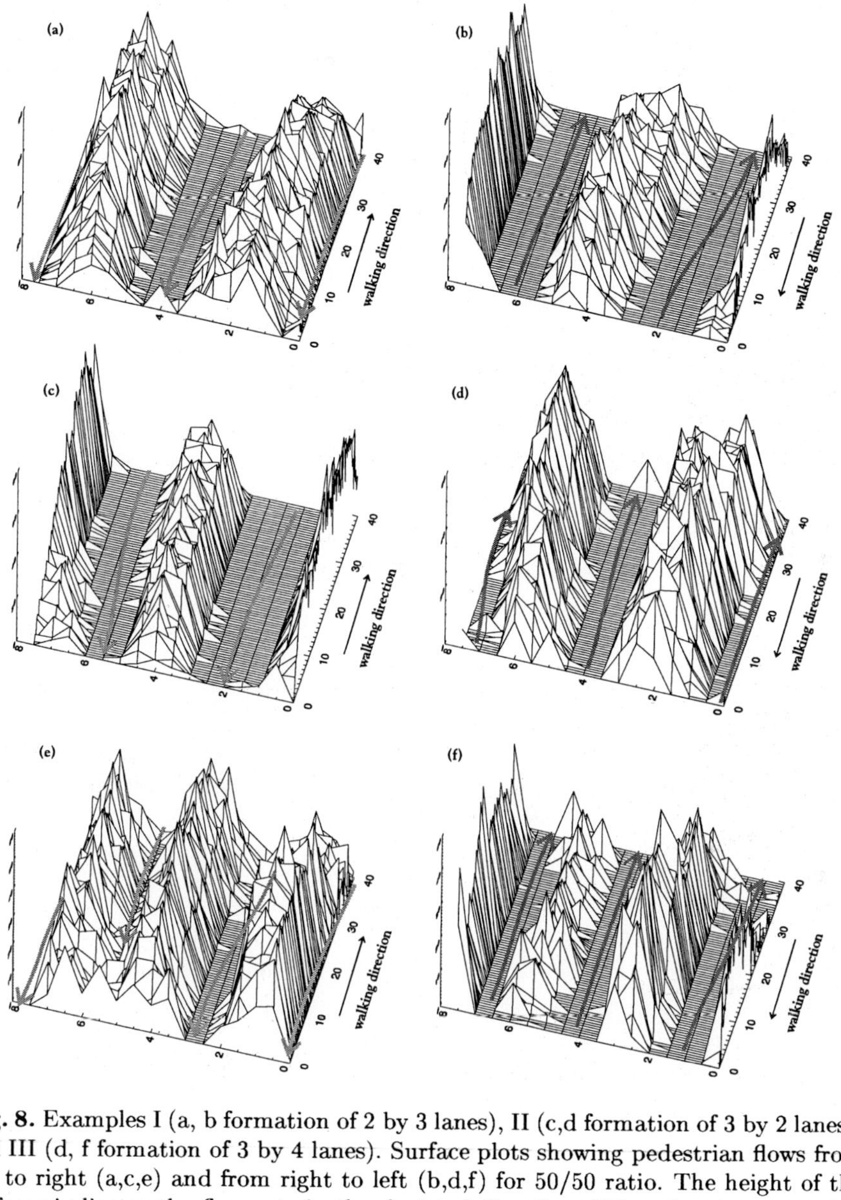

Fig. 8. Examples I (a, b formation of 2 by 3 lanes), II (c,d formation of 3 by 2 lanes), and III (d, f formation of 3 by 4 lanes). Surface plots showing pedestrian flows from left to right (a,c,e) and from right to left (b,d,f) for 50/50 ratio. The height of the surfaces indicates the flow-rate in the depicted direction. 250 pedestrians occupied the hallway at all times. Note that the arrows indicate the opposing flows.

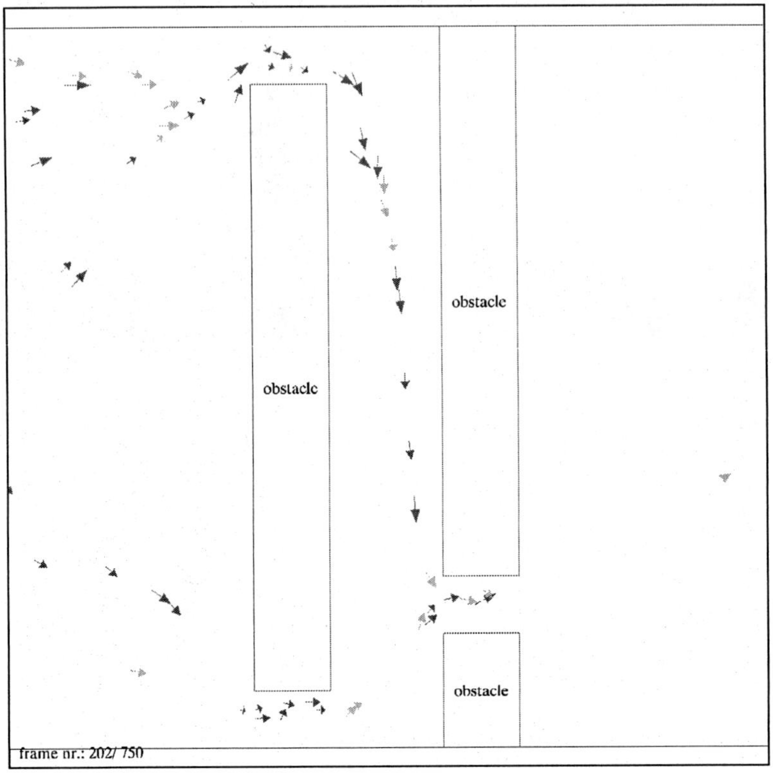

Figure 9. Pedestrian wayfinding for unexperienced pedestrians.

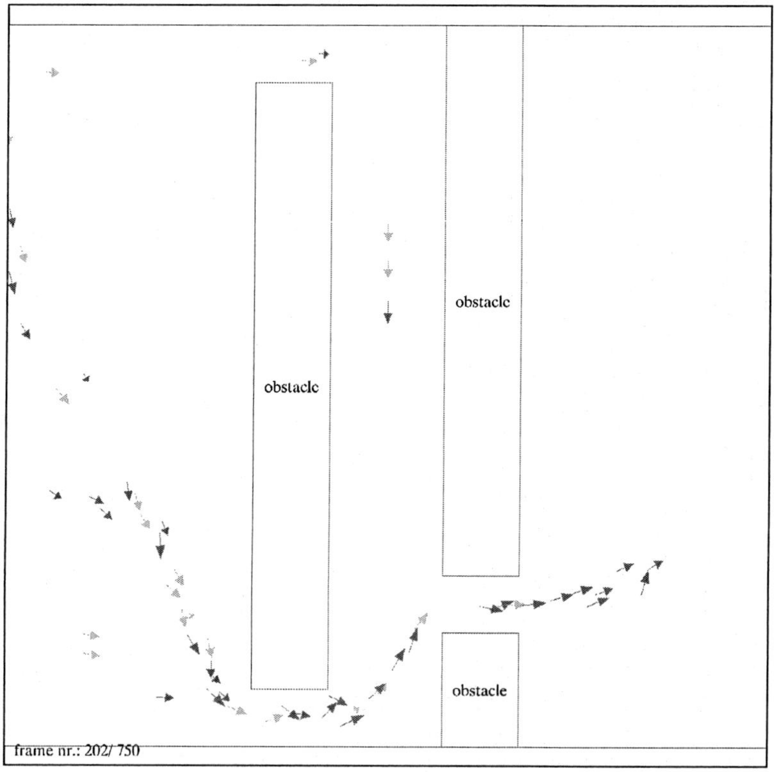

Figure 10. Snapshot from pedestrian simulation in case pedestrian choose the route optimising the expected travel time.

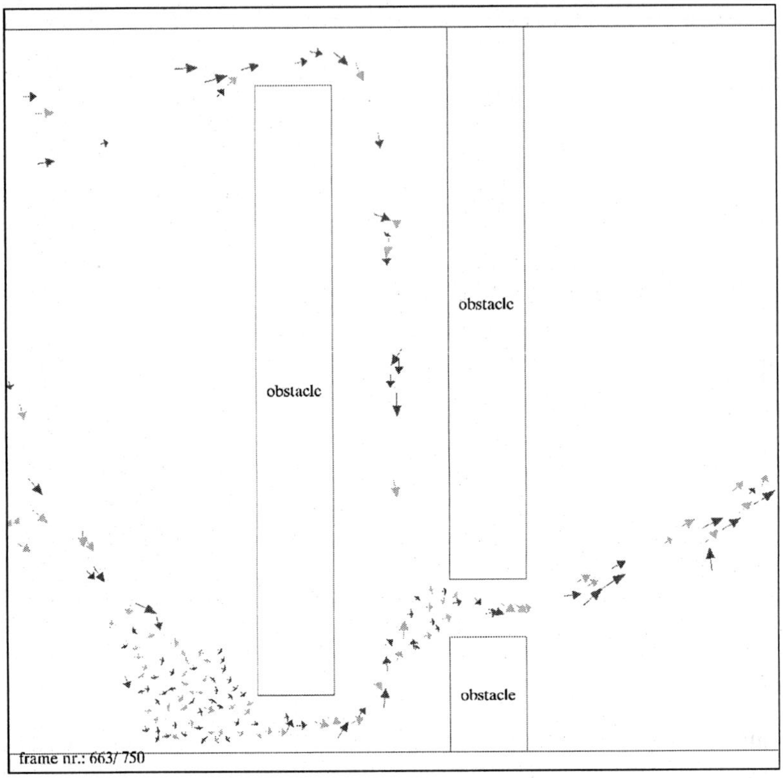

Figure 11. Snapshot of pedestrian simulation showing that once congestion occurs, the increasing travel time on the formerly optimal route yields an increase in pedestrians preferences for the route alternative along the top of the left-hand-side obstacle.

while walking. The developed approach yields the optimal walking direction at each location in the considered area, which can be determined by solving a partial differential equation stemming from dynamic programming (the so-called *Hamilton-Jacobi-Bellmann* equation).

The new model is characterized by among other things the fact that positions and velocities of pedestrians are continuous in space. Compared to other microscopic simulation models, such as the CA-models, space-continuity enables easy handling of pedestrian flows moving in different directions. Moreover, the pedestrian dynamics are governed by some rather simple dynamic processes (compared to the relatively large number of behavioral rules of the recent CA-models). The model is discretized with respect to time using explicit schemes, implying that the state of the system at the next time-step can be determined completely by the state at the previous time step. This is beneficial, since it enables parallel updating (just like the CA-models), enhancing potential computational efficiency of the model.

The dynamics of the pedestrians are governed by moving, longitudinal acceleration towards the pedestrian's free speed, changing the direction towards the destination of the pedestrian, and pedestrian interaction with consequent remedial maneuvering. The latter process is based on the principle of stimulus and response, implying that pedestrians will only consider changing their direction and speed once they are impeded by a slower pedestrian or a pedestrian walking into the opposite or perpendicular direction. If there is a stimulus, the response of the pedestrian depends on the flow conditions in the direct environment. For instance, whether the impeding pedestrian can be passed depends on the available free space (gaps). How the impeding pedestrian is passed, depends on the characteristics of the pedestrians that are present in the target direction of passing. We have assumed that impeded pedestrians make user-optimal choices regarding the direction of passage, given that passing is possible.

Using the developed model, speed-density relations have been established and compared to empirical speed-density relations. It appeared that these are similar, while small differences can be corrected by improving the model parameter settings. Moreover, we can conclude that the model features a very important characteristic of pedestrian flows, namely self-formation of dynamic lanes and clusters, meaning that pedestrians walking into the same direction cluster together, thereby forming (dynamic) lanes. The simulation experiments have also revealed that this process appears chaotic, in that although the final state is stable, it cannot be predicted beforehand, since it critically depends on initial conditions and microscopic behavior. Finally, we have shown how pedestrian wayfinding influences the location of congestion occurrence.

Data collection and empirical analysis methodologies are important issues. It is likely that video observations can provide data that will enable determination of essential model parameters, velocity distributions, and other (microscopic) characteristics of pedestrian flows. Using video, pedestrian interaction behavior can also be studied in the near future.

Acknowledgement

This research is funded by the Social Science Research Council (MaGW) of the Netherlands Organization for Scientific Research (NWO).

References

1. V. Blue and J. Adler, *Bi-Directional Emergent Fundamental Pedestrian Flows from Cellular Automata Microsimulation*, in Proceedings of the 14th International Symposium on Transportation and Traffic Theory. Ceder, A. (ed.), 235-254 (1999).
2. V. Blue and J. Adler *Modeling Four-Directional Pedestrian Movements*, To appear in Transportation Research Records (2001).
3. Chiolek, M.T. *Spatial Behaviour in Pedestrian Areas*, Ekistics **268**, 120-121 (1978).
4. W. Daamen and S.J. Van de Stouwe, *Modelling pedestrians in transfer stations*, Conference Proceedings World Congress on Railway Research 1999, Tokyo, October 1999.
5. P.G. Gipps *Simulation of Pedestrian Traffic in Buildings*, Schriftenreihe des Instituts fuer Verkehrswesen 35, University of Karlsruhe (1986).
6. D. Helbing and P. Molnar, *Social Force Model for Pedestrian Dynamics*, Physical Review E **51**, 4282-4286 (1995).
7. D. Helbing, *Traffic Dynamics: New Physical Modeling Concepts* (Springer, Berlin, 1997).
8. S.P. Hoogendoorn, *Multiclass Continuum Modeling of Multilane Traffic Flow*, TRAIL Dissertation Series, Delft University of Technology, (Delft University Press, Delft, 1999).
9. S.P. Hoogendoorn and P.H.L. Bovy, *Gas-Kinetic Modelling of Pedestrians Flows*, To appear in Transportation Research Records (2001).
10. S.P. Hoogendoorn, S.P. and P.H.L. Bovy *Microscopic Modeling of Pedestrian Flows Development and application results of MiPSim*, TRAIL Studies in Transportation Science (2001).
11. R. Hughes, *The Flow of Large Crowds of Pedestrians*, To appear in Mathematics and Computer Simulation (2000).
12. F.P.D. Navin and R.J. Wheeler *Pedestrian Flow Characteristics*, Traffic Engineering, June 1969, 30-36 (1969).
13. A. Toshiyuki, *Prediction Systems of Passenger Flow*, In *Engineering for Crowd Safety* (R.A. Smith and J.F. Dickie eds.) (Elsevier Amsterdam, 1993), 249-258.
14. U. Weidmann, *Transporttechnik der Fussgänger*, ETH, Schriftenreihe IVT-Berichte 90, Zürich (in German) (1993).

2-Dimensional Optimal Velocity Models for Granular Flow and Pedestrian Dynamics

Y. Sugiyama[1], A. Nakayama[2], and K. Hasebe[3]

[1] City College of Mie, Tsu, Mie 514-0112, Japan
[2] Gifu Keizai University, Ohgaki, Gifu 503-8550, Japan
[3] Aichi University, Miyoshi, Aichi 470-0296, Japan

We extend the 1-dimensional Optimal Velocity (OV) Model, which describes car following behaviour, to a model in 2-dimensional space. We present the possibilities of applying the model to the investigation of several kinds of phenomena of collective behaviors, such as granular flow and pedestrian dynamics.

1 Introduction

The Optimal Velocity Model is one of the most successful models for the description of the dynamics of traffic flow [1,2]. The model is a 1-dimensional car following model governed by the following equation

$$\frac{d^2}{dt^2}x_n(t) = a\left\{V(\Delta x_n) - \frac{d}{dt}x_n(t)\right\}. \tag{1}$$

x_n is the position of the n-th car, and $\Delta x_n = x_{n+1} - x_n$ is the headway. a is a sensitivity constant. $V(\Delta x_n)$ decides the optimal velocity, which should have a form such as $V(x) = \tanh(x - c) + d$.

The OV model can be extended to several directions. Here, we try to develop our model to a 2-dimensional version. We apply the models to the phenomena of several kinds of collective motions of interacting particles, such as granular flow in a pipe, actually the pseudo 1-d model [3] is very similar to a 1-d OV model, or pedestrian dynamics in passage and the bio-motions, etc.

2 Model Description

In the OV model, the interaction of two particles acts not to crash each other. If the interaction is isotropic, the particles always move uniformly distributed in 2-dimensional space. So the interaction should be anisotropic. The interaction is strong between a particle and its neighbour in the direction of motion within the short range. The interaction with the particle moving behind is very weak. On the basis of these assumptions, we present the models in two different ways.

First, we express 2-*d* OV model as the naive extension of the 1-*d* OV model in 2-dimensional space.

- Model-1

$$\frac{d^2}{dt^2}\boldsymbol{x}_n(t) = a\left\{\sum_{m}^{N(n)} V(r_{mn}(t))P(\theta_{mn}(t))\boldsymbol{e}_{mn}(t) - \frac{d}{dt}\boldsymbol{x}_n(t)\right\} \tag{2}$$

$$r_{mn}(t) = |\boldsymbol{x}_n(t) - \boldsymbol{x}_m(t)| \tag{3}$$

$$\boldsymbol{e}_{mn}(t) = \frac{\boldsymbol{x}_n(t) - \boldsymbol{x}_m(t)}{|\boldsymbol{x}_n(t) - \boldsymbol{x}_m(t)|} \tag{4}$$

$$V(r_{mn}) = \tanh(r_{mn} - c) + d \tag{5}$$

where $\boldsymbol{x}_m(t)$ is the 2-dimensional position of the m-th particle. The optimal velocity $V(r_{mn})$ is determined by the distance of each two particles. Here, we chose it as the same as in the 1-*d* OV model. The function $P(\theta_{mn})$ gives the anisotropy of the interaction. We can choose the function $V(r_{mn})$ and $P(\theta_{mn})$ freely according to what kind of phenomena we want to investigate in 2-*d* OV models.

$\sum_{m}^{N(n)}$ denotes the summation with respect to m which satisfies $m \in \{\|\boldsymbol{x}_n - \boldsymbol{x}_m\| < 2r$, s.t. $r = \min_m |\boldsymbol{x}_n - \boldsymbol{x}_m|\}$. $N(n)$ means the number of particles with the condition surrounding the n-th particle. We note the summation is dependent on time according to the positions of all particles. Instead of this 'simple summation' we can define the 'average summation' as $\{1/\sum_{m}^{N(n)} \rho(\theta_{mn})\} \sum_{m}^{N(n)} \rho(\theta_{mn})$, where $\rho(\theta_{mn})$ denotes the weight which depends on the direction to the m-th particle. This means that the particle pays more attention to the particle just in front of it than to the particles in other directions. If we choose $\rho(\theta_{mn}) = 1$, the average summation turns to $\{1/N(n)\} \sum_{m}^{N(n)}$, which can be called 'simple average'. But this summation is not realistic. Actually, the model with this summation results in the trivial motion. We have two possibilities for the choice of summations.

Model 1 is formulated with consciousness of the effect of following behavior of particles. This model may be suitable for the flow of bio-motion or evacuation dynamics.

We present another expression of 2-*d* OV model as

- Model-2

$$\frac{d^2}{dt^2}\boldsymbol{x}_n(t) = a\left\{V_c\boldsymbol{e}_x - \sum_m^{N(n)}[v_{max} - V(r_{mn}(t))]P(\theta_{mn}(t))\boldsymbol{e}_{mn}(t) - \frac{d}{dt}\boldsymbol{x}_n(t)\right\}. \tag{6}$$

In contrast to model-1, model-2 is formulated with consciousness of the exclusive effect. $V_c\boldsymbol{e}_x$ is the homogeneous velocity in the background, which has some constant value. We choose the x-axis in the direction of the flow, which corresponds to the direction of pipe in granular flow or the passage in pedestrian flow.

3 Results

In general, the OV model has two characteristics: the following behavior and the exclusion behavior. The 'attractive' force in the following behavior and the 'repulsive' force in the exclusion behavior are originally different. But they can not be distinguished from each other in 1-dimensional space. The 2-*d* OV model can be expressed in the above two different concepts with different formulas. Actually, if we express the model-1 and 2 in 1-dimensional space, each model is reduced to the same 1-*d* model.

In this report we concentrate on the dynamics of granular flow in a pipe and pedestrian flow in a passage. For this purpose model-2 is adequate and we perform the simulation using model-2.

For a model of granular flow in a pipe, we introduce the anisotropy of interaction as

$$P(\theta_{mn}(t)) = \frac{1}{2}(1 + \cos\theta_{mn}(t)), \tag{7}$$

where

$$\cos\theta_{mn}(t) = \frac{\boldsymbol{x}_m(t)\cdot\boldsymbol{x}_n(t)}{|\boldsymbol{x}_m(t)||\boldsymbol{x}_n(t)|}), \tag{8}$$

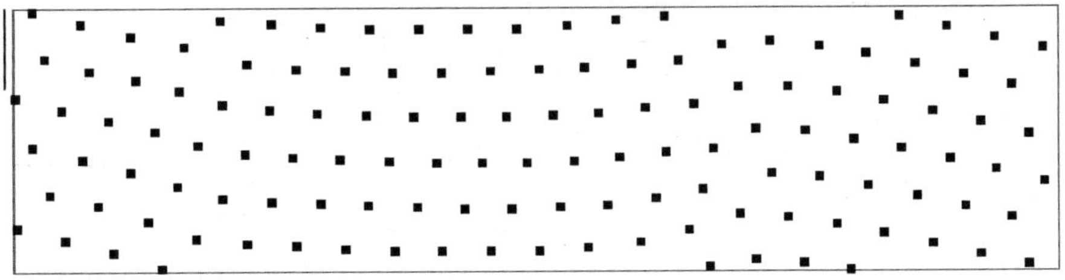

Figure 1. The snapshot of granular flow after optimization. The sensitivity $a = 0.5$ and the number of particles N=130.

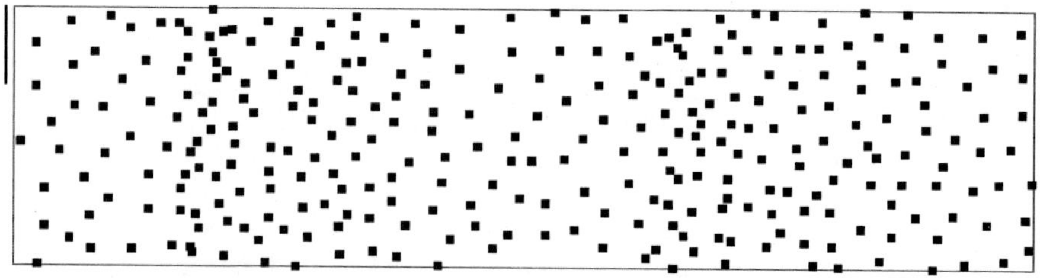

Figure 2. The snapshot of granular flow after optimization. The sensitivity $a = 0.5$ and the number of particles N=300.

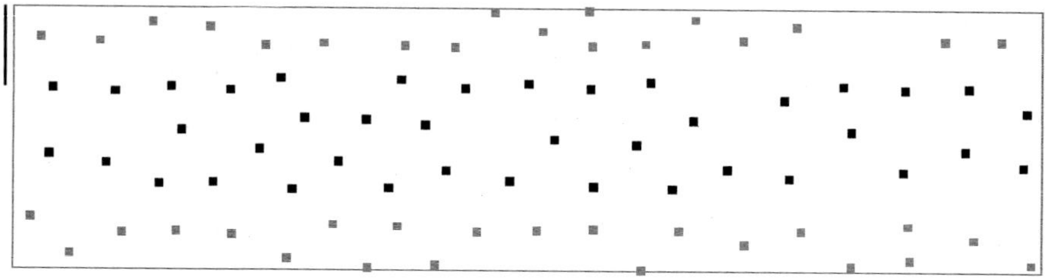

Figure 3. Snapshot of a simulation for pedestrians: fast particles (light) and slow particles (dark). $a = 0.5$ and N=80.

and choose the simple summation. The setting seems reasonable from the physical point of view, but this is one trial among various possibilities. For simplicity, the simulations are performed with periodic boundary conditions in the both directions of 2-dimensional space. The flow moves in the x-direction on the torus.

In the low density of particles, the typical behavior is shown in Fig.1. The homogeneous flow is realized.

In high density we can observe the jam formation on the torus which resembles the slug in pipe, as in Fig.2.

For a model of pedestrians in a passage, we introduce an anisotropy into the interaction in the following way:

$$P(\theta_{mn}(t)) = \cos\theta_{mn}(t), \tag{9}$$

and choose the average summation with the weight function as

$$\rho(\theta_{mn}(t)) = \cos\theta_{mn}(t). \tag{10}$$

For the simulation of pedestrians we introduce two different kinds of particles, which have different OV functions. In this case we can observe the fast particles pass the slow particles. For low density the flows of slow particles and fast particles are sharply separated and each flow moves homogeneously, as in Fig.3.

For intermediate density in Fig.4, an 'island' of jam cluster is formed, which is qualitatively different from the case of granular. The jam of granular forms the 'band' perpendicular in the direction of the flow. The bunching appears, which is similar to the phenomena in traffic flow.

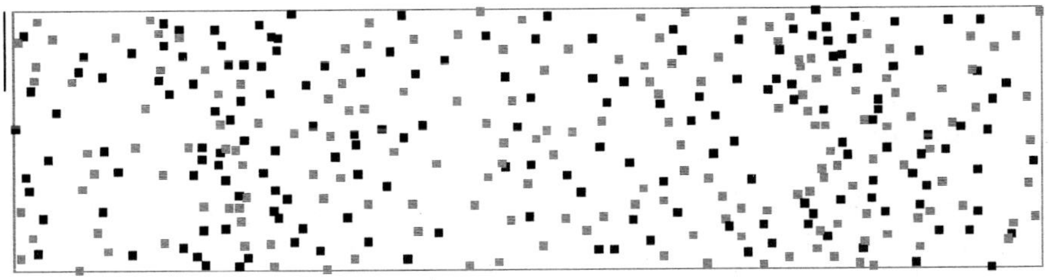

Figure 4. Snapshot of a simulation for pedestrians: fast particles (light) and slow particles (dark). $a = 0.5$ and N=350.

Finally, we show the relation between the average velocity of the flow and the density (the number of particles N) and the Q-k diagrams: the relation between the flow and density, for both models in Fig.5.

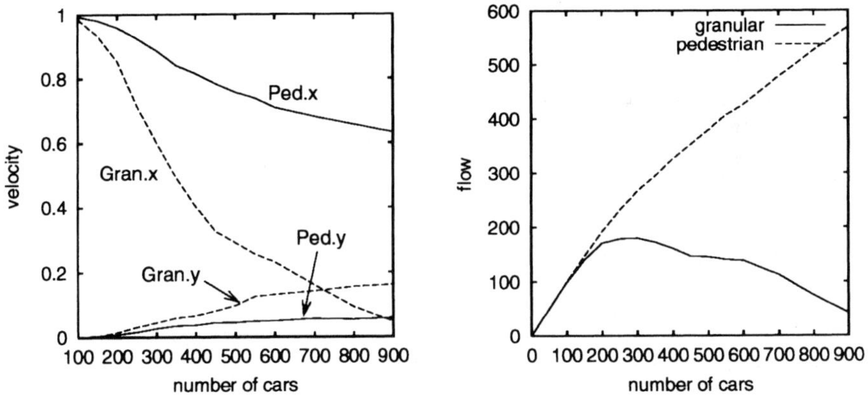

Figure 5. The average velocities in x and y direction and Q-k diagrams, for granular (Gran.) and pedestrian (Ped.) flow. The sensitivity $a = 1$.

4 Summary

The simulation of the 2-d OV model may provide some information about phenomena in multi-lane (including two-lane) traffic flow such as 'synchronization'. The effect of multiple lanes can be investigated by the 2-d OV model. The simulation of two flows moving in opposite directions shows the behavior of pedestrians in a passage [4,5]. We can also apply the 2-d OV models to the phenomena of evacuation and bio-motions.

References

1. M. Bando, K. Hasebe, A. Nakayama, A. Shibata, Y. Sugiyama, Japan J. of Ind. and Appl. Math, **11**, 203 (1994); Phys. Rev. E**51**, 1035 (1995). Y. Sugiyama; D. E. Wolf, M. Schreckenberg A. Bachem (Eds.), Traffic and Granular Flow'(World Scientific) 137 (1996); M. Schreckenberg, D. E. Wolf (Eds.), Traffic and Granular Flow '97 (Springer) 301 (1998).
2. D. Chowdhury, L. Santen, and A. Schadschneider, Physics Reports **329** 4-6, Statistical Physics of Vehicular Traffic and Some Related Systems, 228 (2000).
3. O. Moriyama, N. Kuroiwa, M. Matsushita, and H. Hayakawa, Phys. Rev. Lett. **80**, 2833 (1998).
4. M. Schreckenberg and D. E. Wolf (Eds.), Traffic and Granular Flow '97, (Springer, Singapore, 1998); D. Helbing, I.J. Frakas, T. Vicsek, p 21.
5. D. Helbing, H.J. Herrmann, M. Schreckenberg, and D. E. Wolf (Eds.), Traffic and Granular Flow '99, (Springer, Berlin, 1998); T. Vicsek, A. Czirók, and D. Helbing, p 147, D. Helbing, I.J. Farkas, and T. Vicsek, p 245: and references therein.

Multi-Modal Traffic in TRANSIMS

K. Nagel

Swiss Federal Institute of Technology (ETH) Zürich, Switzerland

Traffic simulations need to include other modes of transportation besides car. This paper explains how this can be implemented in a large scale micro-simulation package. The most important element is to recognize that trips start and end on the walk network, and to explicitely model this mode.

1 Introduction

Transportation planning considers questions of the long-term impact of infrastructure or policy changes – such as an additional subway line or an increase of gasoline costs. Typical questions are: Where does traffic/congestion shift to? Will trips be dropped or added? Might people change where they live or work as a longterm response to changes in the transportation system?

Traditional models for transportation planning typically put most of their emphasis on car traffic. Other modes are often considered via modal split calculations only – that is, once transportation demand has been calculated, some of the demand may decide to take other modes instead of the car, usually via a discrete choice model based on travel times and different values of time on the different modes.

The reason for this is that static assignment models do not have a realistic representation of the dynamics. For example, time-dependent effects of car traffic, such as queue build-up and spill-back during the rush-hour, cannot be represented. Similarly, more complicated effects in other modes, such as congestion of a pedestrian facility, or bus-car interaction, cannot be modeled.

The approach is nevertheless justified as long as the other modes quantitatively do not play a large role, and it may be justified under certain circumstances if the car mode is the only congested mode. In many metropolitan regions, this is not true: other modes besides cars are contributing more than 25% of all trips, and these other modes are heavily congested.

As a result of these and other thoughts, there is currently a push towards microscopic (or agent-based) transportation planning models. In these models, all entities including the travelers themselves are individually resolved. This makes is in principle straightforward to include arbitrary elements of reality. The downside is that it may be hard to implement, it may use considerable computer resources, and data for calibration of these effects may be hard to obtain. Nevertheless, microscopic transportation simulation models are feasible, and several groups are working on them.

This paper will first give an introduction into how such microscopic transportation simulation packages are designed. This is followed by a section about a possible implementation of multiple modes into such a package. The paper is concluded by a discussion and a summary.

2 Large Scale Transportation Simulations

One aspect of any transportation simulation package needs to be the actual traffic dynamics. Ideally, one wants to have travelers walking to the vehicles, entering the bus or the car, then those vehicles drive through the traffic, stop at lights, obey speed limits, and make turns at intersections after getting into the necessary lane. This is the task of the traffic micro-simulation module.

One aspect of such a traffic simulation is that travelers need to know how they navigate through the system. When do they enter a bus, when do they make a turn, etc.? This task is typically solved by a routing module.

Any routing module needs as input the origin and the destination of the trip. Traditionally, one uses origin-destination matrices here – that is, tables which contain information about how many people travel between any pair of destinations in the region. A disadvantage of this method is that all information about the travelers themselves is lost at this level – and in consequence, decisions cannot be coupled to such information any more. For example, a person who arrives late at work because of congestion may skip lunch in order to catch up. An alternative method of demand generation for travel is called "activity-based". This method is closer to how individual humans think and plan – that is, for each individual in the simulation the module generates individual plans for activities.

It is not sufficient to run these three modules in sequence since, for example, plans depend on congestion but congestion depends on plans. In order to find a solution which is consistent between the modules, it is common to run feedback iterations, where the agents slowly adapt to the situations they encounter. – The next sections will describe the modules including feedback in more detail.

2.1 Demand Generation

In order to generate the demand for transportation, many of today's projects are based on activities. That is, instead of using an abstract origin-destination matrix, they look into people's motivation to travel. This is achieved by having the simulation generate sequenced activities for the day, for example: Sleep until 7am. – Be at work at 9am. – Go to lunch at 12:30pm for one hour. – Leave work after 8 hours of work. – Go to the kindergarten to pick up a child. – Go shopping. – Go home and stay there for the rest of the day. These activities come with locations, that is, they generate demand for travel. Two successive activities which take place at two different locations will generate a trip request.

It is clear that one needs to have information about the network impedance (i.e., the travel times between different locations) when making activity plans. We will come back to this in the section about feedback and in the outlook.

Once the simulation "knows" where and when people do their activities, transportation is generated via connecting activities that take place at different locations. This includes modal choice (walking, bicycle, train, car, etc.) and the precise routing.

2.2 Traffic Simulation

So far, we have generated "plans" of the individuals. The travelers in the simulation compute their plans for the following time period, which may for example be a day. Now these plans need to be executed, which is called the traffic micro-simulation. These simulations come at many different levels of resolution and fidelity, reaching from the traditional steady-state flow-based cost function to very detailed micro-simulations.

If one is interested in time-dependent results, as for example the queue build-up during the onset of rush periods, the simulation needs to be sufficiently realistic to contain such dynamics. Traditional flow-based cost functions are *not* able to realistically deal with such dynamical effects, at least not in a straightforward way. Thus, the right traffic simulation has to be chosen according to what aspects of the dynamics one wants to have represented for a given question. There are currently more than hundred traffic microsimulations [1]. However, in most of these travelers do not follow individual plans as was explained above. Examples of plan-following traffic simulations are TRANSIMS [2], DYNAMIT and MITSIM [3], DYNASMART [4], and LEGO [5]. In some of these, travelers do not "know" their full routes but only their destinations (and are routed via the simulation which knows the paths); the practical impact of this difference is not known.

2.3 Feedback

The traffic simulation needs input from the demand generation, since it executes the plans from the demand generation. However, the demand generation depends on the traffic simulation because for example congestion only shows up in the traffic simulation, and demand adjusts to such shortages. In order to deal with this situation, one iterates between demand generation and traffic simulation. For example, the demand generation module is run assuming no congestion, the resulting traffic simulation is run, then the demand simulation is run again now including the congestion from the last traffic simulation run, etc., until a steady state is reached. That is, the system is systematically relaxed towards a consistent state.

An important issue in this context is the question of computational efficiency versus behavioral realism. Traditional static equilibrium assignment has, once the origin-destination matrix and the network including link-cost function are specified, a unique solution (in terms of the paths flows). This allows one to concentrate on the fastest computational algorithm to find this solution. In iterated transportation simulations, this issue is considerably more complicated. Although such iterated simulations relax to some kind of reproducible steady state, this state is stochastic. Also, it is not clear if this state is unique or if it depends on the path of the computation – in dynamic traffic assignment (DTA) computational evidence indicates that it is unique [6,7] but once one includes other aspects such as public transit or activities rescheduling it is easy to construct scenarios where this is no longer true.[1] Another important aspect is that real systems may not be at this relaxed state at all. In that case, transients matter much more, which means that the rules of the synthetic travelers would have to be much more realistic than for just reaching the relaxed state.

2.4 Computing

A metropolitan region easily consists of several millions of travelers. In order to get an idea of the necessary computing let us assume that we want to simulate a scenario of 24 hours (= 86400 secs) with 1 million travelers. Simulations typically do second-by second updates. In each update, several variables such as traveler speed or traveler location are computed. Let us assume that there are ten such variables, and the update of each variable needs 100 basic computer cycles. This is not a lot: For example, a simple algorithm for lane change will consider the following: Do I need to change lanes? Where are my neighbors? Is there enough space for me to change lanes? Also, fetching data from RAM typically takes about 10 cycles. A result of these assumptions is that, with a 1 GHz-CPU, the scenario takes

$$\frac{24 \text{ hours} \times 10^6 \times 10 \times 100}{1 \text{ GHz}} \approx 24 \text{ hours}$$

of computing time. Running 50 iterations thus will take 50 days of computing time on a single CPU. Using parallel computers will reduce this number accordingly; for example, using a so-called Beowulf of 50 Pentium PCs will lead

[1] Proofs of ergodicity [8] for such systems do not always apply for actual implementations, and they are also often not practical since they imply thousands of iterations instead of the usual number of fifty.

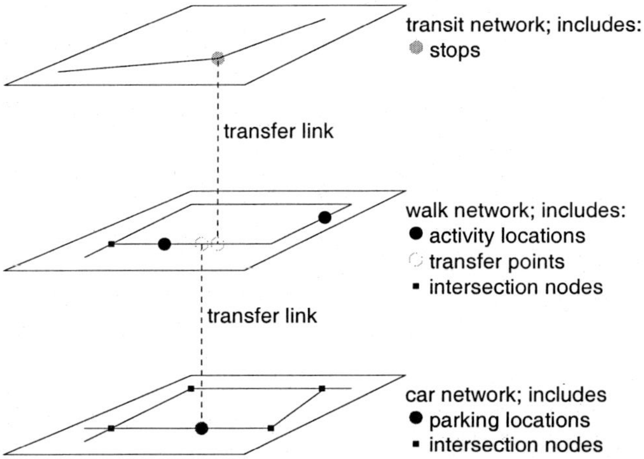

Figure 1. Conceptual representation of network layers

to a computing time of between one and two days. See Refs. [9,10] for more information.

3 Multi-modal Approach

3.1 Conceptual Approach

The arguably best starting point for getting a clean concept of a multi-modal transportation simulation package is to recognize that all activities take place on the walking network. Even if the car is parked in the garage, people have to walk to it. As a result, the simulation needs to have a walking network, on which all activities are located, and which has transfer points to the other modes of transportation (Fig. 1).

In the actual implementation, the walking network may look similar to the car network, and it may use the same nodes and links, but conceptually it is a separate thing. The walking network does not have to be connected; Fig. 2 shows an example of a activity location which can only be reached via a freeway.[2]

[2] Also, in this construction, different lines of public transit are not connected – if one wants to change transit lines, one first has to go, via a transfer link, back to the walking layer. In consequence, the transit layer consists of a number of unconnected lines.

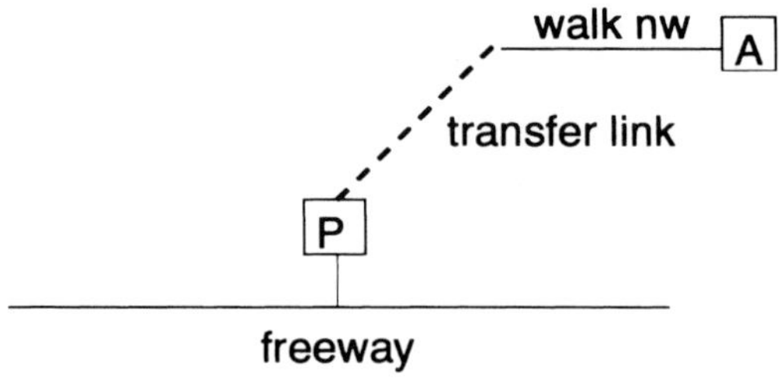

Figure 2. Conceptual representation of an activity location which can only be reached via a freeway.

3.2 Multi-modal Activities

TRANSIMS activities files come with a specification of the desired mode how an activity should be reached. This specification is a mode string – a sequence of letters denoting the different modes. Examples for mode strings are:

- wcw: walk–car–walk
- wbw: walk–bus–walk
- wbwbw: walk–bus–walk–bus
- wtw: walk–transit–walk (transit includes walking)

This mode specification can be changed via the feedback process. That is, a traveler can try out different modes, and eventually settle on one according to arbitrary performance criteria. This will be described in more detail on a section on feedback.

3.3 Multi-modal Router

It is possible to construct a search graph such that a generalized Dijkstra algorithm can be used to find a "best" multi-modal path through the network [11]. The construction assumes that the mode string is given, for example "wbwbw", where again "w" stands for walk and "b" stands for bus. This example would involve a route which uses two different bus lines.

The search graph for the algorithm is obtained as follows (see Fig. 3): For each additional entry in the mode string, a separate network layer is assumed. That is, for "wbwbw" we have, from bottom to top, first a walking layer, then a bus layer, then again a walking layer, etc. The layers are connected via the transfer links as was already discussed above. The main difference is that for the algorithm, network levels are replicated if they appear at more than one position in the mode string.

When consulting Fig. 3, it becomes clear that a path through the search graph can be found in an organized way via a Dijkstra algorithm: Assume that the starting location is on the left, and the traveler wants to reach a destination on the right. Since we have "wbwbw" as mode sequence given, the algorithm will construct five levels in exactly that sequence, as can be seen in the figure.

The Dijkstra algorithm will then start in the bottom left, and expand systematically in its usual way through the search graph. Transfer links are logically treated the same way as links for other modes, that is, they have a link travel time and possibly other attributes. In this way, the algorithm will deterministically find the fastest path through the expanded network, which will correspond to a path of the desired mode sequence. Note that it will be possible to stop the algorithm if the destination can be reached faster without changing bus lines. For further information, see Ref. [11].

The router also keeps track of car availability. It will plan car trips only with available cars (which are usually parked close to home), and it will make sure that the car is returned at the end of the day. In complicated situations (such as a different family member returning the car), it is however the microsimulation which will find errors and notify the selector to correct them. For example, the microsimulation will, during a 48h simulation, notice if a car was not returned the previous evening since it will not be available for the trip.

3.4 Router Implementation

The above description is conceptual. For efficiency, the implementation does several simplifications and modifications:

- The walk network and the street network share the same nodes. This makes sense since most of them are the same anyway. By setting the links accordingly, it will still be possible to have nodes which can be reached by car only or by walking only.
- Activity locations and transfer points are made the same on the walk network.
- There is only at most one activity location and one parking location on each link. This means, for example, that all parking along a link is aggregated into a single location.
- In fact, the walking and street levels are not replicated at all in the actual implementation. Instead, a counter along the "fringe" nodes of the Dijkstra algorithm keeps track of where in the mode string sequence each fringe node is during the algorithm.

3.5 Multi-modal Traffic Micro-simulation

As said above, the micro-simulation executes all plans simultaneously in a realistic representation of the traffic system, and it is here that interactions between travelers, for example causing jams or missed busses, are computed.

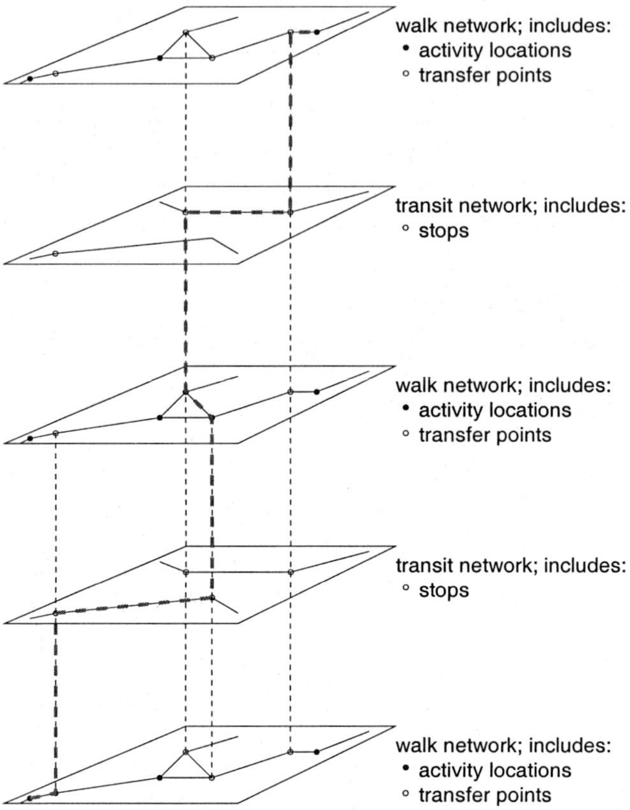

Figure 3. Conceptual representation of product construction, including one path.

As also said above, TRANSIMS, in its current implementation, assumes that all plans (including routes) are pre-calculated before the simulation starts.

In principle, the micro-simulation is just a good representation of reality. With respect to multi-modal travel, this means that again all travel of persons begins at an activities location, which is located on the walking network. If the travelers use additional modes besides walking, the simulation moves the travelers to, say, the bus stop or the parking lot, where they enter a car.

That means, for example, that traffic lights follow schedules or adaptive procedures as they do in reality, and drivers change lanes according to traffic conditions or in order to be in correct lanes for intended turns. It also means that the simulation logically moves the travelers from their starting locations (usually home) to the bus stop or the parking lot, then lets them enter their vehicles, etc.

A convenient feature of the design as it is described here is that the simulation can actually deal with multi-modal plans even if the corresponding mode of travel is not implemented in the micro-simulation. For example, walking may not explicitely be implemented. In such a case, the micro-simulation will just assume that the plan contains completely correct information, and execute that. For example, the plan may say that the traveler leaves home at time t_1 and arrives at the bus stop at time t_2. If walking is not explicitely implemented, the micro-simulation will just assume that the travelers leaves home at t_1 and arrives at the bus stop at time t_2. Conveniently, this works for arbitrary modes which are not implemented, so that the router does not have to take into account possible limitations of the micro-simulation, and also micro-simulations with different capabilities can run on the same set of plans. It should be clear that for such modes no interactions can be modeled. For example, when walking is not modeled, missing a train because of crowding at the subway station will not happen.

A different aspect of the multi-modal implementation is the actual representation of public transit schedules. In the TRANSIMS traffic micro-simulation, public transit service is generated the same way car traffic is generated, i.e., via drivers picking up their transit vehicle at the depot and following a prescribed route with prescribed stops, a prescribed schedule, and possibly prescribed procedures for the case of schedule deviations. This is similar to the modeling of a shared ride, where a driver in one car has to pick up one or several passengers and drop them off later. In that way, effects like the bunching of busses or schedule delays because of congestion will come out naturally. Also, buses have a capacity limit, possibly leaving passengers behind at the bus stops, and it is possible to make the loading/unloading times a function of the crowdedness. If the actual routes of drivers and transit vehicles are not available, then synthetic routes can be generated from the schedule. Such synthetic routes will however obviously not include certain effects of reality, as for example the effect that a vehicle which switches lines may not be available because of a delay.

3.6 Explicit Pedestrian Traffic in TRANSIMS?

Currently, TRANSIMS does not explicitely implement pedestrians; they are treated as any other non-implemented mode as described above. It should be clear that TRANSIMS could benefit from the addition of a realistic pedestrian mode, in particular in cities where pedestrian facilities suffer from congestion. It should also be clear that including pedestrian traffic into TRANSIMS would be relatively straightforward, and could be done via arbitrary models as long as they follow the same plans format. This, however, puts constraints on the pedestrian model design, namely that is has to follow the graph-oriented structure of TRANSIMS. In particular, it must be possible to define paths from one activity location to another, plus from activity locations to transfer points and vice versa. Agents need to be able to follow such paths, and the information must be structured in a way so that the router can work with it. This excludes, for example, random wandering across a wide plaza or through an old city where a person only approximately knows where he or she is. Future versions of TRANSIMS may eventually be extended in this regard. Also note that aspects of this can be included via putting delays on the transfer links.

3.7 Feedback

No computational module will be able to anticipate everything which can happen in other modules. For example, strongly fluctuating congestion could lead to missing the train although according to mean travel times everything looks fine. Similarly, the shared ride may depart without me if I am too late, or the parking lot at the park-and-ride station may be full. In many cases, researchers may want to define non-optimal behavioral decision models for humans, which would exclude from the beginning that travelers take, say, the fastest path.

For all these elements, TRANSIMS provides its feedback mechanism (also called "selector"). Instead of (computationally) dwelling forever on a decision between two different modes, or a range of possible starting times, the simulation can just decide "to try it out". For example, a transit and a car option can be tried by an individual, and the performance of both be evaluated. In that performance evaluation, aspects like the number of stop signs, or the predictability of the respective options can be taken into account. This allows both the modeling of non-optimizing behavioral models, and the inclusion of non-linear costs which cannot be picked up by Dijkstra-type routers (also see below). In the second case, an "optimal" solution probably cannot be guaranteed, but an individual improvement procedure can be implemented.

In many cases, this also solves the question of "where to put" certain decisions. For example, a mode choice decision could be made on the level of the router, via putting values of time and other things on different modes and then finding the optimal mode for the trip. It however also makes sense to look at this during the activities planning, since mode choice and activities chaining are intimately related. Via the feedback mechanism, neither of these modules has to make the decision alone – instead, in critical cases where several solutions are plausible, the selector will be able to make the individuals try out different solutions and then pick a good one according to heuristic (or optimizing) rules. Clearly, there is a tradeoff between modeling effort for the trip planning and the number of iterations one may have to run.

4 Discussion and Outlook

Any design has shortcomings. Although the TRANSIMS design seems robust in the sense that design decisions from many years ago have not gotten in the way of newer issues, there are always elements which are easier to model with a certain design and others which are harder to model. Here are some issues:

- The router will find a good or the best transit line for a trip. In many cases, however, several alternative lines could be taken, for example at typical transit malls in American Central Business Districts. In such cases, the router will pick one line which seems convenient from the schedule; but it may happen that, when the traveler arrives at the stop, another line might be better.
  Including these effects into the router is possible, but only by defining additional "virtual" transit lines for each segment of the transit network which is served by several lines. For example, if a transit mall is served along its entire length by bus lines 1, 3, and 9, then a new line with a not yet existing number could be defined along the length of the path where the three lines overlap. Unfortunately, this yields a possibly high number of such "virtual" lines, since all combinatorial combinations of lines and segments could become new virtual lines. Also, the micro-simulation would have to pick it up in some way, that is, the route plan would have to state which bus lines are equivalent. – What this still would not capture is if two different paths via different transit lines could bring the traveler to the next activity.
- Dijkstra-type routers cannot deal with non-additive costs, which are used by most public transit fare systems or some road pricing schemes. The TRANSIMS router cannot deal with this directly, but at least for simple cases the feedback mechanism can pick it up. For example, it can force the simulation to try out both the transit and the car option, and add the monetary costs only at the point when the relative performance between both is evaluated.

5 Summary

This paper explains the concept of the implementation of multi-modal traffic in TRANSIMS. For this, the paper has two parts: A first, explaining the TRANSIMS design using examples from car-only runs, and a second, which explains the provisions for multi-modal trips. The emphasis of the paper is on the second aspect. TRANSIMS starts by assuming that all activities are located on the walk network, and in consequence any trip connecting two activities at different locations will involve walking at the beginning and at the end of the trip. Transitions to other modes are made via transfer points and transfer links. The activities generator of TRANSIMS provides the preferred mode to reach an activity, and the multi-modal router routes these trips. The traffic microsimulation reads these plans, executes them in detail if the corresponding mode is implemented, and "believes" them if the mode is not implemented. The latter means that for a non-implemented mode the traveler is just moved to her/his destination according to the timing provided in the plan; no interaction between travelers is computed in this case. This also gives a clear implementation path to the addition of realistic pedestrian modules. The TRANSIMS feedback mechanism provides an answer to some of the usual conceptual questions, for example of where to place the mode decision (into the activities planning or into the routing) or what to do about non-additive link costs.

Acknowledgments

The multi-modal implementation of TRANSIMS is documented in [2]. The work leading to this was started when the author was still at Los Alamos, as a collaboration of many including Chris Barrett, Dick Beckman, Madhav Marathe, Riko Jacob, Goran Konjevod, and Stephen Eubank. Many others have contributed to the current implementation.

References

1. SMARTEST (Simulation Modelling Applied to Road Transportation European Scheme Test. See www.its.leeds.ac.uk/smartest/.
2. TRANSIMS, TRansportation ANalysis and SIMulation System, since 1992. See transims.tsasa.lanl.gov.
3. DYNAMIT/MITSIM, 1999. Massachusetts Institute of Technology, Cambridge, Massachusetts. See its.mit.edu.
4. H.S. Mahmassani, T. Hu, and R. Jayakrishnan. *Dynamic traffic assignment and simulation for advanced network informatics (DYNASMART).* In N.H. Gartner and G. Improta, editors, *Urban traffic networks: Dynamic flow modeling and control* (Springer, Berlin/New York, 1995).
5. In *Verkehr und Mobilität*, number 66 in *Stadt Region Land. Institut für Stadtbauwesen* (Technical University, Aachen, Germany, 1998).
6. M. Rickert and K. Nagel, *Issues of simulation-based route assignment.* Presented at the International Symposium on Traffic and Transportation Theory (ISTTT) in Jerusalem, 1999. See www.inf.ethz.ch/~nagel/papers.
7. J. A. Bottom, *Consistent anticipatory route guidance.* PhD thesis, Massachusetts Institute of Technology, Cambridge, MA, 2000.
8. C. Cantarella and E. Cascetta, *Dynamic process and equilibrium in transportation network: Towards a unifying theory*, Transportation Science A, **25**, 305–329 (1995).
9. M. Rickert and K. Nagel, *Dynamic traffic assignment on parallel computers in TRANSIMS,* Future generation computer systems, **17**, 637–648 (2001).
10. K. Nagel and M. Rickert, *Parallel implementation of the TRANSIMS micro-simulation*, submitted. See www.inf.ethz.ch/~nagel/papers.
11. C. L. Barrett, R. Jacob, and M. V. Marathe, *Formal language constrained path problems*, Los Alamos Unclassified Report (LA-UR) 98-1739, see transims.tsasa.lanl.gov, 1997.

A Multi-Agent Cellular Automata Model of Pedestrian Movement

J. Dijkstra[1], J. Jessurun[1], and H. Timmermans[2]

[1]Eindhoven University of Technology: Faculty of Architecture, Building and Planning – Design Systems Group, The Netherlands

[2]Eindhoven University of Technology: Faculty of Architecture, Building and Planning – Urban Planning Group, The Netherlands

This paper reports on the development of a multi-agent cellular automata model of pedestrian movement and decision making within the built environment. The approach will lead to a system that may serve as a tool in the design process to better understand how a design will influence user behaviour.

1 Introduction

The problem of simulating traffic flows is one of the topics that traditionally has played a central role in transportation research. It has received renewed interest, however, due to the popularity across disciplines of complexity theory. As a result, there is a growing interest to apply agent technology to describe driver-vehicle objects [1] and to describe the behaviour of drivers as a mean to achieve anticipatory traffic forecasts [2].

This paper reports on a research project that uses a multi-agent model, which received the acronym Amanda, in which a cellular automata is embedded for network decision analysis. In the specific application described in this contribution, agents represent pedestrians, moving over a pedestrian network. Each agent is located in a simulated space, based on the cellular automata grid. Each iteration of the simulation is based on a parallel update of the agents conforming local rules. In Dijkstra, *et al.* [3] a framework for the agent model part of the system is given to describe the behaviour of pedestrians.

Point of departure is pedestrian movement within a shopping mall. The focus is on simulating pedestrian dynamics in a shopping mall, the choice mechanisms that are involved and the desire to realise particular goals by conducting particular activities. Not all these aspects are discussed in this paper. This paper explores the use of cellular automata for modelling pedestrian dynamics.

2 Approaches to Traffic and Pedestrian Dynamics

Cellular automata (CA) are discrete dynamical systems whose behaviour is completely specified in terms of a local relation. They are mathematical models of spatially distributed processes. The purpose of the model is to simulate dynamic processes. CA models have also been used to model transportation systems. For instance, CA models for road traffic have received a great deal of interest during the 1990s. Nagel and Schreckenberg [4] have analysed vehicular movements with a CA car-following model. Based on the Nagel-Schreckenberg model the dynamics of CA models are investigated. An example may be the so-called 'velocity dependent rules'-model, where the focus is on the occurrence of meta-stable states or the synchronised traffic [5]. This simulation model can be used for large-scale networks and because of the speed capability of the model even for traffic assignment and traffic forecast purposes [6].

Renewed interest in the gas-kinetic description of freeway traffic flow has led to a multilane gas-kinetic model of freeway traffic [7]. Hoogendoorn and Bovy [8] have extended the gas-kinetic modelling approach to describe generic traffic flows in an n-dimensional space and present a new gas-kinetic model for two-dimensional pedestrian flows.

In recent years, there has been growing interest in understanding traveller behaviour, including that of pedestrians. Pedestrians are an integral part of the transportation system. Blue and Adler [9] have applied CA to model pedestrian flows and demonstrated that these models produce acceptable fundamental flow patterns. Klüpfel, *et al.* [10] presented a simple CA-model for the description of crowd motion in the microscopic simulation of evacuation processes on passenger ships.

Although pedestrian flows are an important consideration in transportation research and these studies represent important progress, there are still few microscopic models for studying the movement and activity of pedestrians. We think, that the principles of CA for traffic flows are also applicable to the modelling and simulation of user behaviour in buildings and public spaces, although the specific mechanisms need to be changed.

3 A Cellular Automaton Model of Pedestrian Movement

In our model approach, a CA model is used to simulate pedestrian movement along the network of the system. The network is the three-dimensional CA model represented by the graphical representation of a state at a certain time. The refinement of the network will be expressed by the format of the network, for example a cube of size 0.4-meter. If an agent moves over the network, the left-behind occupied cell changes into an empty cell and another cell will be occupied. The navigation speed v is the number of cells that are crossed during an update time-step Δt. The speed can be influenced by the occupation of other cells in the network and by the behaviour of an agent. The population of occupied cells could result in emergent behaviour.

We define a regular lattice of some width W and length L. Cells within the lattice are given the notation c_{ij}, where $1 \leq i \leq W$ and $1 \leq j \leq L$, and the type of cell $\in$ {*empty*, *decision*, *wall*}. *Empty* means the cell belongs to the walkway. *Decision* means the cell belongs to those cells (at the decision node) where agents have to decide where to go. *Wall* means that the cell is part of a wall, shop, etc.

We define the state s of a cell x (actually c_{ij}) in the pedestrian walkway to be *on* ($s(x)=1$) if it is occupied by pedestrian p_n, otherwise it is *off* ($s(x)=0$). Also we define a density $\delta_r(x)$ which shows the activity intensity around cell x. More specifically, it represents the number of neighbours of an *on* state in relation to the total number of neighbours in the Moore neighbourhood of cell x with radius r. In a d-dimensional grid with a Moore radius r, the number of cells n in the adjacent neighbourhood of cell x (x included) = $(2r+1)^d$. The updated cell x depends on the *on* states of all cells in the neighbourhood N of x ($N(x)$). To summarise the transition of a state of a cell from time t to $t+1$,

$$s^t(x) = 0 \rightarrow s^{t+1}(x) = 1, \text{ if hit by an object or a user-agent}$$

$$s^t(x) = 1 \rightarrow s^{t+1}(x) = \begin{cases} 1 & \text{if cell activity remains} \\ 0 & \text{if a user-agent left cell behind} \end{cases}$$

$$\delta_r(s^{t+1}(x)) = \sum_{i=0}^{n-1} s^{t+1}(x+y_i)/n$$

$$\text{where } y_i \in N(x),\ y_0 = 0,\ n = (2r+1)^d$$

The CA micro-simulation proceeds in time-step-slices. Let v_{max} the maximum speed and Δt the time-step; a time-step-slice is defined as $\Delta t_s = \Delta t \ / \ v_{max}$. For example, a maximum speed of 5 cells per second (2m/s) means a time-step-slice of 0.2 second. In each time-step-slice lane assignment changes the positions of all pedestrians according to local rules applied to each pedestrian. During a movement of a pedestrian based on the original Speed, the subsequent steps are ruled as follows:

Step1: check decision-point
  If decision-point is passed Then goto Step3.

Step2: check celltype
  If cell-type='decision'
  Then examine the behaviour of the pedestrian, which direction the pedestrian eventually wants followed by a turn in that direction; decision-point will be passed .

Step3: check cell
  If cell isn't occupied by a pedestrian and cell-type='empty' or cell-type='decision' Then Walk .

Step4: check adjacent cells
  If the cell over to the left (right) isn't occupied by a pedestrian
  Then Move to the left (right) cell Else Wait.

The movement during an update time-slice-step involves a non-deterministic crossing of at most one cell. Visualisation of pedestrian movement is organised in terms of single time steps. For example, a speed of 3 cells during a time-step of one second and a time-step-slice of 0.2 second implies a movement structure of *walk-wait-walk-wait-walk.*

4 Simulation Model of Pedestrian Movement

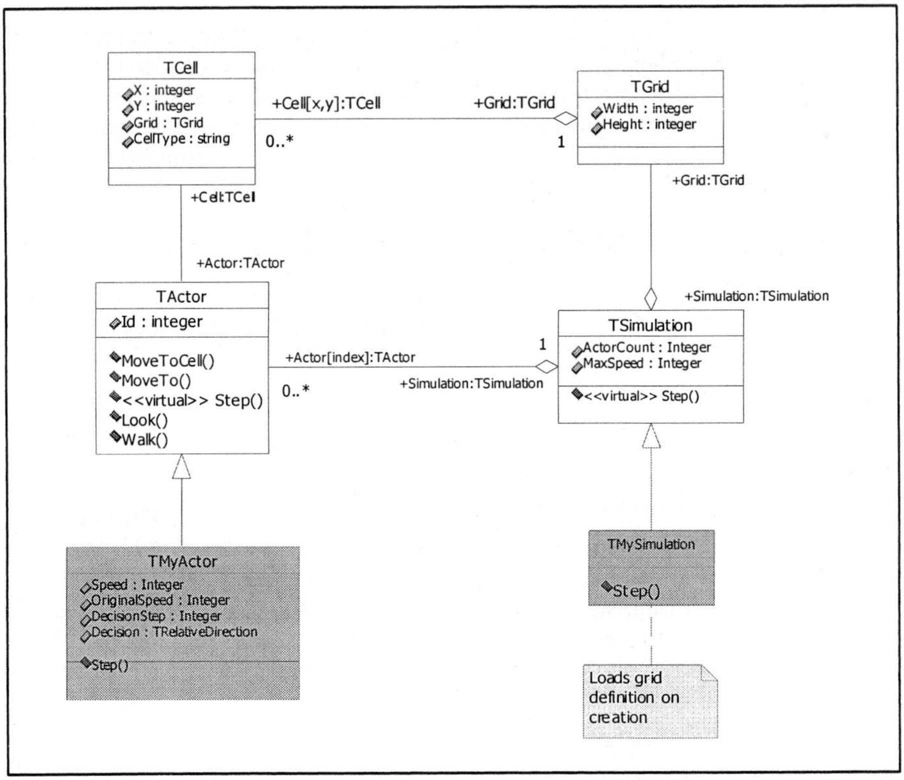

Figure 1: Graphical notation of the simulation process.

Consider a shopping mall or shopping environment with shopping pedestrians. This environment consists of a street network, which can be presented by a network consisting of N nodes and L links and a set of establishments, consisting of S stores, restaurants, etc. A subset E of these N nodes represents the entry/departure points of the model. In the multi-agent model approach, agents represent pedestrians. Each pedestrian p_n possesses a scenario, which includes among others behaviour-, intention- and activity- properties. A pedestrian p_n is supposed to carry out a set of activities. That is, pedestrians are supposed to purchase a set of goods, become involved in window-shopping, and possibly conduct other, some related activities. We assume that the completion of an activity is a key *decision point*, where pedestrians will adjust, if necessary their activity agenda and anticipated time allocation to the activities not yet completed. Another decision point is every node of the network where pedestrians may decide to take another route, changing the anticipated duration of the overall visit to the shopping centre.

The aim of our simulation model is to gain insight into pedestrian activity behaviour and movement. Fig. 1 is the graphical notation that expresses the simulation process where actors represent pedestrians.

Initially, we will realise different graphic representations of our simulation: an actor-based view to represent the pedestrians in the simulation, a network-based view to represent the density size of occupied cells in the network, and a node-based view to represent the decision points in the network.

For a first impression, a T-junction walkway is considered in a simulation experiment where pedestrians enter at random.

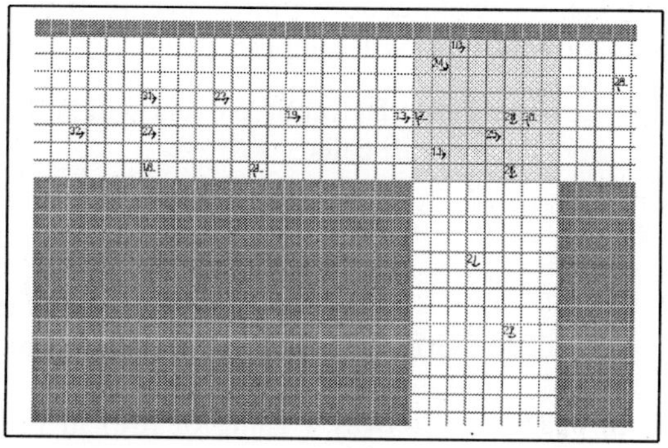

Figure 2: Cellular automata grid

Figure 2 gives an impression of the cellular automata grid with pedestrians, represented by arrows and actor-number, moving along the network.

Figure 3 shows the first ideas for a generalised cellular automata object model. This model includes a TGrid and a TCell class definition. Herein, a TGrid object can contain multiple cells and where the position of a cell is expressed by an (x,y,z)- coordinate.

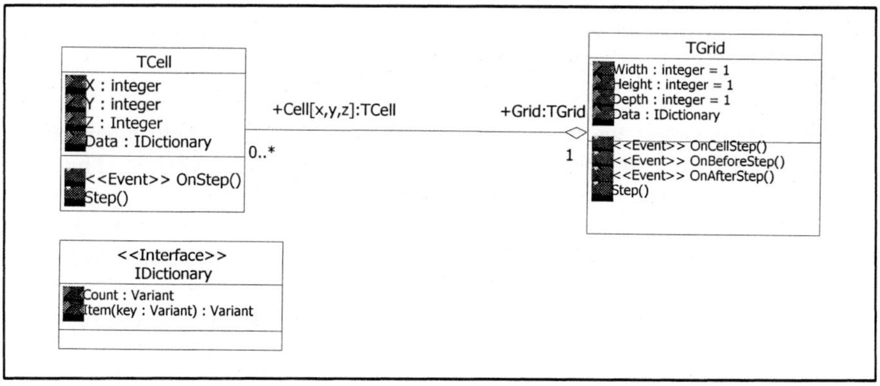

Figure 3: First cellular automata object model set-up

By setting the grids width, height and depth properties, the grid object will create the accompanying cells on the correct positions. The 'Data' property is another significant property, which both the TGrid and TCell classes have. With this 'Data' property, the objects of those classes can keep track of any needed data. For example, the alive state for a cell x of type Tcell can be set by the assignment x.Data.Item("Alive")=True. By calling the method Step of TGrid, one simulation step will be executed.

In the TCell class definition, the methods with <<stereotype>> event are events; you can associate code for those events of the objects of that type. When calling Step the following will happen:

1. The TGrid.BeforeStep event will be fired.
2. Then for all cells:
    a. The OneCellStep event will be fired. This event will get a reference to that cell.
    b. The Step method of the cell will be called causing the Tcell.OnStep method of that object to be fired.
3. The TGrid.AfterStep event will be fired.

An example (Conway's game of life) in Pseudo Visual Basic with this object model is given:

```
Sub Grid_OnCellStep(cell:Tcell)
    count=CountLivingNeighbours(cell)
    If cell.Data("Living")=True Then
      If count<2 Or count>3 Then cell.Data("Living")=False
    Else cell.Data("Living")=False
      If count=2 Then cell.Data("Living")=True
    End If
  End Sub
```

Besides this cellular automata part of the model, an agents part will be designed. These two parts of the model will be combined to let agents occupy a cell and give it the possibility to see the CA as its environment. So, the agent should be able to walk, look, or sense other information contained in the grid and their cells.

5 Discussion

In the present paper, we have discussed the concept of a multi-agent cellular automata model for simulating and visualising pedestrian movement. The development of a multi-agent system offers the promise of simulating autonomous individuals and the interaction between them. Given limited allowed space, we have described some principles underlying a simulation of pedestrian movement to illustrate our approach. In addition, we have briefly discussed some ideas for a generalised object model of the cellular automata model. In the near future, aspects concerning the simulation of movement and behavioural aspects such as visual range, attractors, avoidance behaviour, and navigation will be further elaborated.

References

1. P. Hidas, *How Can Autonomous Agents Help Microscopic Traffic Simulation?*, Workshop: *Agents in Traffic Modelling*, The Fourth International Conference on Autonomous Agents, (Barcelona, 2000).

2. J. Wahle, A. Bazzan, F. Klügl, and M. Schreckenberg, *Anticipatory Traffic Forecast Using Multi-Agent Techniques,* in: *Traffic and Granular Flow '99*, D. Helbing, H.J. Hermann, M. Schreckenberg, and D.E. Wolf, (Eds.), pp. 87-92 (Springer, 2000).

3. J. Dijkstra, H.Timmermans, and B. de Vries, *Towards a Multi-Agent System for Visualising Simulated Behaviour within the Built Environment*, in: *Design and Decision Support Systems in Urban Planning*, Proceedings of the 5th International Conference, pp. 101-117 (Maastricht, 2000).

4. K. Nagel and M. Schreckenberg, *A Cellular Automaton Model for Freeway Traffic*, Journal de Physique I **2**, 2221-2229 (1992).

5. R. Barlovic, A. Schadschneider, and M. Schreckenberg, *Metastable States in Cellular Automata for Traffic Flow*, European Physical Journal B **5**, 793-800 (1998).

6. J. Esser and M. Schreckenberg, *Microscopic Simulation of Urban Traffic: Based on Cellular Automata*, International Journal of Modern Physics C **8**, 1025-1036 (1997).

7. D. Helbing, *Verkehrsdynamik*, (Springer-Verlag, Berlin, 1997).

8. S. Hoogendoorn and P. Bovy, *Gas-Kinetic Modelling and Simulation of Pedestrian Flows*, 79th Transportation Research Board Meeting & Conference, (Washington D. C., 2000).

9. V. Blue and J. Adler, *Emergent Fundamental Pedestrian Flows from Cellular Automata Microsimulation*, Transportation Research Board, Record **1644**, 29-36 (1998).

10. H. Klüpfel, T. Meyer-König, J. Wahle, and M. Schreckenberg, *Microscopic Simulation of Evacuation Processes on Passenger Ships*, in: *Theoretical and Practical Issues on Cellular Automata*, Proceedings of the Fourth International Conference on Celullar Automata for Research and Industry, Karlsruhe, S. Badini and T. Worsch, (Eds.), pp. 63-71 (Springer-Verlag, London, 2000).

Laying the Foundations: The Use of Video Footage to Explore Pedestrian Dynamics in PEDFLOW

A. Willis, R. Kukla, J. Kerridge, and J. Hine

Transport Research Institute, Napier University, Redwood House, 66 Spylaw Road, Edinburgh, EH10 5BR, United Kingdom

This paper is concerned with how video footage of pedestrian activity in various urban environments can be used to inform the development of PEDFLOW - a microscopic, agent-based model of pedestrian flow that can be used to simulate the effects of an environmental change (e.g., widening a pavement) on how pedestrians negotiate urban space. The process can be summarized as follows: (1) video footage of pedestrians is digitized and calibrated according to real-world measurements of the filming area; (2) an interesting scenario is identified (e.g. two pedestrians approaching each other on a collision course) and all cases of this scenario are selected from the video for analysis; (3) for each case, qualitative information (e.g. pedestrian group size, or final action) and quantitative data (e.g. walking speed, gap size) are recorded. For the latter, the frame-by-frame position of selected pedestrians is plotted using commercially-available image analysis software and related to key features of the environment (such as kerbs and obstacles); (4) statistical analyses are performed to extract patterns of behaviour which can be translated into algorithms that form the basis of the PEDFLOW model. In this way, the model can be developed according to realistic approximations of how pedestrians negotiate urban space.

1 Background

Many governments now seem committed to reversing the decline in walking that has characterized towns and cities across Europe over the last 20 years. In the UK, for example, regional governments have been given responsibility for developing Local Walking Strategies that encourage walking in urban areas. In order to increase the number of journeys made on foot, environments must be made more attractive to pedestrians: however, remarkably little is known about which measures (such as pedestrianization, or traffic-calming) are successful in achieving the desired changes in pedestrian mobility, walking behaviour, and perceptions of safety, attractiveness and convenience. Here, we describe the development of a new, agent-based model of pedestrian movement (PEDFLOW) which aims to simulate the effects of environmental layout on pedestrian movement. Unlike other models of pedestrian flow, PEDFLOW is being developed according to empirical studies of how pedestrians actually negotiate urban space. The result, it is hoped, will be a powerful, flexible design tool that can be used by urban planners to evaluate potential changes to the layout of urban space before they are implemented.

2 Modelling Pedestrians

If one were to predict the impact of an infrastructural change on the movement behaviour of pedestrians, one would need a way to simulate a group of people and put them in a virtual environment where they can be observed.

A look at a crowd of pedestrians might give the impression that the movement is random, but closer observation reveals certain patterns. Pedestrians with similar goals organize themselves in lanes, certain parts of the observed area are overcrowded and others are nearly unused. These macroscopic features however are not generated on purpose but are rather a result of many little individual decisions that people make. Thus, in order to achieve the same level of self-organization [2] in a software model, it is necessary to represent pedestrians as individual software objects as well.

Since it is infeasible and also unnecessary to represent specific, "real" people, one needs to abstract the "typical" behaviour – mechanisms that are identical for all people or at least common to a group of people. A similar generalization can be performed for entities that pedestrians encounter during their journey. Does it make a difference if a person faces a lamp-post or a group of stationary pedestrians, or can both be merged into one general concept (such as "*blockage*") for the purpose of the model?

What are the 'typical' parameters that describe a pedestrian and how can they be determined? The most obvious is speed. We assume that the speed of a person in an unperturbed environment is constant. Under certain condition, that speed will be adjusted. A second assumption is that each person has a goal which it wants to get to as quickly as possible, *i.e.* by moving in a straight line. However the ideal trajectory might be impossible, because of obstructions, resulting in a divergence from the optimal path. People differ in how soon they diverge when faced with an obstruction. The distance can depend on the type of obstruction but also on the type of pedestrian. The choice of which side to use when passing an entity is another decision that can be observed and analysed.

With these considerations in mind, the PEDFLOW group decided to go for an autonomous agent approach to modelling pedestrians. An autonomous agent is a software object which works independently towards a goal using local knowledge and algorithms: *i.e.* "*a system situated within and part of an environment that senses that environment and acts on it, over time, in pursuit of its own agenda and so as to effect what it senses in the future*" [2].

This now well-established programming paradigm is supported by a number of development environments and programming languages. However most of them are unsuitable for this particular purpose (as shown in [3]). The PEDFLOW model was developed using JAVA, which allowed us to streamline and optimize certain computational aspects of agents that are not or only minimally applicable to pedestrians. The model software supports real-time visualization and features an atypical representation of time and space. All random shapes - that is pedestrians as well as buildings and pieces of street furniture - are transformed into rectangles of unified side length and aligned to a grid. Pedestrians are thus modelled as squares that can move between adjacent grid elements or diagonally. The size of a pedestrian defines the grid size. Movement speed is represented as the time delay between movements (delay = grid size / speed). The latter is justifiable as pedestrians move in a succession of steps; in PEDFLOW the step size is unified and positions are aligned to simplify computation with respect to collision detection.

The movement of each agent is determined by rules. The inputs for them are the distances and properties of entities in the viewing area, which is divided into three zones: left, straight, right. Entities can appear within critical distance-zones, which are dependent on their type. The decision is then made using rules that work on the observation of the virtual environment adjusted by the agent's internal parameters.

To summarize: the PEDFLOW modelling environment provides a means to set up a generalized, virtual environment containing representations of buildings, street furniture and pedestrians which are visualized on-screen. Pedestrians are implemented as autonomous agents which have the ability to move at certain speeds and directions. These agents need to be calibrated with parameters that are derived from the empirical study of "real" pedestrians, including, for example:

- desired walking speed,
- conditions for speed changes,
- conditions for direction changes,
- preferred direction,
- divergence distance from obstructions,
- passing distance to obstructions.

where obstructions can be stationary or mobile (other pedestrians). The following sections deal with how this information can be obtained from reality by observation, measurement and analysis.

3 The Empirical Study of Pedestrian Movement

3.1 Rationale for Using Video-Based Observational Techniques

In order to gain a comprehensive and reasoned understanding of how pedestrians negotiate the walking environment, a combination of observational and interview methodologies is desirable. Observational methods are invaluable in exploring the movement patterns of individual pedestrians – in particular, to examine how people negotiate obstacles in their path, and how they position themselves within their environment more broadly. It is important, however, not to rely solely on these techniques: first, observational methods are unable, on their own, to elucidate the decision-making processes underpinning walking behaviour. Although the decision rules in the PEDFLOW model do not need to be identical to those underpinning human behaviour, some understanding of why pedestrians react to their environment the way they do can only enhance the development of the behavioural rules in our model. Second, observational studies reflect suppressed rather than actual demand for pedestrian facilities - in other words, they are only able to provide an insight into the behaviour of individuals already choosing to walk a particular journey.

The behavioural research being undertaken as part of the PEDFLOW project combines interview and observational techniques to explore the relationship between pedestrians' perceptions of the environment and their behaviour within it. As well as observing pedestrian movement within a range of different environments, we are undertaking in-depth interviews with volunteers to explore the decision-making processes of individuals in some detail, and also large-scale questionnaire-based surveys to uncover generalities in pedestrian decision-making within various sections of the pedestrian population as a whole. This paper, however, is concerned with the observational studies - in particular, the methods we have devised to track the movement patterns of pedestrians in a fast and accurate way.

Our observational studies are based on the premise that the optimal method of exploring how pedestrians negotiate urban space is through the use of unobtrusive, video-based surveys. Observation must be unobtrusive in order to avoid the well-known "observer effect" of individuals changing their behaviour when they know they are being watched [4]. If we want to capture naturalistic movement patterns, then, we must ensure that pedestrians are unaware they are being observed. In order to address the obvious ethical issues associated with such methods, all studies are carried out in accordance with a strict Code of Conduct.

Observational studies can take a number of forms. The great power of the video survey over other methods lies in its ability to provide a permanent record of the environment under observation: extremely accurate measurements of both time and space can be derived from this record – providing, of course, it is calibrated effectively with respect to the real-world measures of each. Two main drawbacks with this technique, however, help explain why no large-scale, systematic exploration of pedestrian movement at a microscopic level has been attempted to date: first, the problem of calibrating a foreshortened video image (which arises when the camera is not positioned perpendicular to the ground) has typically restricted video surveys to locations where bulky scaffolds can be erected to hold the camera directly overhead the observation area [5]. Such techniques profoundly restrict the range of environments that can be surveyed. Second, plotting the microscopic trajectory of pedestrians as they navigate urban space has so far proven immensely laborious and time-consuming.

Recent developments in digital image processing provide the potential to overcome both these difficulties. The rest of this paper outlines the process through which footage of pedestrian environments is collected, calibrated and analyzed, and a picture of how pedestrians behave at a microscopic level developed.

3.2 Data Collection: Filming, Sampling and Calibration

Once the survey location is selected, video footage of pedestrians is collected using either a standard hand-held digital camcorder mounted on a tripod, or CCTV cameras operated by collaborators within local City Councils. Private survey companies are contracted to film areas that cannot be accessed using either of these methods. Between one and twelve hours' footage are collected in any single session.

Video recordings are converted into digital format (.avi) using a standard video capture card housed in a PC. For each of the empirical questions identified during the development of the prototype PEDFLOW model [see section 2] all instances of the appropriate scenario in the digital movie files are selected according to predefined, objective sampling criteria, then saved as short clips (typically a few seconds long) for subsequent analysis. If, for example, we were interested in comparing the deviation distance associated with an item of street furniture (such as a lamp post) with that for a cluster of stationary pedestrians, all instances in which a pedestrian walks within a certain trajectory of the object (lamp post / cluster of pedestrians), and in which no other confounding variables enters the frame, would be clipped for inclusion in the sample.

Each clip is then imported into a commercially-available motion analysis package. This software contains a method of calibrating images for distance using measurements we make of the area under investigation, thus overcoming the problem of perspective foreshortening.

3.3 Data Analysis 1: Tracking Individual Pedestrians

Motion analysis software is used to plot the frame-by-frame position of selected objects (in this case, a pedestrian approaching an obstruction) for the duration of the clip. In order to provide an effective environmental context for pedestrian movement, we use the tracking function in the program to generate a "static make up" (.smu) file, which marks out the location of salient, static features of the environment (such as street furniture or pavement boundaries).

The software outputs the real-world x and y positions of each tracked object separately as a function of time (t) in frames. Unfortunately, this format does not allow the subsequent measurement of distances (and thus speeds); what is required is an x versus y plot, which essentially represents the pedestrian's trajectory in real-world space. In order to rectify this, we have developed a custom software tool that takes the output of the motion analysis program, collects the *.smu* file associated with the clip, and generates a visualization of the x versus y trajectory of the pedestrian under examination within its environmental context. The tool provides a range of functions enabling the accurate measurement of distances between any two points, and, if applicable, the associated speed and angular separation. In our original example, the deviation distance of pedestrians with respect to any object in its environment (either stationary, or moving) can be plotted with an accuracy as great as the calibration and the tracking processes allow – typically in the range of a few centimetres.

3.4 Data Analysis 2: Informing the PEDFLOW Model

By focussing on a single example of an event (in this case, a pedestrian approaching an obstruction), the technique outlined here allows a truly global analysis of movement behaviour in a given situation. Not only can individuals' movement behaviour be described – in terms of deviation distances, gap sizes, walking speeds, *etc.* – but factors that may affect the decision-making process (such as age, or group size, or type of environment) can be explored. In our study, any such factors that may be observed from the video footage are recorded in a spreadsheet alongside the descriptions of the behaviour for each individual clip. This allows for the statistical analysis of whole data sets to uncover general differences in behaviour between certain sections of the population as a whole, or according to certain environmental features – allowing, for example, the comparison of behaviour in pedestrianized shopping areas versus a suburban "route to school"). Such findings should, in conjunction with those from the interview-based findings, help inform the development of the behavioural rules that determine the movement decisions of individual agents in the PEDFLOW model.

References

1. D. Helbing, *A mathematical model for the behavior of pedestrians,* Behavioral Science **36**, 298-310 (1991).

2. S. Franklin and A. Graesser, *Is it an agent, or just a program?: a taxonomy for autonomous agents*, in: *Proceedings of the Third International Workshop on Agent Theories, Architectures, and Languages*, (Springer-Verlag, 1996).

3. R. Kukla, J. Kerridge, A. Willis, and J. Hine, *PEDFLOW: development of an autonomous agent model of pedestrian flow*, presented at the 8th Annual Meeting of the Transport Research Board, Washington, (2001).

4. R. Bechtel, *Human movement in architecture.* in: *Environmental Psychology,* H.M. Proshansky et al. (Eds.), (Rinehart & Winston; New York, 1970).

5. B. Pushkarev and J. M. Zupan, *Urban space for pedestrians*, (The MIT Press; Cambridge, Mass, 1975).

Problems of Human Attention Decrease Detection

T. Tichý

Laboratory of System Reliability, Czech Technical University, Faculty of Transportation Sciences Konviktská 20, Prague 1, Czech Republic

This paper contains the discussion of the decrease of attention of a human operator in course of his/her long term interaction with technical systems. Primarily it concerns possible creation of microsleep and its detection by means of elektroencephalography (EEG) followed by evaluation by means of mathematical methods. Secondly this paper deals with prevention of attention decrease by means of biofeedback or possibly with scope of warnings of technical system for human operator evaluated by means of principle „fail safe“.

1 Introduction

Problems of reliability interaction between artificial and human being is still unsatisfactory solved. Implementation of new technologies with elements of intelligence decreases probability of defects and increases a service life of such device, but at the same time it is more demanding on reliability of human operator.

Operators who work with complex and efficient artificial systems such as transport systems (aircraft, train expresses, trucks, ships) have to make fast and correct decisions with maintaining maximum focus and concentration on the specific task. These complex, demanding tasks on human operator are jointed by many other internal and external factors such as lenght of working hours, mental and physical state, extreme climatic conditions, quality of working environment, but also monotonous scene or view which operator must watch.

In present time tendency of our research is to recognize decrease of attention and thereby decrease of reaction on created stimulation. With almost constant scene or view seen by operator and optional personal indisposition can result in significant decrease of his attention, which can possibly lead into micro sleep (momentary drowsiness). This state of human brain is very dangerous and might lead to huge material and financial damages but more importantly to losses of human lives.

2 Material and Methods

To be able to find out decrease of operator's attention in interaction with a system we have to monitor most significant parameters, which will unambiguously identify decrease of attention and possible accession of microsleep. Most significant parameters are following: electric activity of brain; frequency of breath; heart beat frequency; eyes movement (or lid movement if needed) and others.

Attention of human operator constantly fluctuates, these changes can be considerable in course of the activity of an operator. Most often it is monotony and fatigue that leads into decrease of attention and micro sleep.

Micro sleep is a brief, light sleep, which can be shallow with open eyes or deep with closed eyes and possible conversion to first phase of synchronic sleep.

Our workstation specializes in measuring electrical brain activity. Electronic brain activity rises by synchronization of large population of neuron cortex near thalami-cortical oscillation. Brain activity, is scanned by electrodes, that are placed on a surface of a skull and potentials transcribe in time EEG curve.

EEG is measured according to international scheme designed by H. Jasper, so-called 10/20 system, see Fig. 1. Individual brain areas are termed font: Fp - pre-frontal, F -frontal, C - central, P -parietal, O - occipital, T - temporal. Odd numbers denote left hemisphere and even numbers denote right hemisphere, character **z** (zero) means center between two hemispheres. Mode scan of EEG is different, we use joint referential electrodes most often connected to both earlobes (A1, A2).

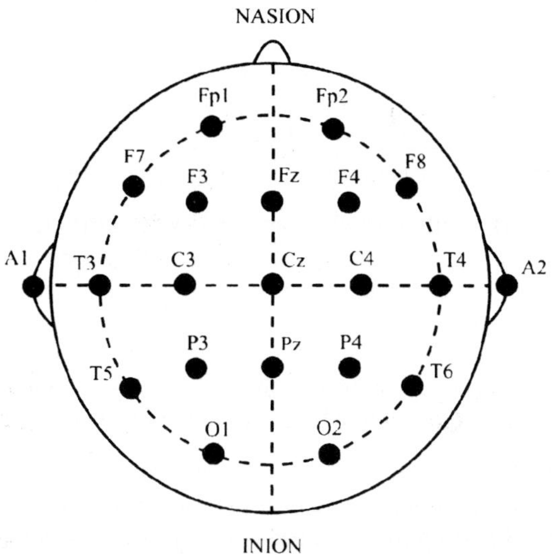

Figure 1: Allocation of electrodes according to international system "10/20".

Resulting EEG record measured on the skull of human operator has a very small energy and so electric potential is on order of only tens of μ V. Therefore we are using electrodes incased in plastic/synthetic and transmission from surface of the scull to the electrode is provided by conductible gel (decreases impedance) or we use electrodes Ag/AgCl, which also use conductible gel as a means of transport.

Resulting EEG record is processed with help of Fast Fourier Transformation (FFT) and single spectrums are created from them. Fundamental brain frequency is: delta (0,5 - 3,5 Hz), theta (4 - 7 Hz), alpha (8 - 13 Hz), and beta (14 - 30 Hz).

By comparing spectra of different states of human brain (vigilance, mental activity – reading, counting, drowsing, sleep) we can determine characteristic distribution and representation of single frequency for existing state.

3 Results

As noted above, attention of human operator constantly changes over the time. In order to determine subjective limit of decrease of attention of the operator we stem from course of his reactionary time.

To ascertain the beginning of micro sleep we measure reactionary time – length of sound or optical stimulation – based on research of their length we ascertain intensity of micro sleep till the first phase of synchronic sleep begins. This measurement is done by external module with exact time and connected to the computer, which records precise reactions of human operator directly into EEG curve on the second computer, see Fig. 2.

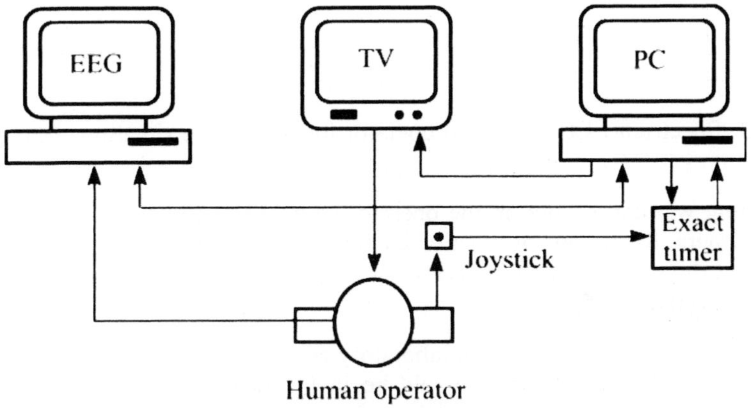

Figure 2: Diagram of EEG recording with reactionary time.

To simulate conditions of human operator in traffic, we show to a tested person a TV screen with a video showing similar view that he sees during his long hours of duty. At the same time we force the operator to react to several visual or sound cues.

After recording EEG with reaction times, with accuracy of milliseconds, we process this information with FFT and obtain spectral curve. Human operator is faced with three situations: a) vigilance (180 – 400 ms); b) relaxation (400 – 800 ms); c) sleep and micro sleep (800 – 1000, 1200 ms); see Fig. 3.

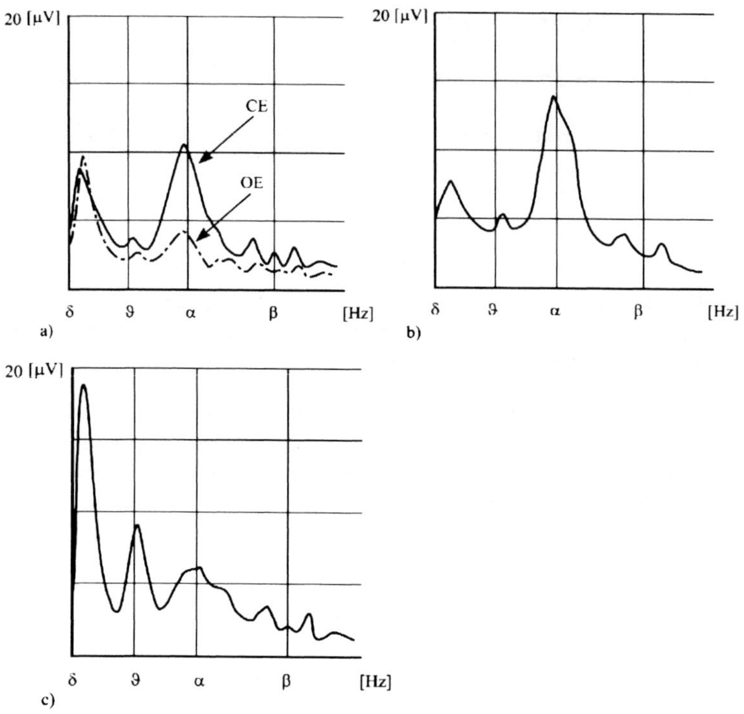

Figure 3: Spectrum EEG of human operator found in three situations near meters reaction times (CE – close eyes, OE – open eyes).

Recording was carried out on the operator who was sleepy and shortly after his working hours. Recording of reactionary time was accompanied by a sound cue assigned at the even and odd intervals when operator was falling asleep the EEG curve started to disintegrate.

Since recording has been done in laboratory conditions reactionary times are disproportionately lower than they would be in practical application. Later on we have used sound cue only, because human brain responses faster to the sound cue then to the visual cue whose path to the consciousness is longer.

If the operator gets into micro sleep sound cue will wake him up much faster than visual cue, for limen of excitation in optic system is comparison to acoustic system increased, therefore realization of visual cue can take quite a long time which might lead to a catastrophe.

4 Conclusion

At first glance a decrease of attention of human operators might not necessarily be a big problem when we deal with computer-operated systems with multi-controlled protection. But if larger failure on computer operated system occurs it is matter of seconds when human operator has to decide what is the most important thing to do.

I think I do not have to mention mistakes that happen during critical situation in the field of transportation and energetic. One of possible causes is carelessness of the operator that comes out of monotony of work, from fatigue or even from neglecting regulations.

Possible prevention of dangerous situations is multiple-protection of appliances beginning with visual cue affecting the operator and ending with several sound cues of different intensity with possibility of self-turn-off system. There are cases when appliances cannot be cut off from service for several different reasons, which might lead into material and financial damages. Therefore we need an attentive operator who will recognize such problems in time and will be able to solve them.

One of the possible methods that help to key up optimal affection of human brain and ensure the highest possible efficiency of the operator and at the same time to increase resistance against micro sleep is exploitation of biofeedback.

Our workstation owns device that exploits biofeedback to affect treatment of several brain dysfunctions and enables to increase efficiency of human brain by his auto-stimulation without use of medications or physical intervention. Biofeedback with help of specially modified software transform electric signal to the brain on basis of simple training program, which is controlled by his brain activity, see Fig. 4.

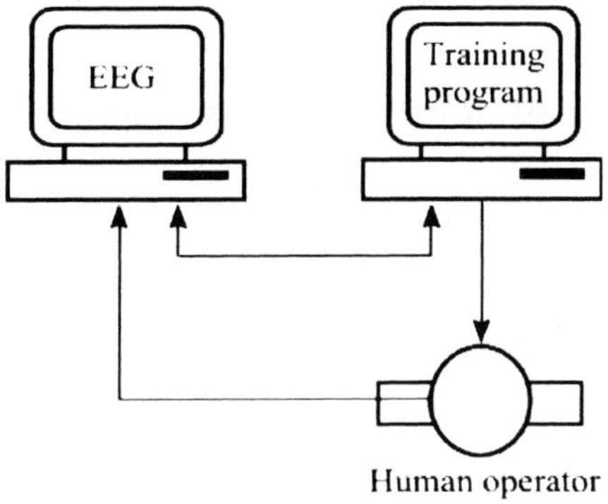

Figure 4: Diagram biofeedback.

References

1. E. Bena, J. Hoskovec, and J. Štikar, *Psychology and Physiology of Driver* (in Czech), (Prague, Publisher Transporting and Communication, 1968).

2. J. Faber, *Electroencephalography* (in Czech), (Prague, UK, 1992).

3. J. Faber, M. Novák, V. Prenosil, and L. Vydra, *Diagnostic, Detection and Predicate micro-sleep*, Development No. LSS 58/99 (in Czech), Prague, (1999).

4. A.S. Gevins and A. Rémond, *Methods of analysis of brain electrical and magnetic signals,*(Handbook of electroencephalography and clinical neurophysiology, (Amsterdam-New York-Oxford, Elsevier, 1987).

5. M. Novák, a *collective: Artificial neuron network, theory and application* (in Czech), (Prague, C. H. Beck, 1998).

6. J. Svatoš, *Biological signals I - Genesis, processing and analysis* (in Czech), (Prague, CTU, 1995).

7. J. Tuma, *When with tunnel turn to fiery hell, Technical weekly 49/99* (in Czech), 16.

8. L. Vydra, *Analysis EEG marker micro-sleep,* Development No. LSS 59/99 (in Czech), Prague, (1999).

9. T. Tichý, M. Leso, J. Faber, and M. Novák, *Problems of human operator vigilance decrease - detection and warning,* Development No. LSS 73/00 (in Czech), Prague, (2000).

10. http://teach.bhs.mq.edu.au/~tbates/imaging_techniques/EEG/EEG.html.

Microscopic Simulation of Pedestrian Crowd Motion

A. Keßel, H. Klüpfel, J. Wahle, and M. Schreckenberg

Physik von Transport und Verkehr, Gerhard-Mercator-Universität, Duisburg, Germany

We present an investigation of a driven random-walk model which is used to simulate pedestrian motion. We analyse the dependency of the global density and the mean velocity on the boundary density for open and periodic boundary conditions. Global and local fundamental diagrams are presented for a random sequential and parallel update. In addition, an analytical solution for the relation between mean velocity and density is presented, too.

1 Introduction

The simulation of traffic and pedestrian motion has gained much attention in recent years [8,15]. The increasing size of recreation facilities makes it necessary to investigate the motion of pedestrians and different properties of human motion. For example it is useful to know how to lead persons through a shopping mall or how to increase the efficiency of an evacuation process onboard a cruise ship [16,17]. The latter point is interesting from the view of ship designers to create corridors and stairs which are able to provide the possibility for a fast evacuation on the one hand and to provide as many cabins as possible on the other hand. It is also interesting for insurance companies to evaluate the risk for passengers on such a ship.

For modelling of pedestrian motion different types of simulations are used. Macroscopic models [9] that focus on the optimisation of pedestrian flows or different types of microscopic models: continuos in space [5] utilising a so called social-force model and models which are discrete in space. The latter are called cellular automaton models if they are also discrete in time and the state variable of the cell [3]. They are as well concerned with fundamental model characteristics [2,11,12,13,19] and their application [10].

There is, however, a considerable lack of empirical data and only a limited number of studies have been performed [4,6,7]. For an overview see [5,14,20,21]. Especially with regard to single-person data like walking speed the data basis is scarce. However, there are investigations under way as shown in other articles of these proceedings.

The outline of this article is as follows. Next we describe our model. After that we present results from simulations with a random sequential and parallel update under open and periodic boundary conditions. The last but one chapter presents an analytical approach for the relation between the mean velocity and the density. The final chapter gives an outlook to outstanding tasks.

1.1 Model

In 1999 Muramatsu et. al. [1] presented a driven random walk model for pedestrian motion. The pedestrians move on a square lattice. The size of the cells is supposed to be 40 cm. The pedestrians are allowed to move one cell per time step.

The decision for the new position of a pedestrian is made as follows. First, the pedestrian counts the neighbouring empty cells. From that number a probability for each unoccupied cell is computed and by means of a random number the new position is chosen. The formulas to calculate the probabilities are listed in table 1. A backstep is not allowed. In case of an occupied cell the probability for that direction is set to zero.

Direction	**Probability**
forward	$D+\frac{1-D}{\#\,\text{unoccupied cells}}$
up	$\frac{1-D}{\#\,\text{unoccupied cells}}$
down	$\frac{1-D}{\#\,\text{unoccupied cells}}$

Table 1: Probabilities for the three possible walking directions.

Figure 1 shows the layout of the corridor. The arrows mark the possible moves of the pedestrians. The amount of possible moves depends on the number of unoccupied cells around each pedestrian.

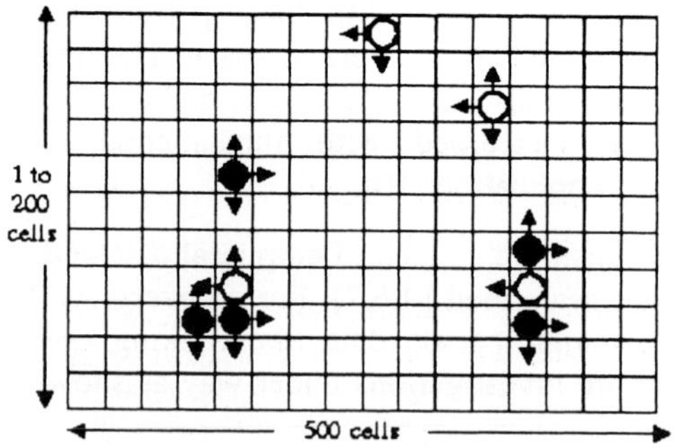

Figure 1: Layout of the corridor and possible moves of the pedestrians.

Under open boundary conditions the pedestrians were added at the left or right boundary, depending on the walking direction. The density at the boundaries was fixed at simulation startup.

The pedestrians were not able to cross the upper or lower boundary.

1.2 Update Procedures

We used two different update procedures for the simulations. The random sequential update (rsu) moves the pedestrians in a random order but every pedestrian only once in a timestep. After the move of a pedestrian the next pedestrian reacts on the new distribution. Because of this there is no problem with multiple occupations of a single cell.

The parallel update moves the pedestrians in the first step only virtually. Due to this there is the possibility for a multiple occupation of single cells because the moving pedestrian does not know if the cell he has chosen is still empty or occupied by a virtually moved pedestrian. Because of this after the virtual move of all pedestrians a conflict solution is performed to resolve the problems with more than one pedestrian per cell. A cell which is occupied by more than one pedestrian is drawn under the concerned pedestrians. The winner gets the cell, the losers move back to their old positions.

2 Simulation Results

The simulations were carried out with the parameters mentioned in table 2. For a width of w = 100 cells the fraction between the two walking directions was additionally set to c = 1/3 and c = 1/5 to study the dependency of the density for the transition from moving to jammed pedestrians from the fraction.

Width	Drift	Fraction	Simulated timesteps	Update	Boundary conditions
10, 20, 50, 200	*0.0, 0.1, 0.4, 0.7*	*0.5*	*10000*	*rsu*	*open*
100	*0.0, 0.1, 0.4, 0.7*	*0.2, 0.3, 0.5*	*10000*	*rsu*	*open*
1, 2, 5, 10	*0.0, 0.25, 1.0*	*0.0, 0.2, 0.5*	*50000*	*rsu, parallel*	*open, periodic*

Table 2: Parameters for the simulations.

For the corridor with widths smaller than 10 cells a drift D = 0.25 was chosen because in the case of three unoccupied cells in the neighbourhood this value means a probability of $p_{\text{forward}} = 0.5$ for a step in forward direction.

Figure 2 shows the mean velocity versus the density at the boundaries for open boundary conditions simulated with the rsu. There is only a slight dependency of the transition from moving to stopped pedestrians on the drift. A similar picture is recorded for periodic boundary conditions. With the parallel update the transition to the stopped phase occurs at a lower density due to the conflict solution. An increasing drift leads to an increasing mean velocity in the moving phase, up to a value of 1 for drift $D = 1$.

Additionally we recorded global and local fundamental diagrams. The local fundamental diagrams recorded the flow and the density on each lane for each walking direction. The global fundamental diagrams just distinguish between the walking directions and not the different lanes.

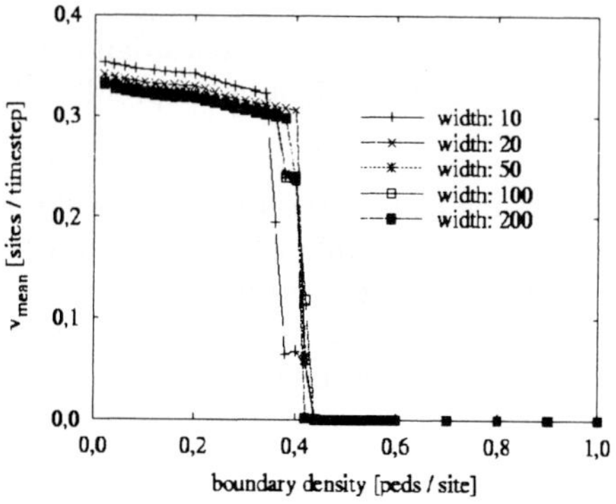

Figure 2: Mean velocity versus boundary density for rsu with open boundaries.

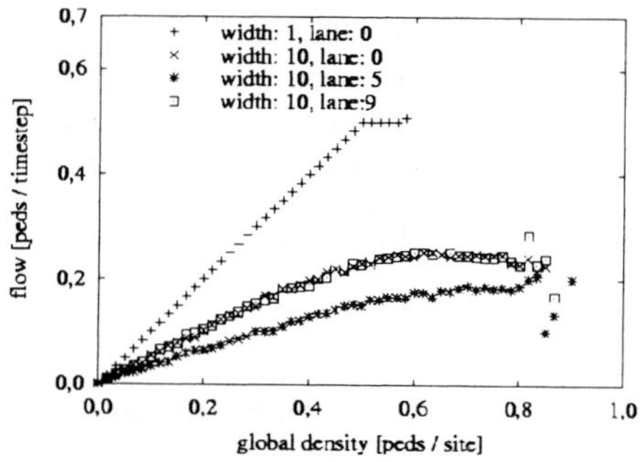

Figure 3: Local fundamental diagrams for the rsu with open boundary conditions.

We found three different local fundamental diagrams. Fig. 3 shows the local fundamental diagrams for corridors with a width of $w = 1$ cell and $w = 10$ cells. The fundamental diagram for the corridor with width $w = 1$ cell is independent from the drift due to the fact that the drift increases the probability for a forward step. Because there is no possibility for another direction than forward in a corridor with a width of $w = 1$ cell, the probability for the forward step is always $p_{forward} = 1$ and so not depending on the drift.

The flow on lanes which are at the upper or lower boundary is higher than the flow on inner lanes due to a missing third direction on a boundary lane. The fact that on boundary lanes the maximum number of unoccupied cells is two increases the probability for these two directions. On inner lanes the maximum number of unoccupied cells is three. This means that in case of a drift $D = 0$ the probability for all three directions is $p = 1/3$. For the same case on a boundary lane the probability for each of the two directions is $p = 1/2$. This results in a higher flow on the boundary lanes.

As a last result we present a graph (Fig. 4) of simulated timesteps until all pedestrians are jammed. We altered the program to simulate either 1.000.000 timesteps or stop the simulation when the mean velocity reached the value 0. We performed 30 recursions for each density for the rsu with periodic boundaries. It was found that for high densities the amount of simulated timesteps was in a range of some hundred timesteps. For medium densities the number of timesteps increased due to an increasing amount of possible configurations.

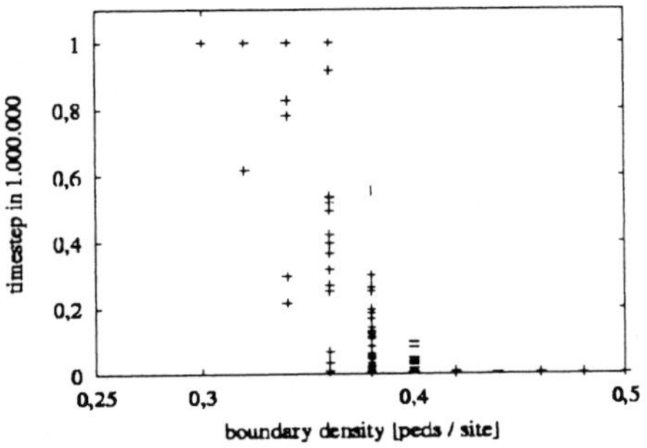

Figure 4: Simulated timesteps until a mean velocity of 0 occurs.

For a density of 0.3 there was no jam found in the system. All 1.000.000 timesteps were performed with still a non-zero mean velocity.

3 Analytical Solution

We tried to formulate an analytical solution for the speed-density relation. The idea is that the mean velocity equals the probability of a certain configuration to arise multiplied with the probability to choose the forward step.

There are three different configurations which provide the possibility for a forward step.

An empty cell occurs with a probability of $(1-\rho)$ and an occupied cell occurs with a probability of ρ.

1. The first configuration is the one with all three neighbouring cells unoccupied. This configuration arises with a probability of $p_{\text{conf.}} = (1-\rho)^3$.

   The probability for the forward step in this case is $p_{\text{forward}} = D + \frac{1-D}{3}$.

2. In the second configuration one of the upper or lower neighbouring cells is occupied. The probability for that configuration is $p_{\text{conf.}} = (1-\rho)^2 \cdot \rho$.

   A forward step is made with a probability of $p_{\text{forward}} = D + \frac{1-D}{2}$.

   This case occurs in two different forms: either the upper neighbouring cell is occupied or the lower neighbouring cell. Due to that, the probability has to be multiplied with a factor 2.

3. The last case is the one with only the front cell unoccupied. The forward step has a probability of $p_{\text{forward}} = 1$ and this configuration arises with a probability of $p_{\text{conf.}} = (1-\rho) \cdot \rho^2$.

   Writing this all together leads to the following expression for the mean velocity:

$$\bar{v} = (1-\rho)^3 \cdot \left(D + \frac{1-D}{3} \right) + 2 \cdot (1-\rho)^2 \cdot \rho \cdot \left(D + \frac{1-D}{2} \right) + (1-\rho) \cdot \rho^2 .$$

Figure 5 shows the analytical solution and some measured graphs for the mean velocity. It is obvious that the analytical solution becomes worthless for narrow corridors. The reason is the effect of the boundary lanes on the mean velocity. For corridors with an infinite width the analytical solution would fit well.

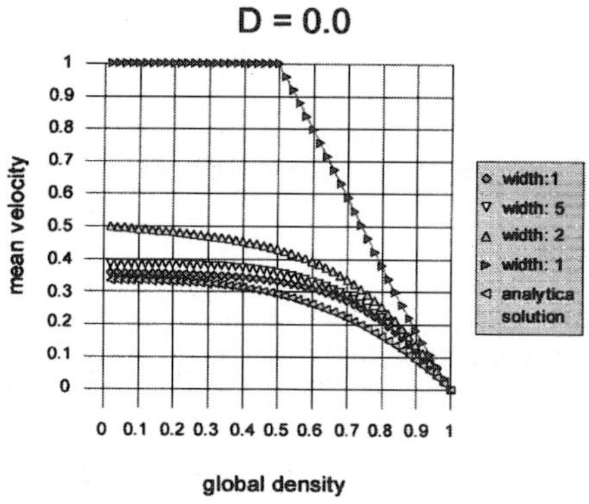

Figure 5: Comparison of the analytical solution with measured values for the mean velocity.

4 Outlook

An improved analytical solution regarding the boundary lanes is in progress. This will lead to a better fit of the analytical solution to the measured data.

A next step for the simulations will be an investigation of the model regarding obstacles and crossing traffic.

A faster implementation of the model is the basis for finding the lowest jam creating density.

References

1. M. Muramatsu, T. Irie, and T. Nagatani, *Jamming transition in pedestrian counter flow*, Physica A **267**, 487-498 (1999).

2. V.J. Blue and J.L. Adler, *Cellular automata microsimulation of bi-directional pedestrian flows,* Journal of the Transportation Research Board **1678,** 135-141 (2000).

3. C. Burstedde, K. Klauck, A. Schadschneider, and J. Zittartz, *Simulation of pedestrian dynamics using a 2-dimensional cellular automaton,* cond-mat/0102397 (2001).

4. J. J. Fruin, *Pedestrian Planning and Design,* (Metropolitan Association of Urban Designers and Environmental Planners, New York, 1971).

5. D. Helbing, *Theoretical foundation of macroscopic traffic models,* Physica A **219**, 375 (1995).

6. L.F. Henderson, *The statistics of crowd fluids,* Nature **229,** 381-383 (1971).

7. L.F. Henderson, *Sexual differences in human crowd motion,* Nature **240,** 353-355 (1972).

8. D. Helbing, H.-J. Hermann, M. Schreckenberg, and D.E. Wolf, (Eds.), *Traffic and Granular Flow '99,* (Springer, Berlin, 2000).

9. H.W. Hamacher and S.A. Tjandra, *Mathematical modelling of evacuation problems - a state of the art,* in: [18], M. Schreckenberg and S.D. Sharma (Eds.), in print.

10. H. Klüpfel, T. Meyer-König, J. Wahle, and M. Schreckenberg, *Microscopic simulation of evacuation processes on passenger ships,* in: *Proceedings of the Fourth International Conference on Cellular Automata for Research and Industry,* pp. 63-71 (Karlsruhe, 2000).

11. M. Muramatsu and T. Nagatani, *Jamming transition in two-dimensional pedestrian traffic,* Physica A **275**, 281-291 (2000).

12. T. Nagatani, *Jamming transition in the traffic-flow model with two level crossing,* Phys. Rev. E **48**, 3290 (1993).

13. K. Nishidate and M. Baba, *Cellular automaton model for random walkers,* Phys. Rev. Lett. **77** (9), 1675-1678 (1996).

14. Di Nello (Ed.), *SFPE Handbook of Fire Protection Engineering,* (National Fire Protection Association, 2nd edition, 1995).

15. K. Nagel and M. Schreckenberg, *A cellular automaton model for freeway traffic,* J. Physique I **2**, 2221-2229 (1992).

16. R.A. Smith and J.F. Dickie (Eds.), *Engineering for crowd safety,* (Elsevier, Amsterdam, 1993).

17. J.J. Sriver and Y.-I. Kwon, *Recent pedestrian planning issues and initiatives in Japan and South Korea,* Transportation Research Board **1695**, 14-18 (1996).

18. M. Schreckenberg and S.D. Sharma (Eds.), *Proceedings of the International Conference on Pedestrian and Evacuation Dynamics,* (Springer, Berlin, 2001), in print.

19. S. Tadaki and M. Kikuchi, *Jam phases in a two-dimensional cellular-automaton model of traffic flow,* Phys. Rev. E **50**, 4564 (1994)

20. Transportation Research Board, *Highway Capacity Manual,* Chapter 13: Pedestrians (1994).

21. U. Weidmann, *Transporttechnik der Fußgänger,* Schriftenreihe des IVT **90**, (ETH Zürich, Januar 1992).

Evacuation Simulation

Simulating Evacuation and Circulation in Planes, Trains, Buildings and Ships Using the EXODUS Software

E.R. Galea

Fire Safety Engineering Group, University of Greenwich, London SE10 9LS, UK

When designing a new building, aircraft, ship or rail carriage, or modifying an existing design, how do we ensure that the proposed design is safe from an evacuation point of view? In the building and aviation industries, computer based evacuation models are being used to tackle these issues. In these industries, the traditional restrictive prescriptive approach to design is making way for performance based design methodologies using risk assessment and computer simulation. In the maritime and rail industries, evacuation models also offer the promise to quickly and efficiently bring these considerations into the design phase, while the ship or carriage is "on the drawing board". This paper describes the development of evacuation models with applications of the EXODUS models to buildings, aircraft, passenger ships and trains and further discusses issues concerning data requirements and validation of these models.

1 Introduction

Evacuating occupants from any inhabited enclosure is a complex and difficult task. There are a multitude of factors that come into play and can influence the nature of the evacuation. However, while buildings, aircraft, ships and trains may be very different types of physical structures, when it comes to evacuation issues, they have a surprising number of features which they share in common, and many features which are unique to the particular type of enclosure.

Following a rail incident, a decision has to be taken as to whether an evacuation is necessary. Similarly to ship board emergencies, but unlike most serious accidents involving aircraft or most buildings (excluding those using the defend in place concept), it is not always prudent to attempt to evacuate passengers from rail vehicles. In many situations the safest course of action is to remain onboard the train or ship as the environment outside can be more perilous than that on board. However, unlike large modern passenger aircraft which carry many highly trained cabin crew that can control each cabin section and each exit, or ships which can have as many as one crew member for every two passengers, trains may carry over 500 passengers yet have only a two or three staff trained to assist or manage an evacuation. Unlike building situations but similar to situations on trains and aircraft, the orientation of the ship may impede easy rapid movement of passengers. However, uniquely to shipping applications, the orientation of the escape paths may be time dependent due to roll. Like modern passenger aircraft, the train environment is a complex congested space not conducive to rapid egress. A train carriage resembles a long tube, fitted with obstacles that impede easy rapid movement, packed with people, with a small number of exits.

Perhaps unique to ship situations, a good deal of time consuming preparatory activities must be completed prior to the actual evacuation. Passengers are instructed to collect life jackets (usually located in cabins or assembly stations) and passengers are usually further required to gather in pre-defined assembly stations prior to attempting to disembark. Unlike the situation in buildings where the way out of a structure is to normally proceed to street level, on board ship, the way off requires passengers to seek the muster station. This could require passengers to travel to an apparently arbitrary location above or below their current position and to the port or starboard. Furthermore, the routes can be complex and confusing to passengers not accustomed to the marine environment, even the term "muster station" may be unfamiliar to most passengers (the industry is replacing the term "muster" with "assembly").

Finally, unlike in buildings, where getting to an external exit usually marks a successful evacuation, passengers in trains, planes and ships are usually faced with the additional ordeal of getting from the exit to a place of safety. This can mean jumping from a train, plane or ship, getting onto an evacuation slide on a plane or ship or getting into life raft/boat on a plane or ship.

One of the issues that separates each of these types of structures is the time available for safe evacuation. Prescriptive regulations in each of the industrial sectors specifies a maximum time for evacuation. In buildings it is 2.5 minutes, in aircraft it is 90 seconds while in passenger ships it can be 30 minutes. However, all of these times are somewhat arbitrary and fail to address the unique situations that can arise even within a particular domain. Performance based regulations used in the building sector attempt to address this issue by specifying not a fixed evacuation time but by specifying the concept that the Available Safe Egress Time (ASET) must be less than the Required Egress Time (RET). The circumstances of M. Flier under consideration dictate the ASET. Computer based models can be used to assist in the determination of both the ASET and the RET.

2 What not to Model

It had been the 'common-sense' view that under extreme conditions, such as evacuations resulting from rail crashes, aircraft crashes, sinking ships or building fires, it was the norm for untrained, inexperienced individuals to 'panic', and therefore act in an irrational and possibly self-destructive manner. In incidents involving fire, this type of non-adaptive behaviour is now considered to represent only a small proportion of behaviour, even under serious and life-threatening conditions [1,2].

Unsuccessful behaviour and behaviour whose motives are not fully understood are often incorrectly labeled as panic. Indeed, reports of panic are now frequently seen more as a description of behaviour by a third party, rather than a useful explanation of the behaviour [3,4]. When the amount of information available to an individual is compiled, and behaviour is then examined, instead of viewing behaviour in a system wide context, behaviour may not seem quite so irrational [5]. The reported panic behaviour, in general, is a rational response to an occurrence about which knowledge is at a premium [6]. Such is the nebulous nature of the term 'panic', that there exists a number of definitions which might be centred on the effect of the behaviour on the individual [7] or the effect on those around the individual [2,3]. All of this adds to the misinterpretation of the frequency of irrational behaviour.

The identification of panic behaviour is also linked to a number of myths attributed to the behaviour and motives of crowds [8]. These suggest that crowds are irrational, suggestible and are incapable of being managed. All of these lead to the conclusion that crowd activities are panic-based. In reality, crowd activities are made up of individual occupant behaviour which is based upon their position in a role structure, which makes the influence of external forces (authority figures, for instance) dependent upon the social filter provided by this role.

People are not indifferent to others, and will in general, attempt not to damage the chances of survival of those around them, except at extreme levels of danger, instead communicating the possible existence of danger, or identifying the best evacuation route [9]. Generally, there exists a level of co-operative behaviour, up until a threshold, after which cues are perceived differently and danger is seen as more immediate.

The presence of a fire will however exert an influence on occupant behaviour through the effect of smoke, heat, irritant and narcotic gases. These fire hazards may have a physiological and/or a psychological effect upon the occupant's behaviour. The perception of fire in an incident is important in the choice of action that the occupant selects in response to the danger. The presence of smoke influences the way in which an occupant decides to act, by both increasing the perceived risk, and through the introduction of movement barriers. If occupants are directly exposed to *flame* it is possible that the self preservation motive takes over and they will attempt to flee. Furthermore, in crowded confined spaces, their behaviour may become increasingly competitive as the situation worsens. A survivor from the Ladbroke Grove train crash stated [10],

"*... There was a flash of fire and the flames came straight down the centre of the carriage. Everybody was screaming. The fireball seemed to pass us by and went straight into the next carriage which burst into flames. I saw people trying to scramble out. They were falling over each other, their faces were blackened...*" [10]

The importance of identifying panic as a relatively rare behaviour is apparent, as it justifies a more diverse behavioural model, which would have to deal with more than instinctive, irrational or even random actions. This model would require intelligent occupants making decisions based upon the events around them. That is not to say that this form of behaviour should be entirely neglected as it can have a significant effect upon the outcome of an evacuation, especially if it occurs at inopportune locations, resulting in the blockage of a busy passageway, for instance.

3 Factors Influencing Evacuation

In order to fully assess the potential evacuation efficiency of an enclosure (i.e. ship, building, aircraft, train carriage), it is essential to address the configurational, environmental, behavioural and procedural aspects of the evacuation process [11].

Configurational considerations are those generally covered by conventional methods and involve enclosure layout, number of exits, exit type, corridor width, travel distance etc. In the event of fire, ***environmental*** aspects need to be considered. These include the likely debilitating effects on the passengers of heat, toxic and irritant gases and the impact of increasing smoke density on travel speeds and way-finding abilities. In addition, for ship based environments the sea conditions can impact on the environment causing heal, list or roll conditions making egress more difficult. ***Procedural*** aspects cover the actions of staff, occupant prior knowledge of the enclosure interior, emergency signage etc. Finally, and possibly most importantly, the likely ***behavioural*** responses of the occupants must be considered. These include aspects such as the occupants' initial response to the call to evacuate, likely travel directions, family/group interactions etc.

Each of these might have several separate effects on an individuals actions, involving physical, psychological, and sociological aspects, and indeed many factors will fall under more than one subject heading. Traditional methods for evacuation design fail to address all these issues in a quantitative manner preferring to rely almost totally on judgement and a set of "prescriptive rules". As these "prescriptive rules" have an almost total reliance on ***configurational*** considerations such as travel-distance and exit width they can prove to be too restrictive. Furthermore, as these traditional methods are insensitive to human behaviour or likely emergency scenarios, it is unclear if they indeed offer the optimal solution in terms of evacuation efficiency. As an alternative to the prescriptive approach, simplistic "hand calculations" are often used. These calculation methods tend to concentrate solely on the load carrying capacity of the various components of the structures interior and as a result this methodology is often referred to as a "hydraulic" model. As an example of one of these sets of criteria, consider a recent marine example. Some of the assumptions listed in the interim guidelines for evacuation analysis of Ro-Ro passenger ships include [12]:

- All passengers and crew begin the evacuation at the same time and will not hinder each other.
- Passengers and crew will evacuate via the primary escape route.
- Flow is only in the direction of the escape route and there is no overtaking.

- No passengers and crew have disabilities or medical conditions that will severely hamper their ability to keep up with the flow.
- Full availability of escape arrangements is considered.
- People can move unhindered.

As with the prescriptive rule approach, these assumptions concentrate on the configurational aspects of evacuation, essentially ignoring the other vital factors. In an apparently more sophisticated yet still crude approach, computer models in which the ***behaviour*** of individuals is essentially ignored have been developed. The direction and speed of egress is determined by physical considerations only (e.g. population densities, exit capacity, etc.). This methodology is often referred to as a "ball-bearing" model as individuals are treated as unthinking objects that automatically respond to external stimuli.

4 Evacuation Models and Data Requirements

Sophisticated evacuation models [13] are being developed that attempt to address each of the identified factors that influence evacuation. Some of these models are being used in a routine manner in the design of buildings and in the evaluation of aircraft concepts. For example, the buildingEXODUS evacuation model [13-16] is routinely used in 18 countries for the design of airport terminals, shopping centres, museums, hospitals, etc. Similarly, the airEXODUS evacuation mode [17,18] has been used in design evaluation projects for Airbus, Boeing, British Aerospace and Bombardier. If evacuation models are to fulfil their promise, they must be capable of addressing the configurational, environmental, behavioural and procedural aspects of the evacuation process. Computer based mathematical models describing the evacuation process have the potential of addressing all the shortfalls listed above and in addition have a role to play in the following areas:

- Design and development of enclosure interiors bringing safety matters to the design phase while the proposed structure is still on the drawing board.
- Implementation of safer and more rigorous certification criteria.
- Development of improved and more efficient staff operating procedures.
- Improved staff training.
- Accident investigation.

Associated with the development of computer based evacuation models is the need for comprehensive data collection/generation related to human performance under evacuation conditions. The nature of the particular "what if" conditions that are to be simulated will dictate the type of data required and the capabilities the model will require. Factual data regarding the evacuation process is essential to the development of computer egress models. Egress models have a high reliance on factual data regarding the evacuation process in order to:

(a) Identify the physical, physiological and psychological processes that contribute to, and influence the evacuation process and hence formulate appropriate models.

(b) Quantify attributes/variables associated with the identified processes.

(c) Provide data for model validation purposes.

Three forms of data are useful in providing the required information. Accident investigation reports that contain human factors analysis and survivor interview accounts are vital in providing information to identify the human element (i.e. item (a) above) that needs to be simulated.

Once identified, the behaviours and occupant performance attributes must be quantified (i.e. item (b) above). Data from full-scale and component tests can be used for this purpose. When undertaking experimentation, it is essential to ensure that a representative population is used for data collection. Often in performing evacuation experimentation specialist sub-populations such as students or military/police personnel or single gender groups are used. The results generated from such experiments are not likely to reflect the performance capabilities of the intended target population and as such are of questionable value. Furthermore, it is important to make use of data that already exists in other sectors whenever possible. However, care must be taken to use appropriate data. For example, while many experiments have been undertaken to assess the speed at which males and females of various ages groups can travel up and down stairs, this is restricted to stairs of design found in buildings [14,19,20]. If the tread depth, raiser height or gradient of the stairs found on ships are significantly different to that found in buildings then this data cannot be used with much confidence in marine applications. Finally, data is required for validation purposes. There are at least four forms of validation/testing that evacuation models should undergo [21]. These are, (i) component testing; (ii) functional validation; (iii) qualitative validation and (iv) quantitative validation.

While each of these four components are essential to the validation process, quantitative validation relies heavily on the availability of considerable amounts of data. Quantitative validation involves comparing model predictions with reliable data generated from full-scale evacuation demonstrations. Here it is essential to emphasis the reliability of the data namely the integrity of the data, the suitability of the experiment and the repeatability of the experiment.

5 The EXODUS Suite of Evacuation Models

EXODUS is a suite of software tools designed to simulate the evacuation of large numbers of people from a variety of enclosures. The family of models currently consists of buildingEXODUS [13-17,21] for applications in the built environment and airEXODUS [17,18,21] for aviation applications. The maritimeEXODUS [22-24] model is currently under development, with prototype versions of the model being used for various marine applications including an application involving a High Speed Craft evacuation [23]. The railEXODUS model is still undergoing development.

The EXODUS model comprises five core interacting sub-models, these are the Occupant, Movement, Behaviour, Toxicity and Hazard sub-models (see Fig. 1). The software describing these sub-models is rule-based, the progressive motion and behaviour of each individual being determined by a set of heuristics or rules.

The spatial and temporal dimensions within EXODUS are spanned by a two-dimensional spatial grid and a simulation clock (SC). The spatial grid maps out the geometry of the structure, locating exits, internal compartments, obstacles, etc. Geometries with multiple floors can be made up of multiple grids connected by staircases, with each floor being allocated a separate window. The structure layout can be specified using either a DXF file produced by a CAD package, or the interactive tools provided, and may then be stored in a geometry library for later use. The grid is made up of nodes and arcs with each node representing a small region of space and each arc representing the distance between each node. Individuals travel from node to node along the arcs.

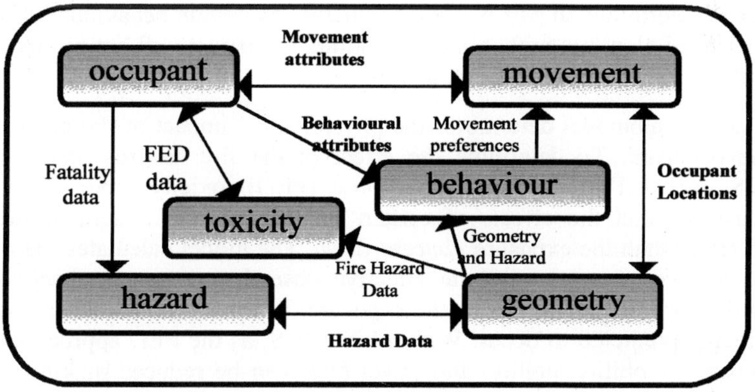

Figure 1: Interaction of EXODUS modules.

The ***Population*** Sub-model allows the nature of the occupant population to be specified. The population can consist of a range of people with different movement abilities, reflecting age, gender and physical disabilities as well as different levels of knowledge of the enclosure layout, response times etc. On the basis of an individual's personal attributes, the ***Behaviour*** Sub-model determines the occupant's response to the current situation, and passes its decision on to the ***Movement*** Sub-model. EXODUS has the capability of implementing two behavioural regimes: NORMAL and EXTREME behaviour. Under NORMAL behaviour, an occupant is prepared to wait until it is possible to make the desired move. Under the EXTREME regime, occupants will patiently queue until a period of time equal to their Patience has been reached (a user defined occupant attribute) before taking actions that result in them jostling around or possibly recommitting to another course of action.

The ***Behaviour*** sub-model incorporates adaptive capabilities that include, structural knowledge, reaction to communication, affiliative behaviour, occupant motivation and a stochastic element to the queuing recommitment behaviour. The behaviour model functions on two levels. These levels are known as GLOBAL and LOCAL behaviour. GLOBAL behaviour involves implementing an escape strategy that may lead an occupant to exit via their nearest serviceable exit or most familiar exit. The occupants' familiarity with a particular enclosure layout may be determined by the user prior to commencing the simulation. It is also possible to assign individuals with an itinerary of tasks – such as visit a pre-defined location - that must be completed prior to evacuation. This is an essential feature if attempting to simulate staff who must take up duty stations during an evacuation or occupants attempting to perform a task such as return to an office prior to evacuating.

The desired GLOBAL behaviour is set by the user, but may be modified or overridden through the dictates of LOCAL behaviour. The LOCAL behaviour includes such considerations as determining the occupants initial response to the call to evacuate i.e. will the occupant react immediately or after a short period of time or display behavioural inaction, conflict resolution, overtaking and the selection of possible detouring routes. The manner in which an occupant will react to local situations is determined in part by their attributes. As certain behaviour rules, such as conflict resolution, are probabilistic in nature, the model will not produce identical results if a simulation is repeated.

The Toxicity submodel determines the physiological impact of the environment upon the occupant. To determine the effect of the fire hazards on occupants, EXODUS uses a Fractional Effective Dose (FED) toxicity model [25]. FED models assume that the effects of certain fire hazards are related to the *dose* received rather than the exposure *concentration.* The model calculates the ratio of the dose received over time to the effective dose that causes incapacitation or death, and sums these ratios during the exposure. When the total reaches unity, the toxic effect is predicted to occur. Within EXODUS, as the FED approaches unity the occupant's mobility, agility, and travel rates can be reduced making it more difficult for the affected occupant to escape. The core toxicity model implemented within EXODUS is the FED model of Purser [25]. This model considers the toxic and physical hazards associated with elevated temperature, thermal radiation, HCN, CO, CO_2 and low O_2 and estimates the time to incapacitation. In addition to this behaviour, the occupant is allowed to stagger through smoke filled environments and is given the ability to select another exit path when faced with a smoke barrier based on their familiarity with the structure. Furthermore, as the smoke concentration increases and visibility decreases, the travel speed of the occupants is reduced according to experimental data [25].

The thermal and toxic environment is determined by the Hazard submodel. This distributes hazards throughout the environment as a function of time and location. EXODUS does not predict these hazards but can accept experimental data or numerical data from other models. A software link has been established between the EXODUS and the CFAST zone model [26]. This allows CFAST history files to be automatically passed to the EXODUS model, thereby enabling the EXODUS and CFAST models to interact in a relatively straight forward manner.

EXODUS produces a range of output, both graphical and textual. Interactive two-dimensional animated graphics are generated as the software is running. This allows the user to observe the evacuation as it takes place. The graphics are interactive allowing the user to interrogate occupants and events. In addition, a data output file is produced containing all the relevant information generated by the simulation, including a copy of the input data. To aid in the interpretation of the results produced by EXODUS several data analysis tools have been developed. These are intended to be used once a simulation has been completed and enable large data output files to be searched and specific data selectively and efficiently extracted (askEXODUS). In addition, a post-processor virtual-reality graphics environment has been developed (vrEXODUS), providing an animated three-dimensional representation of the evacuation (see Fig. 2).

Figure 2: vrEXODUS generated representation of a buildingEXODUS simulation.

Finally, the introduction of the 'Itinerary list' capability extends the applications of buildingEXODUS beyond emergency scenarios to normal circulation patterns in structures. Using this capability, it is possible to investigate how a structure and its users will interact in normal usage such as at peak congestion periods in transport terminals, shopping malls and sports, conference and exhibition venues.

6 The buildingEXODUS Evacuation Model

In this section some of the advanced adaptive behavioural features found in buildingEXODUS are demonstrated in a practical example involving a complex building geometry subject to fire and a range of occupants with varying awareness of the enclosure layout. In this case the NORMAL behavioural regime is used. This allows the potential redirective behaviour of the occupants to be better understood, as they will initially only use the exits that are in general use. For a fuller account of this example and the techniques used, the reader is referred to [27].

The geometry consists of four exits. The localised occupant familiarity system is used to model the evacuation and so it is necessary to indicate the level of occupant exit familiarity. The main exit is 2m wide and is assumed to be familiar to all of the occupants. The remaining three exits are assumed to be emergency exits measuring 1m in width. As these are emergency exits they are not as familiar to the occupant population. The degree of occupant familiarity with these exits is indicated in Fig. 3. The geometry is populated with 150 male and 150 female occupants, randomly located throughout the structure. All of these occupants are fully mobile. The occupant population is assumed to require between 90 and 180 seconds to respond to the call to evacuate. An arbitrary fire atmosphere is imposed on the simulation. The fire atmosphere is represented by a series of zone averages imposed by the authors. The fire atmosphere, while arbitrary is intended to represent an evolving fire hazard and contains the key components of heat (convective and radiative), smoke concentration, and the narcotic gases CO, CO_2, HCN and reduced O_2 (see [27]).

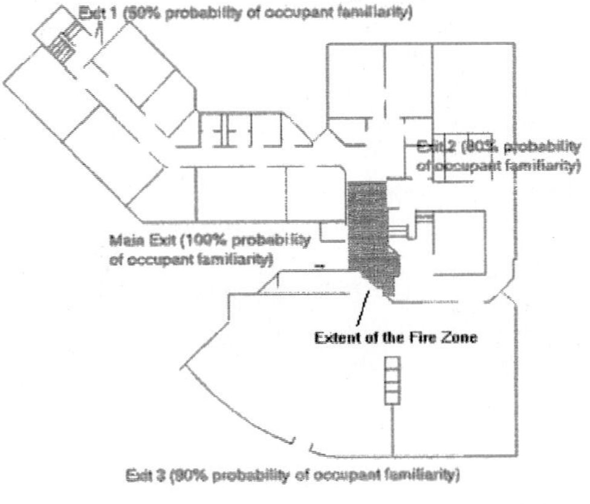

Figure 3: Geometry used in buildingEXODUS demonstration.

For simplicity, the fire is represented by a single spatial zone of fixed size that covers the main entrance and the approach to the entrance (see Fig. 3). Furthermore, the hazard values within the zone are kept constant over the duration of each time step of the fire evolution. The buildingEXODUS toxicity model is activated and so FED calculations are made for each occupant during the simulation. The scenario is repeated 10 times to generate a distribution of results. When the occupants eventually responded (after between 90-180 seconds), the majority of the occupants moved towards the main exit. This is due to the nature of the geometry and the fact that all occupants were familiar with the main exit. However, their path was blocked by the developing fire hazard that was seated in the corridor directly ahead of the main exit. On average 42 occupants redirected towards the other available exits, avoiding any interaction with the hazard entirely (see table 1 and Fig. 4). Had more occupants been familiar with the other exits, a greater number would have elected to redirect.

From table 1, the majority of redirecting occupants selected exit 1, even though this represented the exit that was least likely to be known to the occupants. This demonstrates that the number of occupants that elect to redirect towards an exit is influenced by a number of factors and is not simply a reflection of occupant familiarity. The complexity of exit usage can arise through a combination of influential factors such as the occupant's initial position, the probability of occupants redirecting, the occupant's familiarity with the structure and the peculiarities of the scenario. In this instance, the initial location of the occupants (who were randomly distributed throughout the geometry), the shape of the geometry and the location of the smoke barrier had an important impact upon exit selection.

Exits 1, 2 and 3 were emergency exits and as such would not have been used in everyday movement behaviour. In the simulation, the occupants only considered using the emergency exits once they had encountered a smoke barrier and realised that they had to redirect. Thus all of the occupants initially moved for the main entrance. Once they encountered the smoke barrier the redirective behaviour was initiated.

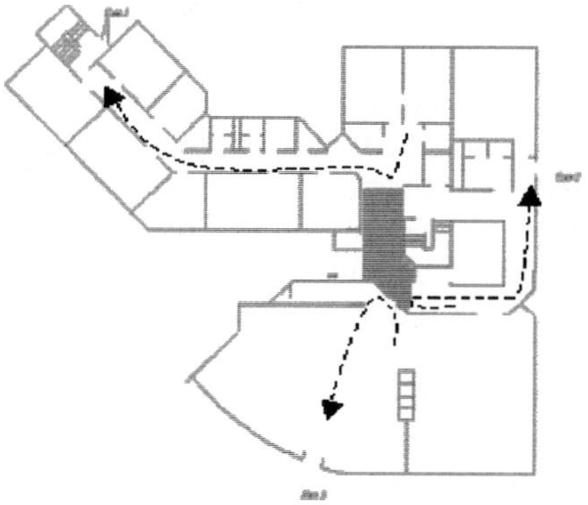

Figure 4: Redirection routes that may have been adopted during evacuation.

Occupants approaching the main entrance from the north part of the building (i.e. the top of Fig. 4) had only one viable alternative exit, namely exit 1. Due to the attractiveness of exit 1, only 50% of the occupants originally located in the north part of the building would have been aware of this exit.

The remaining occupants travelling from the north side of the smoke barrier were forced to proceed through the smoke barrier. The other redirecting occupants had two alternative exits, exit 2 and 3. Thus, these redirecting occupants were split between exits 2 and 3, reducing the potential number of occupants using either of these two exits. The precise choice of these occupants was dependent on the relative attractiveness of the two exits. This explains why exit 1 attracted approximately twice as many people as either exit 2 or exit 3. As can be seen from table 2, *irrespective of occupant awareness*, no occupants redirected towards the main exit, as this was situated behind the seat of the fire.

No of Occ. That redirected towards					
Exit 1	Exit 2	Exit 3	Exit 4	Evac. Time (sec)	Fatalities
20 [14-26]	10 [5-13]	12 [8-16]	-	221.53 [205.5-240.1]	162 [154-172]

Table 1: Results from evacuation simulation (numbers in brackets represent minimum and maximum values observed over the 10 repeat simulations).

7 The airEXODUS Evacuation Model

Consider the next generation Very Large Transport Aircraft (VLTA). Designs currently being considered are capable of carrying 800+ passengers with interiors consisting of two aisles and possessing two full-length passenger decks. Other more radical concepts consist of a Blended Wing Body (BWB) design, involving one or two decks but with possibly five or six aisles. Questions concerning seating arrangement, nature and design of recreational space, the number, design and location of internal staircases, the number, location and type of exits, the number of cabin crew required and the nature of the cabin crew emergency procedures are just some of the issues that need to be addressed.

The massive increase in passenger capacity and aircraft size being suggested also challenge some of our preconceptions in equipment design and operating procedures. For instance, in order to efficiently complete an evacuation, will it be necessary to extend emergency procedures to the marshalling of those passengers evacuated to the ground? Imagine a situation with 800 passengers on the ground, possibly on one side of the aircraft. What impact will they have on fire fighting and rescue operations? Who should take responsibility for the grounded passengers? Should evacuation procedures allow passengers to travel between decks before exiting the aircraft? Do we need to consider new concepts in exit design, perhaps introducing three or four lane exits? How efficient can a three or four lane exit be in evacuating passengers? Will the proximity of multiple emergency slides have a detrimental effect on evacuation efficiency and safety? Can exits be safely spaced further apart than the current arbitrary 60 foot limit? What impact will this have on evacuation times and survivability? How much time is actually required for safe egress? Does the 90 seconds concept have any relevance to these aircraft? If BWB aircraft become a reality should designs incorporate continuous solid cabin partitions along the length of the aircraft? Should these cabins have cross aisles linking each cabin section? Will it be sufficient to simply have exits in the forward and aft sections of the aircraft?

Quite apart from questions of emergency evacuation, issues concerning the appropriateness of VLTA and BWB designs in allowing the rapid and efficient movement of passengers during boarding and disembarkation are an additional essential design consideration. Furthermore, these requirements may potentially conflict with the requirements for emergency egress. Ultimately, the practical limits on passenger capacity are not based on technological constraints concerned with aircraft aerodynamics but on the ability to evacuate the entire complement of passengers within agreed safety limits.

The airEXODUS evacuation model is being used to address these and related issues. Associated with the development of airEXODUS is the need for comprehensive data collection/generation related to human performance under evacuation conditions. Three forms of existing data are being used as the source of the required information. Aircraft accident human factors reports produced by for example the NTSB and the AAIB [28], 90-second certification data held by the aircraft manufacturers [29], and large-scale experimentation devised to answer operational questions [30]. All three have extensively been used in the development of airEXODUS. airEXODUS makes use of 90-second certification data to specify certain key parameters [29]. In particular, data concerning the *EXIT READY TIME, PASSENGER EXIT DELAY TIMES and OFF-TIME,* and parameters for all the exits must be specified. These are described briefly as follows:

(a) Exit Ready Time

Exits are opened by cabin crew or passengers. Within airEXODUS it is possible to specify the time require to open the exit. In addition, the exit ready time can include the time required for the slide to deploy. In effect the exit ready time specifies the time from the start of the simulation to when the exit is ready to allow passengers to pass through the exit. Exit ready times for aircraft involved in past certification trials have been analysed as part of the data extraction exercise reported in [29].

(b) Passenger Exit Delay Time

One the most important parameters in airEXODUS is the pax exit delay time. This time represents two stages of the exiting process, the exit hesitation time and the exit negotiation time. In virtually all cases, the paxs exhibit a hesitation at the exit, before negotiating it. Typically, this starts when an outstretched hand first touches the exit. The latter time considers the amount of time taken to pass through the exit. The precise points at which hesitation begins and ends is based on a somewhat subjective decision and sometimes impossible to judge due to crowding, camera angle, light intensity etc. In general, the exit delay time is due in main to paxs either waiting at the exit for the path to clear and/or contemplating how to negotiate the exit. In either case, the negotiation stage does not usually start until there is space for it to commence. Furthermore, the process of passing through the exit and travelling from the exit to the ground are considered as separate events (controlled by exit delay time and Off-Time respectively within airEXODUS) which can occur in parallel.

Within airEXODUS, each passenger is randomly assigned a delay time as they pass through the exit. The delay time is assigned using a probability distribution derived from past certification trials [29]. The delay time is dependent upon a number of factors. The following list represents the most prevalent of these factors:

- **exit type** - the exit type (thus size) causes different kinds of exiting technique for each exit type, for example passengers tend to crouch and climb out of Type-III exits, and jump out of Type-A exits.
- **exiting behaviour** - different behaviour traits may be exhibited by different passengers, even on the same exit type. For example, some passengers jump through Type-A exits, whereas others sit on the sill and push off.
- **passenger physical attributes** - The gender, age and physical size of the passengers has also been found to have an impact on the hesitation time. However, there is currently insufficient data available to perform a meaningful analysis on all exit types.
- **presence of cabin attendants** - the presence (or absence) of cabin attendants at exits can enormously influence the behaviour exhibited by passengers at exits. Undirected passengers tend to take more time deciding how to use the exit, and indeed, which exit to use.
- **behaviour of cabin attendants** - when cabin attendants are present at an exit, the degree of assertiveness they display also influences the hesitation times. As the level of assertiveness increases, the range of slower hesitation times decreases, thus increasing the overall flow throughput of the exit.

(c) Off Time

In airEXODUS, when a passenger has reached an exit, a further time penalty is added to reflect the time spent either travelling down a slide or across a wing. The time penalty added is randomly assigned between user specified limits, similar to the passenger exit hesitation time. In reality, this time will depend upon a number of factors, such as slide sill height and slide travel technique. When added to the exit time of a passenger, this produces the *on-ground* time. Off Times for aircraft involved in past certification trials have been analysed as part of the data extraction exercise reported in [29].

In order to model the above scenario, airEXODUS was run with the competitive behaviour options deactivated. The population used within the airEXODUS simulations consisted of 38 males and 22 females. The exit hesitation time distribution used in these simulations was derived from 90 second trial data for Type-I exits with assertive cabin crew. The extremes in evacuation performance were used to define the numerical window of results using an identical process to that described for the experimental results. A total of twelve airEXODUS simulations were conducted involving four repeats of three different seating arrangements. Fig. 5 depicts the window of results for both the numerical simulations and the experimental trials. As can be seen there is quite good agreement between the numerical and experimental results.

While airEXODUS has successfully been used to predict the certification performance of several aircraft, the outcome of these trials was known to the authors before the simulations were performed. A more challenging validation exercise was requested by the UK CAA, requiring airEXODUS to predict the performance of a modified Boeing B767 aircraft, (designated the B767-304ER), prior to the actual test, in order to establish the predictive capabilities of airEXODUS for 90 second certification trials. A confidential report [31] containing details of the model formulation and results of the simulations was produced by FSEG and distributed to the UK CAA and US FAA prior to the trial, and Boeing after the trial. The geometry of the B767-304ER was constructed within airEXODUS using the interactive tools available, based on plans supplied by Boeing. The aircraft seats 351 passengers and has four pairs of exits arranged with two pairs of Type-A exits forward of the wing, a pair of Type-I exits just aft of the wing and a pair of Type-A exits in the rear.

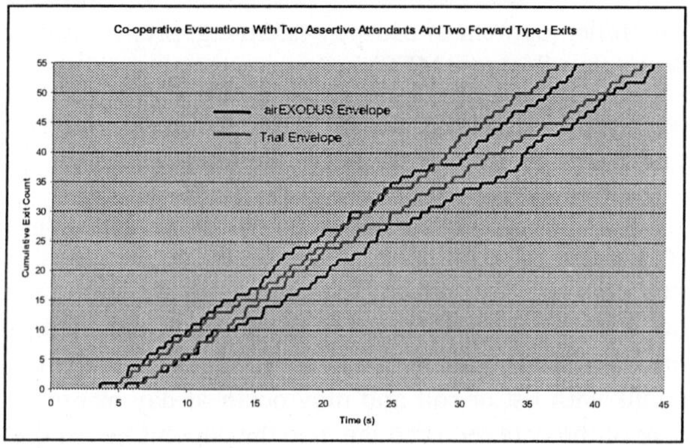

Figure 5: Window of results for numerical simulations (dark lines) and experimental trials (light lines)

The model results suggest that the B767-304ER is capable of producing evacuation times in the range from 69.5 to 74.4 seconds with an average of 71.8 seconds. In order to make a direct comparison with the trial results, the time for the crew to leave the aircraft and the slide times must be subtracted from the actual measured trial times. This comparison reveals that the average evacuation time predicted by airEXODUS is within approximately 2% of the actual recorded time. Furthermore, general trends in passenger flow behaviour predicted by airEXODUS appear to have been corroborated by actual events, for instance, the passenger split within the cabin predicted by airEXODUS was achieved in the actual trial.

The airEXODUS evacuation model has obvious applications to the design and certification of aircraft as well as the training of cabin crew. In addition, the model can be used to assist in the interpretation and understanding of prescriptive rules – such as the "60 foot" rule. In this way, evacuation models can be used to help frame rules, removing some of the magic from the "magic numbers" that abound in the aviation industry.

8 The maritimeEXODUS Evacuation Model

From the modelling perspective, there are at least five components to a ship evacuation:

(1) Recognition by passengers of the need for emergency action i.e. response to alarm.

(2) Preparatory actions e.g. collect life jacket, reunite family, deploy crew, etc.

(3) Progressive evacuation to place of relative safety i.e. refuge or assembly station.

(4) Preparation/deployment of escape system e.g. prepare lifeboats, deploy Marine Evacuation System (MES).

(5) Abandoning the vessel e.g. boarding and lowering lifeboat, going down an MES and boarding life boat, climbing down a ladder, etc.

Depending on the nature of the "what if" conditions being simulated, all of the above aspects may need to be modelled, while other applications may only require some of these components. To perform the required simulation reliably requires an evacuation model with the appropriate set of capabilities and access to the necessary data. Furthermore, the scenario under consideration may be under conditions of calm or involve situations with list or roll and may occur at day or night. This will affect not only the nature of the data required but also the capabilities of the model.

The current prototype version of maritimeEXODUS does not posses marine specific data for the performance of passengers under conditions of list and roll. However, when this data is available it can easily be incorporated into the model framework. As the model has the flexibility to allow the user to alter all of the preset default values, it is easily adaptable when new data becomes available. Furthermore, the current version of maritimeEXODUS does not include reliable data to represent the preparation/deployment of the escape system or the actual abandonment of the vessel. Again, these aspects can be included when data becomes available. In its current form the maritimeEXODUS model is suitable for applications involving the first three components of the identified ship evacuation process.

A simple demonstration of the current capabilities of the maritimeEXODUS prototype is presented using a section of a hypothetical passenger ship. The section under investigation consists of two decks containing passenger cabins and public rooms (see Fig. 6 bottom right). In the section shown, two main stairways link the two decks, one on the port and the other on the starboard side of the ship. A total of 200 passengers are located in the section under consideration, most of which are positioned in their cabins. In the scenario under consideration, the assembly stations are located on the lower deck. In the simulation passengers must retrieve their life jackets from their cabins before moving to their particular assembly stations. In this particular "what if" situation, the emergency occurs during daylight hours and in calm conditions. The following types of behaviours are included in the scenario:

- Not all the passengers react initially to the command to gather at the assembly stations. Some of the passengers require time to react.
- The majority of passengers return to their cabins and pick up their life jackets.
- Some passengers go in search of their group members prior to retrieving life jackets. Thus passengers from the lower deck go to the upper deck which creates contra-flow situations in the corridors and more importantly on the staircases.
- Some passengers undertake a room to room search for lost group members.
- The majority of passengers move to their muster stations once their life jackets have been retrieved. Fig. 6 (top right) shows a scene from the lower deck with the corridors filling with people as they make their way to their muster stations. A small group can be seen already located at a muster station. Fig. 6 (top left) depicts a scene from the upper deck. As can be seen, some passengers are moving along the corridors to the staircases, some passengers are still in their cabins while other passengers are still gathered in the public spaces.

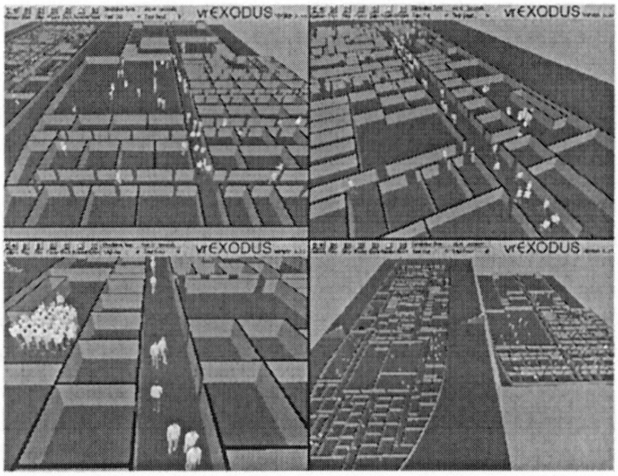

Figure 6: vrEXODUS representation of martimeEXODUS predicted ship evacuation.

- Some passengers go to the wrong muster station and are redirected to their correct muster station, again creating contra-flow situations in the corridors.
- Crew members are moving towards various duty stations, this also creates contra-flow situations within the corridors and staircases.
- Once at the muster stations, passengers gather in groups awaiting the command to evacuate. Depicted in Fig. 6 (bottom left) is a large group of passen-

gers gathered at a muster station however, several passengers are still moving along the corridor heading towards their muster stations.

- Once the command to abandon is given, passengers exit the vessel. While data relating to the time required to undertake the abandon process is not included in the current prototype version of martimeEXODUS, an arbitrary delay time representing the abandonment phase included and assigned randomly to each passenger as they abandon the vessel.

Research and development work for the maritimeEXODUS model is continuing. In collaboration with the Canadian company FLEET TECHNOLOGY and the Canadian Government, a detailed abandonment model is currently under development and a series of trials is being developed in order to collect human performance data under different conditions of list. This information will be incorporated within martimeEXODUS along with appropriate modifications to the behaviour model.

9 The railEXODUS Evacuation Model

While the prevention of rail accidents should be the primary goal of the rail industry, improving the survivability of passengers involved in such accidents, in terms of the crash worthiness of carriages and escape provision, remains a high priority. Though incidents requiring the rapid egress of passengers from trains are thankfully rare, there is an obvious requirement to ensure that the system design - the design of the rail vehicles and signage/communications, through to crew operating procedures - is adequate to allow safe and efficient egress of passengers under a range of conditions. The complexity of rail networks (a mix of underground, light rapid transit, commuter, intercity, and international systems) and the diversity of the rolling stock (type and age), makes generalising about safety difficult. Evacuation models can assist in the design of rail carriages by ensuring that safe egress is possible from all proposed rolling stock designs.

The simplest form of train evacuation involves the egress of passengers through the doors on the side of the train onto a platform. While this type of evacuation can be very quick (see for example [32]), it can only occur when the train has not been disabled and is capable of making its way to a station. In some circumstances it may be necessary to evacuate passengers from one carriage to another through the doors at the end of the carriage. While superficially "simple", these evacuation scenarios, like evacuation from any structure, is a complex process dependent on a number of configurational, procedural, environmental and behavioural interacting factors. The speed and efficiency of evacuation in these circumstances will be dependent on amongst other factors, the; number of passengers; level of overcrowding; nature of the behaviour of the passengers i.e. co-operative or competitive; initial makeup of the passenger population i.e. age, gender, social affiliation, physical ability levels, time of day, etc; physical, psychological and physiological condition of the passengers during evacuation; presence and behaviour of controlling staff; level of lighting; internal layout of the carriage; presence of debris in the carriage; size of exits, etc.

In cases where the evacuation is onto a platform additional factors such as the number and location of exits; size of the gap between carriage and platform; height of step down onto the platform; size of available platform area; egress paths leading away from the platform area, etc. must be considered. For the most part, the highlighted factors that influence evacuation from rail vehicles has not been the subject of rigorous scientific study, and as such are not fully understood or quantified.

In more severe yet still favourable circumstances, it may be necessary to evacuate passengers from an upright train onto the track level. Under these conditions, the passengers may use the side doors as in a platform evacuation or the end door as in a carriage to carriage evacuation. In the latter case, passengers move through the train, passing from carriage to carriage, until a suitable useable exit is found. In some circumstances an emergency ladder (at least one is carried on most U.K. trains) may be used to assist in detraining. Evacuations of this type can take a considerable amount of time, for instance, it is reported by a railway official that an actual non-emergency but necessary detraining of approximately 400 passengers required approximately one hour and twenty minutes for the passengers to descend to track level [33]. This equates to an average of 12 seconds for a passenger to negotiate the exit and ladder (producing a flow rate of 5 passengers/minute). In this incident, the passenger population included elderly and infirm people and passengers attempting to evacuate with their luggage.

Perhaps the most challenging scenario for passengers involves the evacuation from an overturned carriage subjected to fire. As part of our effort to develop the railEXODUS evacuation model, a study of human behaviour exhibited during rail crash incidents through the study of past accidents was undertaken [34,35]. In addition a series of experiments were undertaken in an attempt to estimate the flow rate capacity of an overturned rail carriage end exit [34,35]. This is achieved through two full-scale evacuation experiments, in one of which the participants were subjected to non-toxic smoke. The experiments were conducted as part of a pilot study into evacuation from rail carriages. Details of the experiment may be found on our web site at http://fseg.gre.ac.uk/fire/net20.html. In reviewing the experimental results, it should be noted that only a single run of each trial was undertaken with a limited – though varied – population. The carriage used in the experiments is a standard class Mark IID, which while an old carriage design, shares many features with those carriages commonly found on the British rail network.

In the evacuation involving smoke, the carriage end exit was found to achieve an average flow rate capacity of approximately 5.0 persons/minute. The average flow rate capacity of the exit without smoke was found to be approximately 9.2 persons/minute. It is noted that the presence of smoke tends to reduce significantly the exit flow rate. Due to the nature of the experimental conditions, these flow rates are considered *optimistic*. Finally, the passenger behaviour observed in these trials and reported by survivors of actual accidents does not suggest any behaviour that could not be represented within computer based models of evacuation. As a result no fundamental impediment is envisaged in the development of railEXODUS. These models could be used to investigate the evacuation efficiency of both proposed new design and in service rolling stock. The models could also be used to evaluate the potential benefit of design modifications and assist in the development of evacuation procedures. The models also have a role in accident investigation. However, further experimental work is essential to generate the data required by these models to characterise passenger performance under a variety of rail specific conditions.

10 Concluding Comments

Evacuation models are an established reality in the building and aviation communities. In both these industries considerable effort and resources has been invested in establishing the knowledge base required to develop evacuation modelling tools and the data required to confidently use them. In these industries, these models are becoming a cost effective way to achieving safer and more functional design solutions, allowing designers to seek answers to questions that can not be addressed using conventional approaches. Evacuation models are also appearing in the maritime and rail industries.

Evacuation models are increasing in sophistication through the incorporation of more complex behaviourial capabilities. The first generation hydraulic and the second generation ball-bearing models are making way for behavioural models with adaptive capabilities that include, structural knowledge, reaction to communication, affiliative behaviour, occupant motivation and a stochastic element to the queuing recommitment behaviour. However, for this development to occur and to establish an acceptable level of confidence in the results produced, more data is required concerning the decision-making process. In order for evacuation models to achieve their full potential, it is essential that industry work together to provide the data essential for the further development and validation of these models.

Furthermore, the challenge facing regulators and approval authorities is to develop an understanding of the modelling technology being developed and with that understanding specify relevant design standards and protocols. It is essential that a minimum range of relevant design scenarios as well as acceptance criteria become universally accepted. These should encompass the:

- nature of the target population being investigated e.g. movement capabilities, response times, etc,
- emergency conditions under which the evacuation takes place e.g. night/day, fire/no fire, etc, and
- performance requirements that the proposed design much achieve to satisfy a deemed to comply requirement.

It is hoped that industry and regulatory authorities will explore these issues as they hold the potential to make already safe structures safer by design and assist in removing some of the "magic" from the "magic numbers" that plague safety analysis.

11 Acknowledgements

Prof. Galea is indebted to his colleagues in FSEG, without whom this paper would not have been possible. Prof Galea also gratefully acknowledges the UK CAA for their financial support of his chair in Mathematical Modelling.

References

1. P.G. Wood, *Survey Of Behaviour In Fires*, in: *Fires And Human Behaviour*, D. Canter and Fulton, (Eds.), pp. 205-234 (2nd Edition, 1990).

2. J. Bryan, *Behavioural Response To Fire And Smoke*, in: *The SFPE, Handbook Of Fire Protection Engineering*, Dilenno et al., (Eds.), pp. 1-241-1-262 (NFPA, Quincy, MA, 2nd Edition, 1996).

3. J. Sime, *Escape Behaviour In Fires And Evacuations*, in: *Fires And Human Behaviour,* D. Canter and Fulton, (Eds.), pp. 56-87 (2nd Edition, 1990).

4. H.E. Nelson and H.A. Maclennan, *Emergency Movement*, in: *The SFPE, Handbook Of Fire Protection Engineering*, Dilenno et al., (Eds.), pp. 3-286-3-295 (NFPA, Quincy, MA, 2nd Edition, 1996).

5. B.B. Pigott, *Fire Detection And Human Behaviour*, in: *Proceedings Of The Second International Symposium On Fire Safety Science*, T. Wakamatsu, Y. Hasemi, A. Seizawa, et al., (Eds.), pp 573-581(1989), ISBN 0891168648.

6. D. Canter, *Overview Of Human Behaviour,* in: *Fires And Human Behaviour*, D. Canter and Fulton, (Eds.), pp. 205-234 (2nd Edition, 1990).

7. J. Fruin, *Service Pedestrian Planing and Design*, Maudep 1971, reprinted 1987.

8. J.M. Lewis, *Theories of the Crowd: Some Cross Cultural Perspectives*, Easingwold Papers, Prontaprint Commercial, (1990), ISBN 1874321043.

9. J. Sime, *Human Behaviour In Fires Summary Report,*, CFBAC Report No. 450, (Portsmouth Polytechnic,1992).

10. M. Wells and H. Sarah, *50 feared dead: the cost of going through a red signal,* in: *Report of the Ladbroke Grove crash*, p. 1 (The Guardian, 6th October 1999).

11. S. Gwynne, E.R. Galea, M. Owen, and P.J. Lawrence, *Escape as a Social Response*, (Pub by the Society Of Fire Protection Engineers, Boston USA, 1999).

12. Draft MSC Circular, *Interim Guidelines for a Simplified Evacuation Analysis on RO-RO Passenger Ships*, Annex 2, FP 43/18, (1999).

13. S. Gwynne, E.R. Galea, P.J. Lawrence, M. Owen, and L. Filippidis, *A Review of the methodologies used in the computer simulation of evacuation from the built environment*, Building and Environment **34**, 741-749 (1999).

14. E.R. Galea, S. Gwynne, L. Filippidis, and P. Lawrence, *buildingEXODUS Technical Manual V3.0*, (University Of Greenwich, 2000).

15. S. Gwynne, E.R. Galea, M. Owen, and L. Filippidis, *A Systematic Comparison Of Model Predictions Produced By The buildingEXODUS Evacuation Model And The Tsukuba Pavilion Evacuation Data*, Applied Fire Science **7**, No. 3, 235-266 (1998).

16. S. Gwynne, E.R. Galea, P.J. Lawrence, M. Owen, and L. Filippidis, *Adaptive Decision-Making in buildingEXODUS in Response to Exit Congestion*, in: *Fire Safety Science, Proc of the 6th Int Symp*, M. Curtat (Ed.), pp. 1041-1052 (Poiters, France, 2000).

17. E.R. Galea, M. Owen, and P. Lawrence, *Computer Modelling of Human Behaviour in Aircraft Fire Accidents*, Toxicology **115**, Nos. 1-3, 63-78 (1996).

18. M. Owen, E.R. Galea, P. Lawrence, and L. Filippidis, *The Numerical Simulation of Aircraft Evacuation and its Application to Aircraft Design and Certification*, The Aeronautical Journal, 301 – 312 (June/July, 1998).

19. J. Fruin, *Pedestrian planning and design*, (Metropolitan Association of Urban Designers and Environmental Planners, New York, 1971).

20. J.L. Pauls, *Building evacuation:Research findings and recommendations*", in: *chp 14 in Fires and Human Behaviour*, D. Canter (Ed.), pp. 251-275, Wiley, London.

21. E.R. Galea, *A general approach to validating evacuation models with an application to EXODUS*, J. of Fire Sciences **16**, 414-436 (Sept/Oct, 1998).

22. E.R. Galea and M. Owen, *Predicting the Evacuation Performance of Mass Transport Vehicles using Computer Models*, in: *Proceedings, Fire Safety On Ships*, pp. 239-251 (IMAS London, 1994).

23. E.R. Galea, L. Filippidis, P.J. Lawrence, and M. Owen, *An evacuation demonstration of a typical high speed craft using the EXODUS evacuation model*, (Univ of Greenwich report for RINA, G/DG/1998/000116, London, 1998).

24. E.R. Galea, Safer by design: Using computer simulation to predict the evacuation performance of passenger ships, Proc. IMarE Conf. **112**, 2, Safety of Large Passenger Ships, pp. 23-32, (London, 2000), ISBN 1902536312.

25. D.A. Purser, *Toxicity Assessment Of Combustion Products*, in: *The Sfpe, Handbook Of Fire Protection Engineering*, P.J. Dilenno, C.L. Beyer, R.L.P. Custer, W.D. Walton, J.M.W. Watts, D. Drysdale, and J.R. Hall, (Eds.), pp. 2-85-2-146 (National Fire Protection Association, Quincy, Ma, (2nd Edition), 1996).

26. R.D. Peacock, G.P. Forney, P. Reneke, R. Portier, and W.W. Jones, *CFAST, The Consolidated Model Of Fire Growth And Smoke Transport*, NIST Technical Note 1299, U.S. Dept. of Commerce, Technology Administration, (National Institute of Standards and Technology, 1993).

27. S. Gwynne, E.R. Galea, P.J. Lawrence, and L. Filippidis, *Modelling Occupant Interaction with Fire conditions Using The buildingEXODUS Evacuation Model*, to appear in Fire Safety Journal 2001.

28. M. Owen, E.R. Galea, P.J. Lawrence, and L. Fillipidis, 1998, *AASK - Aircraft Accidents Statistics and Knowledge: A Database of Human Experience in Evacuation*, derived from Aviation Accident Reports, The Aeronautical Journal, pp. 301-312 (June/July 1998).

29. M. Owen, A.J.P. Dixon, and E.R. Galea, 90-second Certification Trial Data Archive Report, prepared for the U.K. CAA for project 049/SRG/R&AD, (March 1999).

30. H. Muir and A. Cobbet, 1996, Influence of Cabin Crew During Emergency Evacuations at Floor Level Exits, Technical Report, CAA Paper 95006, Parts A+B, ISBN 0860396495.

31. E.R. Galea and M. Owen, 1996, Initial EXODUS Model Predictions for Certification evacuation Scenarios of the B767-304ER Aircraft, Confidential Report for the U.K. CAA.

32. H. Muir, *Human Behaviour in Emergency Situations*, in: *Collection of conference papers for the confernece "Putting People at the centre of a safer railway"*, Railtrack Plc, The BAFTA Centre London, (May 1999).

33. Letter from D.A.Marshall to Inquiry Chairman Southall Rail Crash, (30 Sept 1999).

34. E.R. Galea and S. Gwynne, *Evacuating an Overturned Smoke Filled rail Carriage*, (CMS Press University of Greenwich, London, Paper No. 00/IM/55, 2000), ISBN 1899991565.

35. E.R. Galea and S. Gwynne, *Estimating the Flow Rate Capacity of an Overturned Rail Carriage End Exit in the Presence of Smoke.*, to appear in Fire and Materials, 2000.

Mathematical Modelling of Evacuation Problems: A State of the Art

H.W. Hamacher[1,2] and S.A. Tjandra[1]

[1] Fraunhofer Institute Techno-und Wirtschaftsmathematik (ITWM), 67663 Kaiserslautern, Germany
[2] Fachbereich Mathematik, Universität Kaiserslautern, Germany

This paper details models and algorithms which can be applied to evacuation problems. While it concentrates on building evacuation many of the results are applicable also to regional evacuation. All models consider the time as main parameter, where the travel time between components of the building is part of the input and the overall evacuation time is the output. The paper distinguishes between macroscopic and microscopic evacuation models both of which are able to capture the evacuees' movement over time. Macroscopic models are mainly used to produce good lower bounds for the evacuation time and do not consider any individual behavior during the emergency situation. These bounds can be used to analyze existing buildings or help in the design phase of planning a building. Macroscopic approaches which are based on dynamic network flow models (minimum cost dynamic flow, maximum dynamic flow, universal maximum flow, quickest path and quickest flow) are described. A special feature of the presented approach is the fact, that travel times of evacuees are not restricted to be constant, but may be density dependent. Using multicriteria optimization priority regions and blockage due to fire or smoke may be considered. It is shown how the modelling can be done using time parameter either as discrete or continuous parameter. Microscopic models are able to model the individual evacuee's characteristics and the interaction among evacuees which influence their movement. Due to the corresponding huge amount of data one uses simulation approaches. Some probabilistic laws for individual evacuee's movement are presented. Moreover ideas to model the evacuee's movement using cellular automata (CA) and resulting software are presented. In this paper we will focuss on macroscopic models and only summarize some of the results of the microscopic approach. While most of the results are applicable to general evacuation situations, we concentrate on building evacuation.

1 Assumptions and Goals of Evacuation Models

Evacuation as one aspect of emergency processes can be simply defined as the removal of residents from a given area that has been considered as a danger zone to safety as quickly as possible and with utmost reliability. Two different evacuation scenarios can be considered.

(i) Precautionary
In this type of evacuation the estimation of the evacuation time compared to the hazard propagation time and the estimation of the risk can be done a priori. Hence, time and potential risks are the key components of this type of evacuation.

(ii) Life-saving operations
This type of evacuation occurs when insufficient warning has prevented the organizer from conducting a pre-emergency evacuation planning. Here it is more likely that problems as the rescue of injured evacuees in and around the damaged area, route clearance, etc. have to be dealt with.

Evacuation problems may arise in different types of systems, such as buildings, cities or regions, or transportation carriers (e.g. train, ship and airplane). The system structures (e.g. population and the behavior of people at risk, hazard propagation speed and charateristics) essentially influence the optimal planning in the corresponding system. In general, the following is a list of informations which would be ideal to have in evacuation planning (and needs to be estimated somehow, if not available) :

- Type of system defined by layout/geographic information and familiarity, for example : office building, shopping mal and airport
- Behaviour estimation of the occupants under panic situation.
- Occupants distribution (includes age, gender and disability).
- Source and location of hazard, hazard propagation speed/charateristics and factors affecting the hazard propagation.
- Safe destinations (refuge places).
- Availability of emergency service facilities and personel.

These informations are used to define the *dynamic severity matrix* $[\mathbf{r_{it}}]$ where vector $\mathbf{r_{it}}$ states the estimated severity or hazard level of area i at time t. The number of components of vector $\mathbf{r_{it}}$ is determined by the number of informations which are considered. Equation (1) shows an example of dynamic severity matrix.

$$[\mathbf{r_{it}}] = \left(\begin{array}{c} \vdots \\ \ldots \, \mathbf{r_{it}} = \left(\begin{array}{c} \text{fire level of area } i \text{ at time } t \\ \text{temperature level of area } i \text{ at time } t \\ \text{smoke density level of area } i \text{ at time } t \\ \text{toxicity level of area } i \text{ at time } t \end{array} \right) \ldots \\ \vdots \end{array} \right) \qquad (1)$$

By giving appropriate weights to vector components, vector $\mathbf{r_{it}}$ can be converted to a scalar r_{it}. If h is the number of vector components and w_l is the weight of component l, then r_{it} can be scalarized as follows.

$$r_{it} = \sum_{l=1}^{h} w_l r_{itl}$$

This matrix will be used to determine the evacuation priorities that will be discussed in more detail in Section 8.

The *evacuation time*, that is the time needed to complete an evacuation process, basically consists of three main time components ([30], [47]) namely:

- The time evacuees need to recognize a dangerous situation. This time is influenced mainly by the reliability of the alarm system and the familiarity of evacuees with emergency signals.
- The time evacuees need to decide which course of action to take. This time is influenced by the experience of evacuees in facing the emergency situation. This can, for instance, be generated through emergency practice and training.

Behavioural and organizational factors are the main contributors to the duration of these times. The subsequent decisions are made during the movement to the safety, especially when evacuees encounter a "hazard" route, i.e. one which is affected by fire, smoke, etc.

- The time evacuees need to move towards the safety area, which is known as *egress time*. The latter is influenced by the availability of emergency exit signs, well planned evacuation procedures, constructional factors (effective width of walkway, slope of stairs), and human behaviour during panic situations.

Since the behavioural and organizational factors are the main contributors to the first two time components, it is hard to predict analytically the duration of those time components. Therefore, most evacuation models emphasize the calculation of egress time and treat the result as the lower bound of the real evacuation time. This will also be the approach in this paper.

The next section describes the idea of macro- and microscopic models and give a list of classes of existing models which can be used in the evacuation problems. Informations on various specific mathematical models are given in Section 3 - 6, like dynamic flows (Section 3), maximum dynamic flows (Section 4), universal maximum flows (Section 5), quickest path and flows (Section 6). In Section 7 we show the interrelation between these models by establishing a triple optimization result. Section 8 deals with multiple criteria. The fact that the travel time may be dependend on the evacuation density is considered in Section 9. A stronger model, but sounds more difficult to solve is the continuous time dynamic network flow model which is the subject of

Section 10. Before concluding the paper we discuss microscopic models, in particular simulation approaches, in Section 11.

2 Macro- and Microscopic Models

Approaches which are used to model evacuation problems may come from different problem fields, such as, network flow problems, traffic assignment problems, and simulation. Table 2 lists classes of models which can be used in the evacuation problem modeling although some of them may have been originally developed for different purposes. In general, there are two approaches used to model evacuation problems which emphasize on the estimation of the egress time, namely *macro-* and *microscopic models* .

Macroscopic models are mainly based on optimization approaches and do not consider individual differences and decisions for selecting egress routes, i.e. occupants are treated as a homogeneous group where only common characteristics are taken into account. Since the time is a decisive parameter in the evacuation process, most macroscopic approaches are based on dynamic network flow models (see e.g. [14], [18], [19], [41], [22], [12], [43], [50]). The common idea of these models is to represent a building and the attributes of the building's components in a static network G. The modeling of evacuation over time is then done in a dynamic network G_T which is the time expanded version of G and the network flows correspond to evacuation processes (see, for instance, [1] for an overview on network flow theory).

The static network G is used to model supply and demand points, and routes which are used to transfer supplies to demands. These routes may have some intermediate transshipment points. In the static network flow models, supply, demand and transshipment points are modeled by nodes while routes are modeled by paths of the graph. A path of the graph is composed by nodes and arcs, where an arc connects two adjacent nodes. The interrelation between nodes and arcs can, for instance, be described by the *node-arc incidence matrix*. In the representation of a building using a static network, nodes may represent rooms, lobbies or intersection points, while arcs can be used to model corridors, hallways, stairways or a connection between two intersection nodes. Some locations in the building that house a significant number of evacuees are considered to be source nodes in the network. The supply of a source node is given by an estimate of the number of evacuees in the location that the node represents. The building exits or safety locations that might be considered as the final destination of evacuees' movement, are considered as sink nodes. In the evacuation problem we have only one sink node by connecting all the exit nodes to one artificial node and assign the total number of evacuees as the demand value of this node. Hence, evacuation problems can be modeled as multi-source/single sink network flow problem. Each node has a capacity which is the upper bound of the number of evacuees simultaneously allowed to stay in the node. This node capacity can be determined, for instance, by

$$\text{node capacity} := \min \left\{ \frac{\text{floor space area}}{\text{minimum required area per person}}, \frac{\text{maximum allowable weight}}{\text{average weight per person}} \right\}$$

Arcs have other attributes, such as flow capacity and travel time. The arc flow capacity is the upper bound of the number of evacuees per unit time that can traverse the arc. The travel time is the time needed to travel from one node to another. This travel time is one of the important components that must be considered in the modeling of evacuation problems. During the evacuation process, the connection between two positions may only be temporary due to, for instance, blocking by fire or smoke. In this case, the arc that represents the connection must also be temporary, i.e. the arc capacity can be set to zero after some times. These time constraints can not be properly modeled by the static network flow models but the dynamic ones. Moreover, we can formulate the dynamic network flow problem in two ways depending on whether we use a discrete or continuous representation of time.

In the area of dynamic network flow problems, some of the existing models assume constant attributes, e.g constant travel time from one node to another and constant arc flow capacity. The constant travel time might be determined according to some predetermined queuing levels such that the model can be solved efficiently but still able to give quite realistic results. In this paper it will be shown that the models may be extended, thus providing a better estimate of the final evacuation time.

Microscopic models, in which the individual evacuees' movement is emphasized, are based on simulation. These models consider individual parameters (e.g. walking speed, reaction time, physical ability) and interaction of each evacuee with other evacuees during the movement. In recent years there is a growing interest to use cellular automata as the base of microscopic simulation in the field of pedestrians and traffic movement (see for example [8], [42], [51]) which have close interrelations with evacuation problems.

3 Discrete Time Dynamic Network Flow Model

A discrete time dynamic network flow problem is a discrete time expansion of a static network flow problem. In this case we distribute the flow over a set of predetermined time periods $t = 1, 2, \ldots, T$.

Definition 1 Let $G = (N, A)$ be a directed network with N the set of nodes and A the set of arcs (the static network). On each arc $(i, j) \in A$ travel times λ_{ij} are given which are assumed to be constant. The time expansion of G over

Model Class	Evacuation Model	References
Static Network	Shortest path	[22], [67]
	Minimum cost network flow	[67]
	Quickest path	[15], [16], [38], [60]
Discrete Time Dynamic network	Minimum turnstile cost	[14], [18], [41], [50]
	Quickest Flow	[12], [23]
	Universally maximum flow	[34], [48], [66]
	Minimum weight path (multi objectives)	[43]
	Lexicographically minimal cost	[33]
	Flow dependent exit capacity	[18], [19]
Continuous Time Dynamic Network	Constant capacity and travel time	[24]
	Time dependent capacity (maximal flow)	[2], [55]
	Universally maximum flow with zero travel time	[25], [52]
Traffic assignment	Transportation network	[61], [68]
	Density dependent travel time (single objective)	[13], [17], [35], [37], [39], [59]
Simulation	Probabilistic models	[21], [46], [47]
	Cellular Automata	[8], [10], [20], [42], [51]

Table 1. Summary of Existing Evacuation Approaches.

a time horizon T defines the dynamic network $G_T = (N_T, A_T)$ associated with G where

$$N_T := \{i(t) \mid i \in N \; ; \; t = 0, 1, \ldots, T\}$$

and A_T consists of the set of movement arcs A_M

$$A_M := \{(i(t), j(t') \mid (i,j) \in A \; ; \; t' = t + \lambda_{ij} \leq T \; ; \; t = 0, 1, \ldots, T\}$$

and the set of holdover arcs A_H

$$A_H := \{(i(t), i(t+1)) \mid i \in N \; ; \; t = 0, 1, \ldots, T-1\}$$

i.e.

$$A_T := A_M \cup A_H$$

Figure 2 shows a T-time expansion of the static network of Figure 1, with $T = 4$.

The time period t is dependent on the basic unit θ in which travel times are measured. Thus, if we choose 5 seconds as the length of the basic unit (i.e. $\theta = 5$), then specifying three time periods (i.e. $t = 3$) for traversing an arc means we need fifteen seconds to do so. The number of time periods T is obtained by dividing the evacuation planning horizon of interest by the length θ of the basic unit. The smaller θ the more acurately the model represents the actual flow's evolution. Choosing θ too small, however, will result in undesirable size of the network and may have fractional arc capacities which make the problem difficult to solve. Hence, the choice of θ is a compromise between model realism and model complexity.

Since the dynamic network has $(T+1)$ copies of each source node and each sink node, the dynamic network will have multiple sources and multiple sinks. Therefore in order to handle many sources and sinks, one introduces a ***super source*** s and a ***super sink*** d to create a single source/single sink network (see Figures 2 and 3). In evacuation problems, the super sink can be interpreted as a common safety area. How the super source is connected to the source is actually problem-dependent. In the network clearing problem (clearing the network from initial occupancies), the super source is connected only to the time zero copy of the source nodes (see Figure 2). In this case, we may have holdover arcs for source nodes. Arcs from the super source have zero travel time and capacities are equal to initial occupancies. In the maximum dynamic flow problem (see Section 4), the super source is connected to all time-copies of the source nodes. In this case, we do not have holdover arcs for source nodes which do not have predecessors (e.g. node 1 in Figure 1) as shown in Figure 3. Arcs from the super source to other nodes have zero travel time and infinite capacities. On the other hand, generally all copies of every sink node are connected to the super sink and there is no holdover arc for sink nodes. All conections to the super sink have zero travel time and infinite flow capacities.

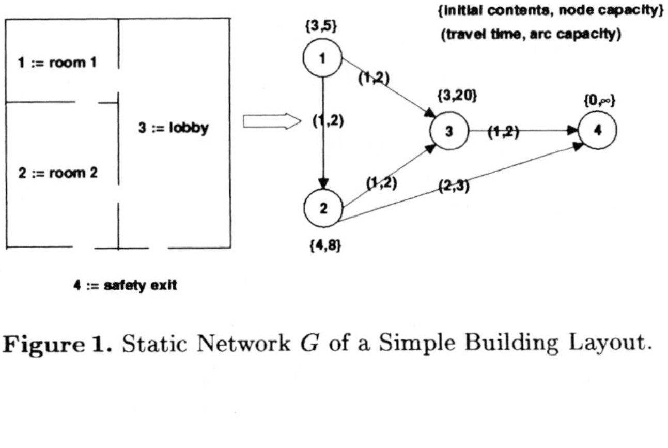

Figure 1. Static Network G of a Simple Building Layout.

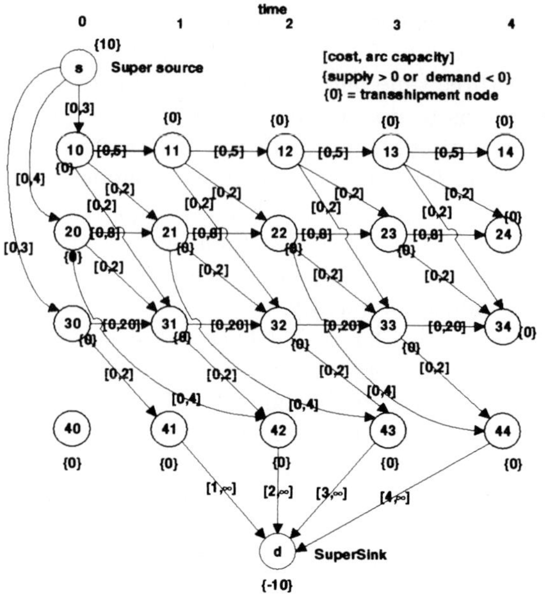

Figure 2. Dynamic Network G_T of the Static Network G of Figure 1, with $T = 4$.

By constructing the dynamic network as defined above, dynamic network flow problems can always be solved as static flow problems in the expanded network. Also, it may be noted that the equivalent static problem does not require keeping arc capacities and travel times fixed over time, as assumed in Definition 1. But these assumptions are essential for building efficient algorithms to solve the problem. The upper bound for the number of nodes and arcs in discrete time dynamic network is described by Proposition 1.

Proposition 1 *If* $n :=| N |$ *and* $m :=| A |$ *then* $n(T+1)$ *and* $(n+m)T + m - \sum_{(i,j)\in A} \lambda_{ij}$ *are the upper bound for the number of nodes and arcs in* G_T *without considering super source and super sink, respectively.*

Since we do not use any arc in the path from the super source to any sink node at time greater than T, we can reduce the size of the time-expanded network by eliminating inessential arcs including the corresponding nodes (see Figure 3).

In the dynamic network flow models, we denote by $x_{ij}(t)$ the flow (e.g. the number of evacuees moving at time t) that leave node i at time t and reach node j at time $t + \lambda_{ij}$. Flows from node i at time t to the same node with travel time $\lambda_{ii} = 1$ represent the number of evacuees who prefer to stay in the building component represented by node i at time t for at least one unit

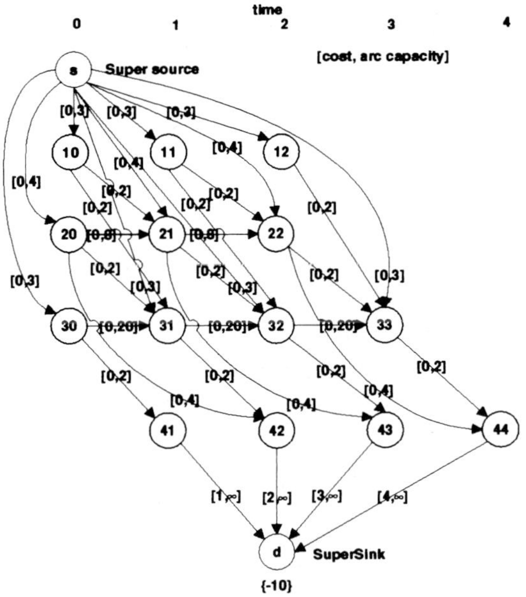

Figure 3. Dynamic Network G_T of the Static Network G of Figure 1, with $T = 4$, without Initial Contents, and by Deleting Inessential Arcs.

time. This flow is denoted by $y_i(t+1)$, i.e.

$$y_i(t+1) := x_{i(t),i(t+1)}$$

The capacity of movement arcs $(i(t), j(t + \lambda_{ij})) \in A_M$ is denoted by $b_{ij}(t)$ where we assume without loss of generality that

$$b_{ij}(t) := \min\{b_{ij}(t') : t' = t, t+1, \ldots, t + \lambda_{ij}\}$$

The capacity of a holdover arc $(i(t), i(t+1)) \in A_H$ is determined by the node capacity $a_i(t)$, and represents how many evacuees can stay in the node at a given time. With $\phi(X, Y)$ as the general objective and with q_i as the initial number of evacuees in any node $i \in N$, the discrete-time dynamic network flow model for evacuation processes can be formulated as follows.

$$\min/\max \quad \sum_{t=0}^{T} \phi(X,Y) \tag{2}$$

$$y_i(t+1) - y_i(t) = \sum_{k \in pred(i)} x_{ki}(t - \lambda_{ki}) - \sum_{j \in succ(i)} x_{ij}(t),$$

$$t = 0, \ldots, T;\ \forall i \in N; \tag{3}$$

$$y_i(0) = q_i, \forall i \in N; \tag{4}$$

$$0 \leq y_i(t) \leq a_i(t), t = 1, \ldots, T-1; \forall i \in N; \tag{5}$$

$$0 \leq x_{ij}(t) \leq b_{ij}(t), t = 0, \ldots, T - \lambda_{ij}; \forall (i,j) \in A \tag{6}$$

where

$$pred(i) := \{j \mid (j,i) \in A\} \ \ ; \ succ(i) := \{j \mid (i,j) \in A\}$$

are the nodes which are predecessors and successors of node i, respectively.

In order to measure the time when evacuees reach their final destinations, so-called *turnstile cost* ([14], [33]) is defined on each arc as follows.

Definition 2 If D is the set of sink nodes of the static network G and d is the super sink node of the associated dynamic network G_T , the (turnstile) cost of any arc $(i(t), j(t' = t + \lambda_{ij})) \in A_T$ is defined different from 0 if and only if $i \in D$ and $j(t') = d$. In this case $c(i(t), d) = t$.

Let us denote $S \subset N$ as the set of source nodes of the static network G. Using the turnstile cost we can define the objective function $\phi(X,Y)$ to model the average evacuation time required by an evacuee to leave the network, i.e.

$$\phi(X,Y) := \frac{\sum_{t=0}^{T} \sum_{i \in D} t x_{id}(t)}{\sum_{i \in S} q_i}$$

Since the denominator is constant and ϕ depends only on the flow variables, one just needs to define the objective function ϕ as

$$\phi(X,Y) := \phi(X) = \sum_{t=0}^{T} \sum_{i \in D} t x_{id}(t)$$

The initial occupancies are modeled by using flow from the super-source s to each source node. Under assumptions of constant capacity (i.e., $b_{ij}(t) = b_{ij}, \forall (i,j) \in A$ and $a_i(t) = a_i, \forall i \in N; \forall t$) and constant travel time, the evacuation model that minimizes the average evacuation time can thus be formulated as follows.

$$\min \quad \sum_{t=0}^{T} \sum_{i \in D} t x_{id}(t) \tag{7}$$

$$x_{si}(0) = q_i, \forall i \in S, \tag{8}$$

$$\sum_{t=0}^{T} \sum_{i \in D} x_{id}(t) = \sum_{j \in S} q_j, \tag{9}$$

$$y_i(t+1) - y_i(t) = \sum_{k \in pred(i)} x_{ki}(t - \lambda_{ki}) - \sum_{j \in succ(i)} x_{ij}(t),$$
$$t = 0, \ldots, T; \ \forall i \in N \tag{10}$$

$$y_i(0) = 0, \forall i \in N, \tag{11}$$

$$y_i(t) = 0, \forall i \in D; t = 0, \ldots, T \tag{12}$$

$$0 \leq y_i(t) \leq a_i, t = 1, \ldots, T; i \in N - D \tag{13}$$

$$0 \leq x_{ij}(t) \leq b_{ij}, t = 0, \ldots, T - \lambda_{ij}; \forall (ij) \in A \tag{14}$$

We can treat the time-expanded network as defined in the Definition 1 as a static network and then apply any minimum cost static network flow algorithm (see, e.g., [1]) to obtain the solution.

Based on this minimum cost dynamic network optimization, Kisko and Francis [41] developed EVACNET+, an evacuation software which can be used to determine the egress time and possible bottleneck locations. The minimum cost dynamic network optimization problem is solved as a static network by using the NETFLO code [40].

In public buildings where the number of evacuees is difficult to estimate, one can model the evacuation problem as a maximum dynamic network flow problem. In the next section we will give a general description of the maximum dynamic network flow problems and the algorithms to solve them.

4 Maximum Dynamic Flows Problem

Given the time horizon T, maximum dynamic flow problems (MDF) maximize the dynamic flows reaching the sink. These problems can be used to model evacuation processes which have no reliable information about the number of evacuees. As already mentioned in Section 3, the super source node in G_T is connected to every time-copy of every source node and there is no holdover arc for source nodes and sink nodes. Arcs from super-source have zero travel time and infinite capacities.

The objective function ϕ of MDF is defined as follows.

$$\phi(X) := \sum_{t=0}^{t=T} \sum_{i \in D} x_{id}(t)$$

Using this definition, MDF with constant arc capacity and non-negative integral travel time can be formulated as follows.

$$\max \quad \sum_{t=0}^{t=T} \sum_{i \in D} x_{id}(t) \tag{15}$$

$$y_i(t+1) - y_i(t) = \sum_{k \in pred(i)} x_{ki}(t - \lambda_{ki}) - \sum_{j \in succ(i)} x_{ij}(t),$$
$$t = 0, \ldots, T; \forall i \in N \tag{16}$$
$$y_i(0) = 0, \ \forall i \in N, \tag{17}$$
$$y_i(t) = 0, \ \forall i \in S \cup D; t = 1, \ldots, T \tag{18}$$
$$0 \leq y_i(t) \leq a_i,$$
$$t = 1, \ldots, T; \forall i \in N - S \cup D; \tag{19}$$
$$0 \leq x_{ij}(t) \leq b_{ij}, \ t = 0, \ldots, T - \lambda_{ij}; \forall (i,j) \in A \tag{20}$$

The solution of MDF defined by Eq. (15) - (20) can be obtained by repeating the feasible flows along some chains (see Definition 3) of the static network from the source to the sink. The flows on these static chains are repeated in the dynamic network for every time period within the time horizon T. This approach is called ***temporally repeated flow technique*** (see Definition 4).

Definition 3 (Chain, chain flow and chain decomposition)
- A chain is a sequence of nodes $P = \{i_1, i_2, \ldots, i_k\}$, $k \geq 2$, such that $(i_j, i_{j+1}) \in A$ and $i_j \neq i_{j'}$ when $j \neq j'$, for $j, j' = 1, \ldots, k-1$, i.e. a chain has no repeated nodes.
- A chain flow $\gamma = <\mid P \mid, P>$ is a static flow of value $\mid P \mid$ along the chain P.
- Let $\Gamma = \{P_1, P_2, \ldots, P_l\}$ be a set of chain flows and let $\mid P_i \mid$ be the chain flow along path P_i. Γ is a chain decomposition of the static flow f if $\sum_{i=1}^{i=l} \mid P_i \mid = f$.

It is well-known that any network flow can be decomposed into chain flows (plus possibly some flows on cycles).

Definition 4 (Temporally repeated flows) Let $\gamma = <\mid P \mid, P>$ be a chain flow. The temporally repeated flow γ^T is a dynamic flow obtained by repeating $(T + 1 - \lambda(P))$ times the chain flow γ , i.e., by sending $\mid P \mid$ units of flow every time period from time zero to time $T + 1 - \lambda(P)$ along the same path (static) P.

The next theorem shows that the maximum dynamic flow problem can be solved as a minimum cost flow problem (MCFP) in the static network. The reader may refer books on network flow theory for more details on MCFP (see for instance [1]).

Proposition 2 ([26]) *Finding a maximum dynamic flow is equivalent to solving a MCFP. In particular, the temporally repeated flow obtained from the chain decomposition of any min cost flow is a maximum dynamic flow.*

In order to solve the maximum dynamic flow problem, and thus find the maximal number of persons, which can be evacuated within T time periods from a given building, we only have to solve a min cost flow problem in the small, static network G.

Algorithm 1

- *step 1* Apply a minimum cost flow algorithm to the original static network G. Let x^* be an optimal solution.
- *step 2* Decompose x^* into k chain flows on $P_1, P_2, \ldots, P_l$ such that

$$x^* = \sum_{i=1}^{i=l} \mid P_i \mid$$

- *step 3* Repeat each chain flow P_i from time 0 till time $T - \lambda(P_i)$.

Example 1
Figure 4 shows a static network of a simple building layout with travel time and capacity parameters attached on every arc. Node 1 and 6 are source and sink nodes, respectively. It is desired to calculate the maximum number of evacuees who can reach the safety during $T = 7$ time units. The optimal solution of MCFP is obtained as shown in table 2 (only the positive flows).

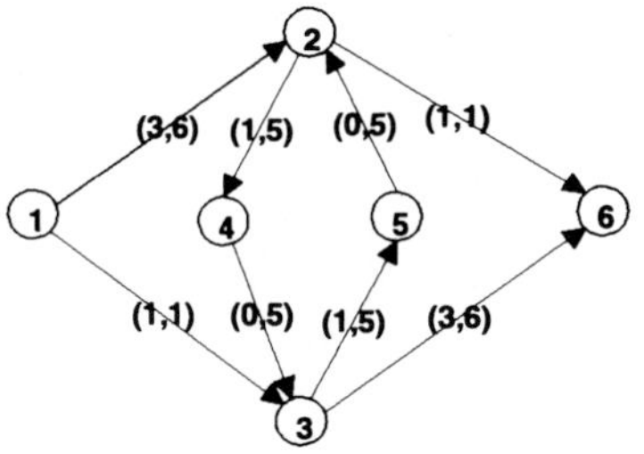

Figure 4. Static Network G for Example 1.

Step 2 of the algorithm gives the following chain flows :

- $P_1 = (1, 2, 6), \lambda(P_1) = 4, \mid P_1 \mid = 1$; P_1 *must be repeated four times for* $t = 0, 1, 2, 3$.

Arc	(1,2)	(1,3)	(2,4)	(2,6)	(3,6)	(4,3)
Flow (f)	6	1	5	1	6	5

Table 2. Optimal Maximum Flows for Static Network in Example 1.

- $P_2 = (1,2,4,3,6), \lambda(P_2) = 7, \mid P_2 \mid = 5$; P_2 *must be repeated only once at time* $t = 0$.
- $P_3 = (1,3,6), \lambda(P_3) = 4, \mid P_3 \mid = 1$; P_3 *must be repeated four times for time* $t = 0, 1, 2, 3$.

The dynamic flows are shown in Figure 5. The total dynamic flow is equal to 13 unit of flow in which 2 units arrive at the sink at time $t = 4, 5, 6$ *and 7 units at time* $t = 7$. *It means that one needs at least 7 time units to evacuate 13 evacuees who are at the location represented by node 1 in the beginning of the evacuation process.*

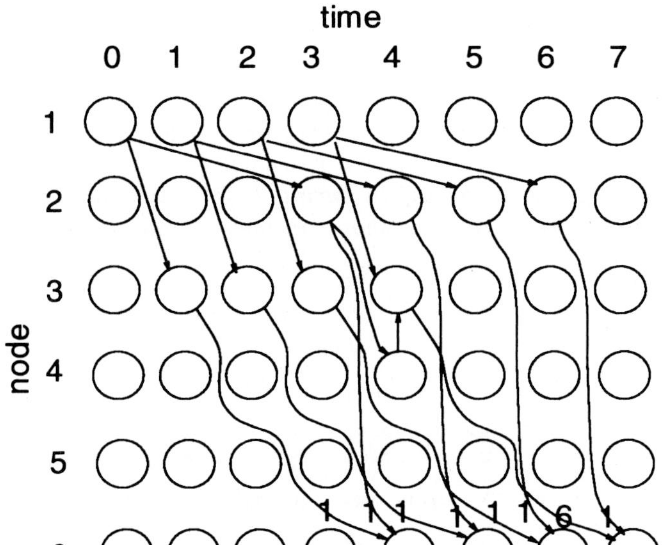

Figure 5. A maximum Dynamic Flow for Example 1.

The solution of the maximum dynamic flow in this example does not use any holdover arcs, i.e. $x_{ii}(t) = y_i(t+1) = 0, \forall i \in N; \forall t = 0, \ldots, T-1$. In fact, it can be shown, that the maximum dynamic flow poblem with constant capacities and travel times never requires hold-over at any nodes [34]. Therefore, variables $y_i(t)$ can be eliminated from the problem formulation.

5 Universal Maximum Flow Problem

The ***(discrete) Universal maximum flow problem*** (UMF) was introduced by Gale [28] as a variant of the maximum dynamic flow problem. UMF is the problem of finding maximum dynamic flows reaching the sink at every time period $t = 1, \ldots, T$. Hence, the optimal solution of UMF is the solution of the maximum dynamic flow problem, not only for the alloted time horizon T, but also for any smaller time horizons. Such a flow is also known in the literature as ***earliest arrival dynamic flow*** (see Hoppe and Tardos [34]). It can be formulated as

$$\max \quad \sum_{t=0}^{t=T'} \sum_{i \in D} x_{id}(t), \forall T' = 1, \ldots, T \tag{21}$$

$$\text{subject to : } \quad (16) - (20)$$

The relevance of UMF for the evacuation problem is obvious. In every time period the maximal number of evacuees is brought to safety, such that an evacuation modeled by universal maximum flows is a very safe one.

By definition, every universal maximum flow is a maximum dynamic flow, but the reverse is not true as is illustrated by the flow distribution in Table 3 using the data of Example 1. In Table 3 the first line of each arc flow shows the optimal solution of maximum dynamic flow problem and the second line shows the one of UMF.

The following algorithm finds a universal maximum flow in the case of static networks with single source and single sink.

Algorithm 2 ([36])

- *step 0 :* Identify K, a lower bound for the value of a maximum dynamic flow with respect to time horizon T, and set K as the total capacity of arcs from super source s to time-copies of source node, i.e.

$$K = \sum_{t=0}^{T-\lambda} b_{si(t)}$$

  with $i(t)$ the t-th time-copy of source node i of the static network and λ the shortest possible time to reach the sink node from the source node. Set all flows to zero and set time counter t to 1.
- *step 1 :* Maximize the flow from s to d via arc $(i(t), d) \in G_T$ by seting temporarily the capacity of arcs $(i(t'), d), i \in D; t' > t$ to zero (i.e., close temporarily those arcs) using the Ford-Fulkerson labeling algorithm([26]).

Time / Arc	0	1	2	3	4	5	6	7
(1,2)	6	1	1	1				
	6	1	1					
(1,3)	1	1	1	1				
	1	1	1	1	1			
(2,4)				5				
				5				
(2,6)				1	1	1	1	
			1	1	1	1	1	
(3,6)		1	1	1	6			
			1	1	6			
(3,5)								
		1				1		
(4,3)					5			
					5			
(5,2)								
			1				1	
arrival at 6					2	2	2	7
				1	1	2	2	7

Table 3. Maximum Dynamic Flows (1-st row) vs Universal Maximum Flows (2-nd row).

– *step 2 :* If $t = T$ stop. Otherwise, increase t by 1, open arc $(i(t), d)$ and go to step 1.

The validity of the algorithm has been proved in Minieka [48]. Note that this algorithm as well as alternative algorithms proposed by Minieka [48] and Wilkinson [66] are non-polynomial (both authors use shortest augmenting path algorithm). Moreover, they all work with the time expanded network, the size of which is dependend on the time horizon T.

Hoppe and Tardos [34] proposed a polynomial approximation algorithm for the discrete UMF by introducing a generalization of the chain decomposition introduced in the previous section, the *non-standard chain decomposition.* It is a chain decomposition of the static flow which can use arcs in the opposite direction of the static flow, i.e. it is allowed to send flows in the direction of arc (j,i) where $(i,j) \in A$ and $(j,i) \notin A$. In this case the travel time along (j,i) is the negative value of the one on $(i,j) \in A$. By using the non-standard chain decomposition, one can again use the idea of temporally repeated flows to produce UMF. Figure 6 shows dynamic flows induced by the non standard chain decomposition. Consider the static network of Example 1. Let $\Gamma = \{P_1, P_2, P_3\}$ be the set of chains, with
$P_1 = \{1,3,5,2,6\}$; $P_2 = \{1,2,5,3,6\}$; $P_3 = \{1,2,4,3,6\}$

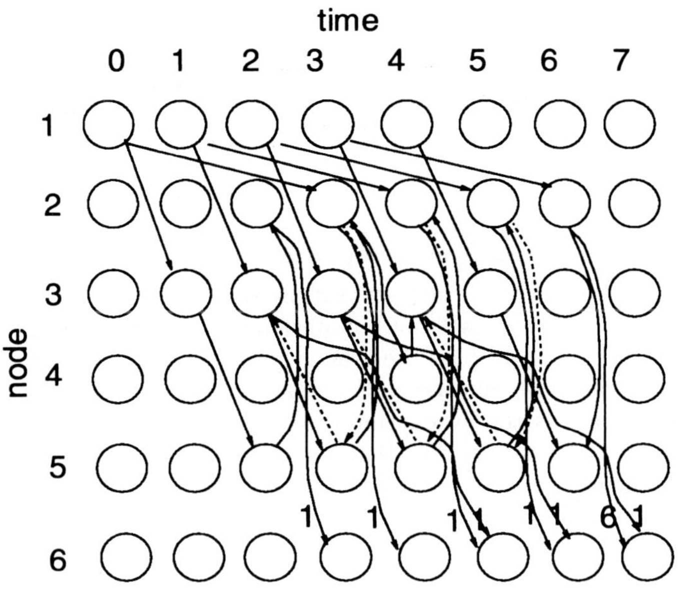

Figure 6. Universal Maximal Flow Induced by Non-standard Chain Decomposition

Γ is a non-standard chain decomposition since $P_2 \in \Gamma$ uses arcs $(5,2)$ and $(3,5)$ in the opposite direction (shown in Figure 6 as dot lines), i.e. uses arc (2,5) and (5,3) that both are not in A.

In general, suppose P is a chain flow that sends flow along arc (i,j) but in the opposite direction of the static flow. In order to keep the feasibility due to capacity constraints, there must be another chain flow P^* that also uses arc (i,j) but sends flow in the opposite direction of P to cancel chain flow P on (i,j). Since the travel time of the opposite arc is nonpositive, if chain flow P arrives at node i at time t, then the chain flow P^* must arrive at i at time $t^* \leq t$. Similarly, if chain flow P stops using arc (i,j) at time t, then P^* must continue sending flow from j until time $t^* \geq t$. In Figure 6, chain flow P_2 starts using arc $(3,5)$ in the opposite direction $(5,3)$ at time $t = 3$ and stops using it at time $t = 4$. On the other hand, chain flow P_1 starts using arc $(3,5)$ at time $t = 1$ and stops using it at time $t = 6$.

By using this non-standard chain decomposition and applying the capacity scaling shortest augmenting path algorithm, Hoppe and Tardos [34] developed the first polynomial approximation algorithm, with time complexity $O(\frac{m}{\epsilon}(m+n\log n)\log U)$, where U is the maximum arc capacity. It is proved to be within $(1+\epsilon)$ of optimality.

6 Quickest Path and Quickest Flow

The quickest path problem as introduced by Chen and Chin [15] is another variant of the shortest path problem. The objective is to send a predetermined number of units from their initial position (i.e. the source node) to the destination (i.e. the sink node) as quickly as possible along a single path. But the notion of "shortest" depends not only on the travel time but also on the number of units that have to be delivered along the path. Flows are sent continuously over the time. The quickest path problem is relevant to a special evacuation problem where evacuees may use only a single path or tunnel from their initial position, that will not be interfered by evacuees from other places. An example of this problem is the evacuation of spectators from a sports stadium. In this case, the quickest path model is applied independently to each network that models the evacuation of each stand in the stadium. Another problem known as quickest flow [12] is similar to the quickest path problem. Here, it is allowed to send flows along multiple paths. The latter problem is known as the minimum time network clearing problem with its obvious relevance for evacuation problems. In this section we first discuss the quickest path and then the quickest flow problem.

6.1 Quickest Path Problems

Definition 5 Given a path $P := (i_1, i_2, \ldots, i_k)$ in the static network G.

- The capacity of path P, $b(P)$ is defined as

$$b(P) = \min_{1 \leq i \leq k-1} b(i_i, i_{i+1})$$

- The length of path P (in time unit) $\lambda(P)$ is defined as

$$\lambda(P) := \sum_{i=1}^{i=k-1} \lambda(i_i, i_{i+1})$$

- The egress time to send out σ units of flow from i_1 to i_k along P is

$$T(\sigma, P) = \lambda(P) + \frac{\sigma}{b(P)}$$

Different from the classical shortest path problem the concatenation property (i.e. the property that every subpath of a shortest path is also a shortest path) is no longer true for the quickest path problem, as is shown by the following example.

Example 2 Refering to the example of Figure 7, we want to send out 20 evacuees from source node s to sink node c. Then the quickest path is $P_1 = (s, b, c)$

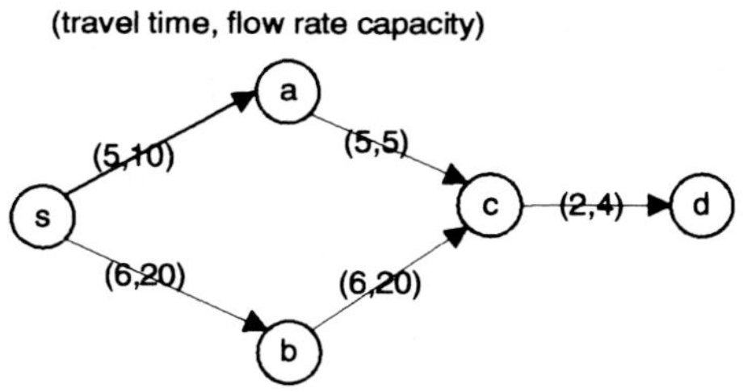

Figure 7. Example for Quickest Path Problem.

with egress time $T(20, P_1) = 12+\frac{20}{20} = 13$. *Suppose instead of c we take d as the final destination, then the quickest path is* $P_2 = (s, a, c, d)$ *with* $T(20, P_2) = 17$. *But* $P_3 = (s, a, c) \subset P_2$ *is not the quickest path from s to c, violating the concatenation property. We see also that* P_1 *is not the (classical) shortest path (with respect to travel time) from s to c. The number of evacuees reaching the destination d over time t is shown as function* $I_d(t)$ *in Figure 8 with*

$$I_d(t) = \begin{cases} 4(t-12) & , t \geq 12 \\ 0 & , \textit{otherwise} \end{cases}$$

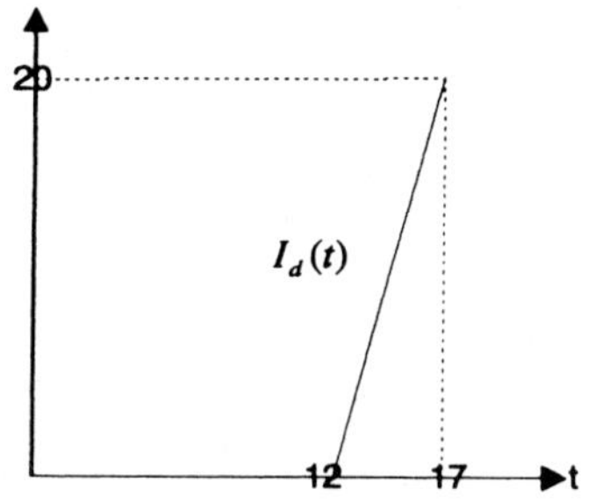

Figure 8. The number of Evacuees Reaching the Destination d.

The following two theorems show the interrelation between quickest and shortest path in G which turn out to be useful in developing an algorithm for solving the quickest path problem. For this purpose we define for given network G and value z the subgraph $G(z) = (N, A(z))$ of G defined by $A(z) := \{(i,j) : (i,j) \in A \text{ and } b_{ij} \geq z\}$

Theorem 1 (Rosen [60]) *If P is the quickest s-d path in the static network G sending* σ *units of flow then*

– *P is a shortest s-d path in $G(b(P))$.*
– *Any subpath of P itself must be a shortest path in $G(b(P))$.*

Theorem 2 (Rosen [60]) *Let r be the number of distinct capacity values and let P_j be a shortest $s-d$ path in $G(b_j), j = 1, \ldots, r$. If*

$$\lambda(P_l) + \frac{\sigma}{b(P_l)} = \min_{1 \le j \le r} \{\lambda(P_j) + \frac{\sigma}{b(P_j)}\},$$

then P_l is the quickest $s-d$ path in G sending σ unit of flows.

Based on these two theorems, Rosen, et al. [60] developed a simple algorithm as follows. In the initial step we compute for each $j = 1, \ldots, r$ a shortest $s-d$ path in $G(b_j)$ and then apply Theorem 2. Using Fredman and Tarjan [27], each computation of P_j requires $O(m + n \log n)$ time such that the overal complexity of the algorithm is $O(rm + rn \log n)$.

This result can be extended by considering the more realistic assumption that the travel time is density dependent (i.e. flow dependent). Here we assume that the travel time is a step function of the flow, which is nondecreasing, and constant in each unit of flow. Let k_{ij} be the number of distinct travel times of arc (i,j) and $k^* := \max_{(i,j)\in A}\{k_{ij}\}$. The travel time of arc (i,j) then can be defined as follows.

$$\lambda_{ij}(x_{ij}) = \begin{cases} \lambda_{ij}^1 & ,0 \le x_{ij} \le b_{ij}^1 \\ \lambda_{ij}^2 & ,0 \le x_{ij} \le b_{ij}^2 \\ \ldots & ,\ldots \\ \lambda_{ij}^{k_{ij}} & ,0 \le x_{ij} \le b_{ij}^{k_{ij}} \end{cases}$$

For each arc (i,j) we create k_{ij} artificial nodes denoted $ij^1, \ldots, ij^{k_{ij}}$. Then, we connect node i to node ij^l and node ij^l to node j for each $l = 1, \ldots, k_{ij}$ as shown in Figure 9. The capacity of arc (i, ij^l) is b_{ij}^l with travel time λ_{ij}^l. The capacity and travel time for arc (ij^l, j) are ∞ and 0 respectively. The modified network will have maximum $(n + mk^*)$ nodes, $2mk^*$ arcs and mk^* number of different capacities. The quickest path then can be obtained by applying the previous algorithm to the modified network.

Proposition 3 ([63]) *Let P is the quickest path with arc $(i,j) \in P$. Then arc (i,j) will use the closest capacity to the $b(P)$, i.e. if $b_{ij}^{l-1} < b(P) \le b_{ij}^l$ then $(i, ij^r) \in P$ for $r = l$ and $(i, ij^r) \notin P$ for $r \ne l$.*

6.2 Quickest Flow Problems

Unlike the quickest path problem, the quickest flow problem (QFP) relaxes the limitation of a single path to multiple paths. QFP is a dynamic network

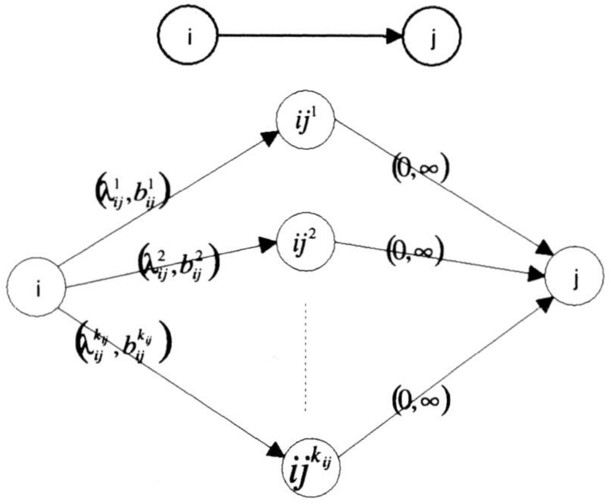

Figure 9. Network Modification for Density Dependent Travel Time.

flow problem with single source and single sink that clear the network in the minimum possible time ([12], [23]). The objective function of QFP can be formulated as minimizing the time horizon $T =: T(v)$ where v is the number of initial occupants of the building. The set of constraints is equal to the one used in the minimum turnstile cost dynamic network flow model discussed in Section 3. The following theorem states properties of the value $v(T)$ of a maximum dynamic flow over time period T, that can be used to derive an algorithm to solve the quickest flow problem.

Theorem 3 ([12])

- *Let T_0 be the length of the shortest path from the source to the sink with respect to the travel time. $v(T)$ is a monotone increasing function and for $T \geq T_0$ it increases strictly.*
- *$\Delta(T) := v(T) - v(T-1)$ is for $T > 0$ monotonously increasing, i.e.*

$$\Delta(T+1) \geq \Delta(T), \forall T > 0$$

- *$\Delta(T)$ attains its value only from the set $\{0, 1, \ldots, \mid x_{max} \mid\}$, where $\mid x_{max} \mid$ is the value of maximum static flow in the network G.*

The interrelations between maximum dynamic flow and quickest flow are described by the following lemma.

Lemma 1 ([12])

- *$T(v) = min\{T \mid v(T) \geq v\}$ where $v(T)$ is a maximum flow over time period T.*

– *Let x be a maximum dynamic flow of value v in the time interval $[0, T], T \geq 0$. If $v(T-1) < v$ then x is a quickest flow of value v, and for the minimum egress time $T(v)$, we get $T(v) = T$.*
– $v(T(v) - 1) < v$

The solution is obtained by applying an iterative process with two main steps [12]. The first step estimates the time period T by utilizing a binary search, Newton or interpolation techniques. The second one solves the minimum cost flow problem with parameter T. The process is repeated until $v(T) \approx v$. Burkard, et. al. [12] developed several polynomial and strongly polynomial algorithms for the quickest flow problem using a continuous version of $v(T)$ in case of single source and single sink.

7 Tripple Optimization Result

Using the turnstile cost definition introduced in Section 3, very interesting properties can be derived interrelating the evacuation time and the maximum number of evacuees that can be sent out to safety in every time period. This interrelation is described in the following theorem known as *tripple optimization theorem.* It says that the flow pattern which minimizes the average evacuation time (see Section 3) also maximizes the number of evacuees reaching the safety at each time period (see Section 5) and minimizes the total time needed to evacuate all evacuees (see Section 6.2).

Theorem 4 ([36]) *Let F_t be the flow vector of arcs connected to the super sink at time t and let c_t is the associated weight (or cost) vector where the weight c is increasing over the time t. Consider three different problems under the assumption that there exists a feasible flow of K units, i.e. the value of the maximum dynamic flow within a time T is not less than K.*
(a) *Universal maximum flow problem :*

$$max \sum_{t=0}^{T'} F_t, \forall T' \leq T$$

(b) *Minimum weighted sum flow problem :*

$$\sum_{t=0}^{T} c_t F_t$$

(c) *Quickest flow problem of initial occupancies K :*

$$min\{T \mid F_{T'} = 0, \forall T' > T\}$$

Then the optimal solution of one of the problems (a) – (c) is also optimal for the other two problems.

As a consequence it suffices to solve either of the three problems using the turnstile cost approach, the UMF approach or the quickest flow approach to obtain a best evacuation plan according to the three goals listed at the beginning of this section!

8 Multiple Objectives

Using the values contained in the dynamic severity matrix (see Section 1), one can divide the evacuation regions into priority levels. For example, locations where hazards initiated, can be classified as regions with the highest priority level. Obviously, evacuees should not move from lower priority regions to regions with higher priority. In the modeling one can enforce this by charging high costs to corresponding arcs. With this priority system, the evacuation process may have several objectives that must be satisfied simultaneously. For example, let $P_1, \dots, P_k$ be a partition of the underlying system into k different parts such that the evacuation of P_1 has the highest, the evacuation of P_2 has the next highest, and finally, the evacuation of P_k has the lowest priority. The objective of the evacuation process is to minimize the evacuation time such that ([33])

(1) The evacuation time of P_1 is as small as possible.

(2) Among all plans optimizing (1), the evacuation time of P_2 is as small as possible.

$\vdots$

(k) Among all plans optimizing (1) until $(k-1)$, the evacuation time of P_k is as small as possible.

Hamacher and Tufekci ([33]) consider this multiple priority level problem as lexicographical minimum cost dynamic flow problem.

The lexicographical ordering is defined as follows.

Definition 6 Suppose we have two vectors with k-components, $\mathbf{c}$ and $\overline{\mathbf{c}}$.

$$\mathbf{c} = \begin{pmatrix} c_1 \\ \vdots \\ c_k \end{pmatrix}, \overline{\mathbf{c}} = \begin{pmatrix} \overline{c}_1 \\ \vdots \\ \overline{c}_k \end{pmatrix}$$

Vector $\mathbf{c}$ is lexicographically smaller than $\overline{\mathbf{c}}$ if and only if c_i is strictly smaller than $\overline{c}_i$ for the first component i, where c_i and $\overline{c}_i$ are different.

Given a set K of vectors with k-components its lexicographic minimum is denoted as *lex min* K.

In the multiple objective evacuation problem, we replace the real valued cost $c_{ij}(t)$ of the single objective problem by the vector

$$\mathbf{c}_{ij}(t) = \begin{pmatrix} c_{1ij}(t) \\ \vdots \\ c_{kij}(t) \end{pmatrix}$$

for each arc $(i,j) \in A$ and for $t = 0, \ldots, T$. The cost value for each priority level $l = 1, \ldots, k$ can be defined by

$$c_{lij}(t) = \begin{cases} t' = t + \lambda_{ij} & , \text{ if } i(t) \in P_l \text{ and } j(t') \in P_{l'},\ l' > l \\ M & , \text{ if } i(t) \in P_l \text{ and } j(t') \in P_{l'},\ l' < l \\ 0 & , \text{ otherwise.} \end{cases}$$

The objective function ϕ can be defined as

$$\phi(X) = \sum_{t=0}^{T} \sum_{l=1}^{k} \sum_{(i,j)\in A} c_{lij}(t) x_{ij}(t)$$

The constraints are (8) - (14). This problem can be solved by using the concept of *lexicographically shortest augmenting path* .

Definition 7 Let $P = \{s = i_0, i_1, \ldots, i_k = d\}$ be a path from super source s to super sink d in the time-expanded network G_T and let

$$P^+ = \{e \in P \mid e = (i_j, i_{j+1}),\ j = 0, \ldots, k-1\}$$

$$P^- = \{e \in P \mid e = (i_{j+1}, i_j),\ j = 0, \ldots, k-1\}$$

If x is a flow in G_T define

$$\epsilon_+(P) = \min \{b_e - x_e \mid e \in P^+\}$$

$$\epsilon_-(P) = \min \{x_e \mid e \in P^-\}$$

$$\epsilon(P) = \min \{\epsilon_+(P), \epsilon_-(P)\}$$

and

$$\mathbf{c}(P) = \mathbf{c}(P^+) - \mathbf{c}(P^-) = \sum_{e\in P^+} \mathbf{c}(e) - \sum_{e \in P^-} \mathbf{c}(e)$$

P is called a lexicographically (lex) shortest augmenting path if $\epsilon(P) > 0$ and

$$\mathbf{c}(P) = \text{lex} \min\{\mathbf{c}(P') \mid \epsilon(P') > 0,\ \forall \text{path } P' \text{in } G_T\}$$

In the definition, the time-expanded network G_T is considered as a static network. The proposed algorithm ([33]) to solve the lexicographically minimum cost flow problem needs the notion of lex extreme flow as explained in Definition 8.

Definition 8 A flow x is a lex extreme flow in G_T if x is a feasible flow with value v and if the cost

$$\mathbf{c}(x) = \sum_{e \in A_T} x_e \mathbf{c}(e)$$

is minimal among all flows with the same flow value v.

The algorithm starts with finding a lexicographically shortest augmenting path and then augments the current flow along this path. Theorem 5 shows that the updated flow after the augmentation is again a lex extreme flow.

Theorem 5 *Let x be a lex extreme flow with flow value v and let P be a lex shortest augmenting path. Then the new flow x' defined by*

$$x'_e = \begin{cases} x_e + \epsilon(P) \ , & \text{if } e \in P^+ \\ x_e - \epsilon(P) \ , & \text{if } e \in P^- \\ x_e \ , & \text{if } e \notin P \end{cases}$$

is a lex extreme flow with respect to flow value $v + \epsilon(P)$.

The search for lex shortest augmenting paths is repeated until no more augmenting path is available. If this situation is reached, the current flow is a lex min cost flow.

Another area, where multi criteria network flows may be used, is when arcs may have several attributes which are time dependent. Travel cost, travel distance and risk are examples of attributes which may be attached to each arc. We may want to consider, for instance, the effect of fire, smoke or toxical chemicals on the availability of arcs over time. If the development is such that evacuees can not walk as fast as usual or even may not be able to pass this arc then the cost change over time in all or some components of the cost vector. If, for instance, the arc (i, j) becomes impassable at time t', then the travel cost of going from node i to node j can be modified to be

$$c_{ij}(t) = \begin{cases} c_{ij}(t) \ , & \text{if } t < t' \\ M \ , & \text{if } t \geq t' \end{cases}$$

with M a large positive number.

Let $\{\mathbf{F}_i\}$ be the set of non-dominated evacuation routes from node i to safety d, and let $\mathbf{c}_{ij}$ be the vector of attributes associated with the arc from node i to node j. Defines also *vmin* as the vector minimization to find the non-dominated evacuation routes. The dynamic programming formulation to find the non-dominated evacuation routes given by ([43], [44]).

$$\{\mathbf{F}_i\} = \text{vmin}_{\ j \neq i, (i,j) \in A} \{\{\mathbf{F}_j\} + \mathbf{c}_{ij}\} \quad , \forall i \in N - \{d\}$$

$$\{\mathbf{F}_d\} = \{\mathbf{0}\}$$

Here, **0** is the null vector. Kostreva and Wiecek ([43]) proposed backward and forward dynamic programming algorithms to solve this problem. They allow discontinuity in the cost functions, an asumption which is certainly relevant for evacuation planning, as seen above.

9 Dynamic Network with Density Dependent Travel Time

So far, we have mainly considered travel times which are either constant or dependent on time. In reality, they are however dependent on the density of the flow. During any kind of movement, and this is in particular true for evacuation movement, the speed (i.e. travel time) will grow with higher density, until we encounter slow down and queuing phenomena at certain degrees of densitiy. In this section we consider therefore travel times $\lambda_e(t) = g_e(x_e(t))$ where g is an appropriately chosen function depending both on time and the flow at this time.

This model is more realistic, but also more difficult to handle from a mathematical point of view. If we consider, for instance, the flow augmentation process, the amount of time a specific arc in a source-sink path is available is depending on the amount of flow sent through this arc. Moreover, since the density dependent travel time is in general nonlinear, the dynamic conservation flow contraints (see constraint (3)) are also nonlinear, which makes the problem much more difficult to solve.

Several references on the dynamic network flow problem with density dependent travel times can be found in the field of traffic assignment (see Table 2). Since many concepts and models of traffic assignment problems can be adapted to tackle the evacuation problems, we review some of these methods in the evacuation context. There are two main objectives in traffic assignment problem, system optimum and user optimum. The former objective tries to minimize the total travel time. It considers the willingnes of the community to share the lateness. The latter tries to optimize every individual's travel time where it is assumed that each individual behaves egotistical, If we assume single destination and allow exogenous flows only at the beginning (i.e. at time zero), the dynamic traffic assignment problem becomes an evacuation problem with known initial occupancy distribution.

In traffic assignment, the nonlinearity of the travel time function is handled by the following approaches.

- Use linear approximation of the travel time by introducing 0-1 decision variables for selecting the appropriate travel time (Kaufmann et al. [39]).
- Use piecewise linear approximation (Carey and Subrahmanian [13]).
- Apply an iterative process where in each iteration the travel time is fixed temporarily according to the current flow (see [37], [59]).

The first approach uses the decision variable δ_{ij}^{ts} with value equal to 1 if the flow entering arc (i, j) at time t needs arc travel time s and zero otherwise. The free-flow travel time gives the lower bound to the parameter s whereas the upper bound is given by the value $(T - t)$. The flow variable x_{ij}^{ts} represents the flows enter arc (i, j) at time t and exits at time $t+s$. The flow value is bounded by the arc capacity corresponds to the selected travel time determined by δ_{ij}^{ts}. The problem is thus formulated as a mixed integer programming problem.

The second approach uses a piecewise linear approximation of the travel times. A time expanded arc is obtained by joining node i at time t with node j at time $(t + k)$ with $k = 0, 1, \ldots, K$ integer breakpoints of the travel time function. The arc flow must lie between at most two neighbouring breakpoints, i.e. the arc flow is represented as a convex combination of two flow values at two neigbouring breakpoints. Using this approach, the problem is formulated as a linear programming problem.

In the last approach, the value of the travel time is approximated using a 2-level iterative process. In the first level, the travel time is temporarily fixed in each iteration. In the second level, a nonlinear optimization problem is solved iteratively as a convex programming problem. This approach is equivalent to the problem of finding the fixed-point of two interdependent algorithmic maps. One algorithmic map is for adjusting the travel time and the other one is for finding the optimal flow under a temporarily fixed travel time.

An even more realistic modelling can be achieved by considering a dependence of the travel time not only on the existing flow, but on three flow components on each arc $(i, j) \in A$ at time t, namely: incoming flow $u_e(t)$, existing flows $x_e(t)$ and outgoing flow $v_e(t)$. Hence, the travel time of any arc e at any time t is given by

$$\lambda : A \times T \to \mathbb{R}$$
$$\lambda_e(t) = g_e(u_e(t), x_e(t), v_e(t)),$$

where g is assumed to be a nondecreasing function, which is convex, continuous and differentiable. This model has been considered by Ran and Boyce [59]. In the evacuation context, the monotonicity assumption mimics properly the situation under congestion. Instead of the node-arc flow formulation used so far, their approach uses an *arc-path flow formulation* in the constraints. Let us define $O(j)$ as the set of arcs leaving node j and $I(j)$ as the set of arcs entering node j. The objective of the model is to minimize the average evacuation time, i.e. we minimize

$$\phi(U) := \sum_{t=0}^{T} \sum_{e \in I(d)} (t + \lambda_e(t)) u_e(t)$$

Constraints in the model include flow conservation, flow propagation, nonnegativity and boundary constraints as follows.

- Let $A(p)$ be the set of all feasible paths from source s to sink d. If ep is an arc e on path $P \in A(p)$, then the interrelation among incoming, existing and outgoing flows of arc e along path p is

$$x_{ep}(t+1) = u_{ep}(t) + x_{ep}(t) - v_{ep}(t), \forall ep \ ; \ \forall t \tag{22}$$

- Evacuees who are in the source s at the beginning can directly move to arcs leaving s or wait in the node if they think those arcs are too crowded. Assuming a single source s and a single sink d, the total movement out of the the source must be equal to the initial contents q. Therefore the supply-demand constraints can be formulated as

$$\sum_{t=0}^{T} \sum_{e \in O(s)} u_e(t) - \sum_{t=0}^{T} \sum_{e \in I(s)} v_e(t) = q \tag{23}$$

$$\sum_{t=0}^{T} \sum_{e \in I(d)} v_e(t) - \sum_{t=0}^{T} \sum_{e \in O(d)} u_e(t) = q \tag{24}$$

- Flows conservation constraints :

$$\sum_{e \in O(i)} u_e(t) - \sum_{e \in I(i)} v_e(t) = 0, \forall i \neq s, d \ ; \ \forall t \tag{25}$$

- Flow propagation constraints :
If platoon dispersion is not allowed then the incoming flow at time t will cause exactly outgoing flow after $\lambda_e(t)$ unit time, i.e.

$$u_{ep}(t) = v_{ep}(t + \lambda_e(t)), \forall e \in A; \forall p \in A(p); \forall t \tag{26}$$

- Boundary conditions :

$$x_e(0) = 0, \forall e \in A \tag{27}$$

- Nonnegativity constraints :

$$u_e(t), x_e(t), v_e(t) \geq 0, \forall e \in A \ ; \ \forall t \tag{28}$$

- Interrelation between node-arc and arc-path formulation:

$$\sum_{p \in A(p)} u_{ep}(t) = u_e(t) \ ; \ \sum_{p \in A(p)} x_{ep}(t) = x_e(t) \ ;$$
$$\sum_{p \in A(p)} x_{ep}(t) = x_e(t), \ \forall e \in A \ ; \ \forall t \tag{29}$$

To handle nonlinearity of the travel time we use again a two level iterative algorithm. Initially the first level (outer level) estimates the travel times using free-flow travel times. The travel times are fixed temporally in order to be used in the second level (inner level). Under fixed travel time, the model has convex objective function but linear constraints. The inner level uses the Frank-Wolfe algorithm (see [40], [59]) to obtain the optimal flows. The resulting optimal flows are used to recalculate travel times and then compare the newest travel times to the previous ones. If the result is not significantly different, then the algorithm stops. Otherwise, the second level is repeated using the new travel times as input. The reader is refered to Tjandra ([63]) for more details on numerical procedures and results.

10 Continuous Time Dynamic Network Flow Model

We have seen in the prevous sections, that discretization plays a vital role in the modeling of evacuation using dynamic network flows. To increase the accuracy of the model one can set the basic time unit θ (see Section 3) very small, but this will enlarge the size of the network and thus the computational complexity of the solutions algorithm. Obviously, a tradeoff is necessary between good accuracy and computational tracktabiliy. Independent of this, the fact that the choice of discretization by choosing a specific basic time unit θ predetermines the possible set of evacuation plans is somewhat unsatisfying.

We therefore discuss in this section a continuous-time approach to evacuation modelling. Continous time dynamic network flow problems have been considered by various authors including Tyndall [64] [65], Grinold [31], Perold [54], Anderson, et al. [3], Philpott [55], Pullan [57], Philpott and Craddock [56], Pullan [58]. Most of existing works emphasize on the analysis of primal-dual relationships and the existence of the optimal solution.

In the continuous model we consider bounded measurable functions $c(t)$ and $b(t)$ where each of the $\mid A \mid$ components assigns the cost of flow and upper bound on the rate of flow in one of the arcs at time t, respectively. Variables $x(t)$ and $y(t)$ define rates of flow in each arc and levels of storage in each node at time t, respectively. The level of storage is bounded above by piecewise linar and continuous function $a(t)$. The *continuous time dynamic (min cost) network flow problem* is formulated as follows (see [4], [55], [6], [58]).

$$\min Z = \int_{t=0}^{T} \sum_{(i,j)\in A} c_{ij}(t) x_{ij}(t) dt \tag{30}$$

$$\text{s.t. } y_j(t) = y_j(0) + \int_0^t [\sum_{(i,j)\in A} x_{ij}(\tau - \lambda_{ij}) - \sum_{(j,i)\in A} x_{ji}(\tau)] d\tau, \; t \in [0,T] \tag{31}$$

$$y_j(t) \le a_j(t), j \in N \; ; \; t \in [0,T], \tag{32}$$

$$x_{ij}(t) \le b_{ij}(t), (i,j) \in A, t \in [0,T] \tag{33}$$

$$x_{ij}(t), y_j(t) \ge 0 \tag{34}$$

Analogous to the three problems in Theorem 4 derived in the discrete dynamic network model context, Philpott [55] formulated three continuous models which can be used for evacuation planning.

(a) Maximize flows into the sink node in the interval $[0, T]$.
$\max \int_0^T [\sum_{(i,d)\in A} x_{id}(\tau - \lambda_{id}) - \sum_{(d,i)\in A} x_{di}(\tau)] d\tau$
subject to (31) - (34)

(b) Minimize the time to clear the network initial occupancies.
$\min T$
subject to (31) - (34)
$y(t) = 0, t \geq T; x_{ij}(t) = 0, t \geq T - \lambda_{ij}$

(c) Minimizing total egress time.
$\min \int_0^T \sum_{(i,j)\in A} c_{ij}(t)x_{ij}(t)dt + \int_0^T \sum_{j\in N-\{d\}} h_j(t)y_j(t)dt$
subject to (31) - (34)
with h_j is holding cost in node j.

Problem (b) is known as the continuous version of the quickest flow problem explained in Section 6. By defining $c_{ij}(t)$ as turnstile cost, problem (c) solves the problem of minimizing the average evacuation time as in Section 3. Relationships among those three problems are explained by the next result.

Theorem 6

– *Any flow which solves (a) will also solves (b) but not the converse*
– *Assuming that*
  - *holding cost for each node is identical and constant,*
  - *arc cost at any time t is defined as*

$$c_{ij}(t) = \begin{cases} \lambda_{ij} & , 0 \leq t < T - \lambda_{ij} \\ T - t & , T - \lambda_{ij} \leq t < T, \end{cases}$$

Problems (a) and (c) are equivalent.

The next result guarantees the existence of a universal maximum flow in the continuous time domain.

Theorem 7 *Under the assumption for holding and arc costs as in Theorem 6, any optimal solution to Problem (c) is also optimal for Problem (a) with respect to all $T' < T$.*

From a model accuracy point of view, modeling evacuation problems using dynamic network problems with continuous time is preferable to discrete time models. But continuous models can to date not be solved satisfactorily for the large scale problems which need to be tackled in the evacuation context. Solution algorithms for continuous dynamic network flow algorithms are due to Anderson and Philpott [5] (continuous-time network simplex algorithm),

Pullan [57], Philpott and Craddock [56] (discretization approaches). Anderson, et.al [2] and Philpott [55] (continuous- time version of Ford-Fulkerson's maximum flow labelling algorithm).

More work is certainly needed in this area. First additional results can be found in Tjandra ([63]). He proposes an algorithm to solve UMF problem with time dependent capacity and utilize the result to solve the quickest flow problem under the same assumption.

11 Microscopic Models

In microscopic models each evacuee is considered as a separate flow object. An evacuee will be exposed to accident effects depending on the route he/she follows and the length of time spent in different locations. An evacuee selects the route 'step by step', which means that the choice of the next piece of the route is decided at every node along this route. The initially selected route might be changed due to some reasons, for instance, blockage by fire or high congestion. Figure 10 shows an example of the evacuation process as it can be modeled for each person. Microscopic models emphasize the modeling of human behavior during an emergency situation. The human model can be provided with some personal attributes, for example, walking speed, personal

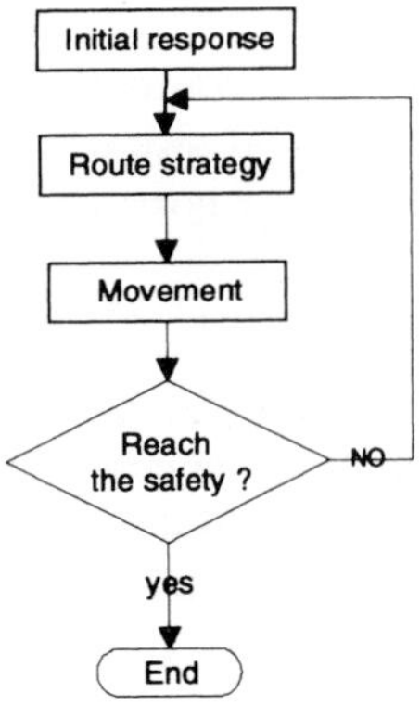

Figure 10. Individual Evacuation Process [46].

memory and psychological condition. These attribute will be used to determine the movement decisions, for example [21], to select the nearest walkway, move on the walkway only when there is no blockage at the end, or change the destination target before reaching it. Løvås [47] proposed some different probability laws for personal movement relative to the route components (nodes and arcs) as follows.

- Random choice. It is applied when the evacuee is not familiar with the surrounding. Lets define X as the state of the underlying system, for instance, the number of evacuees on each arc/node or the hazard (smoke, fire, etc.) level of each arc/node. Moreover, let $p_k(i,j,X)$ be the probability that person k, with his personal attributes, will move from node i to node j when the whole system is in state X. If the number of walkways connected to node i is denoted as δ_i, then

$$p_k(i,j,X) := \frac{1}{\delta_i}$$

- Modified random choice. It includes the possibility that evacuees will return to their previous walkway. Define n_k as the node-number of the last node visited by a person k and introduce a number ω, in which $0 \leq \omega \leq 1$. The parameter ω is a factor showing how a person is willing to return to his/her previous walkway. Using this parameter, the probability $p_k(i,j,X)$ can be redefined as follows.

$$p_k(i,j,X) := \begin{cases} \frac{\omega}{\delta_i} & ,j = n_k \\ \frac{1-\frac{\omega}{\delta_i}}{\delta_i - 1} & ,j \neq n_k \\ 0 & ,otherwise \end{cases}$$

- Suppose there is an evacuation route P determined a priori by the evacuation planner (this is a typical situation, indicated by evacuation sign attached to the wall of rooms or corridors).

$$P := (s = \text{origin}, \ldots, i, j*, \ldots, \text{destination} = d)$$

Suppose that $\delta_i > 1$ for all nodes in this route, except possibly the origin and destination. When an emergency situation arises there is a high tendency that evacuees will use familiar routes that may not follow the route recommended in the evacuation plan. For all $i \in P - \{d\}$ the parameter $\gamma_i \in [0,1]$ indicates the probability that an evacuee will leave the planned route at an intersection node. Then $p_k(i,j,X)$ can be written as

$$p_k(i,j,X) := \begin{cases} 1 - \gamma_i & ,j = j* \\ \frac{\gamma_i}{\delta_i - 1} & ,j \neq n_k \end{cases}$$

Path choice rules like the ones presented above can be used together with attributes of evacuees and building components to design a simulation of a building evacuation (see [62]). Obviously, such simulation is a good tool to model individual behaviour in evacuation planning and can also be used to validate optimization models as presented in the preceding sections.

In recent years, there is an increasing trend to use simulation based on cellular automata (CA). Either deterministic or probabilistic rules can be applied to model the movement patterns between time periods relative to the movement of other persons and/or physical barriers. CA simulation offers the possibility to emulate the essential, diverse movements of evacuees as behavioural responses to varying and uncertain local conditions. A *cellular automata* is defined as a regular n-dimensional lattice partitioned into discrete elements called *cells* or *sites* which has a dicrete step time evolution. Using cellular automata, the space of the evacuation area is divided into accessible and non accessible cells, each of fixed and equal size. The state of each cell is one of finitely many values and has a dynamical behaviour. It is updated simultaneously based on the values of the states in its neighborhood at the preceding time step and according to a specific *local rules* . The set of local rules is defined to control the movement of evacuees, or the state transition of each cell. Since the rules are needed to govern only local relationships among the neighboring cells, CA is considered to be very effective for simulating physical phenomena, the relationships of which over the whole domain are unknown. In the 1-dimensional lattice, a cell has only two neighbourhood cells, the left and right cell adjacent to it. For 2-dimensional lattice, there are some neighbourhood classifications of a cell as shown in Figure 11. The von Neumann neighbourhood consists of 4 cells, the cell above, below, right and left of the reference cell. The radius r of the von Neumann neighborhood is 1, since it considers only the next layer of a cell. The Moore neighborhood is an extension of the von Neumann neighborhood in which the diagonal cells are added. Its radius is still $r = 1$. The expanded Moore neighbourhood further extends the Moore neighborhood by including two layers (i.e., $r = 2$). Another type of neigborhood is the Margolus neigborhood in which 2×2 cells of lattices are considerd at once. The transition from one state to another is arranged by a set of rules according to the neighboring state of the preceding time step. If $x(t)$ is a state of cell x at time t then the state of cell x at time $t + 1$ can be written as

$$x(t+1) = f(x(t), N_x(t))$$

where $N_x(t)$ is state of the neighborhood cells of cell x at time t and where f is the rule. A simple example of a rule can be given as follows. Consider a narrow corridor (assumed wide enough only for one person) with a given length that can be modeled as a one dimensional lattice (see Figure 12). A cell represents a space in the corridor that may be occupied by one person or it may be empty. The walking speed of the ith person $v(i)$ corresponds to the number of cells that a person advances in one iteration (one time step). By defining $x(i)$ as the position of the ith person in the corridor, the distance

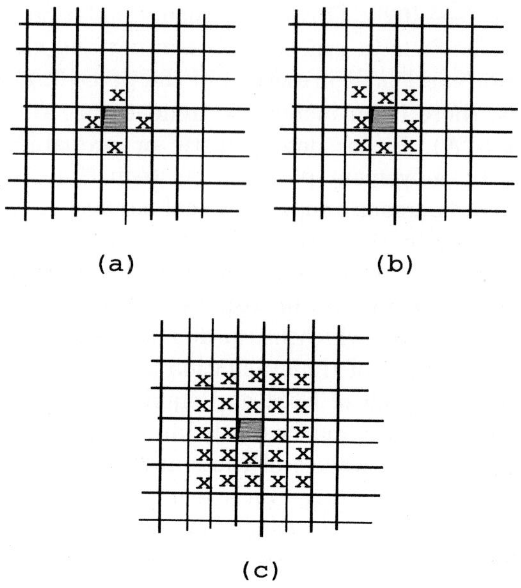

Figure 11. (a) von Neumann Neighborhood (b) Moore Neighborhood (c) Expanded Moore Neighborhood of a Cell (Shaded Cell).

between the ith person and the person immediately ahead is given by

$$g(i) = x(i+1) - x(i) - 1$$

Using these 3 variables the transition rule can be defined as follow :

- Acceleration of free person :
  If $v(i) < v_{max}$ and $g(i) \geq v(i) + 1$ then $v(i) = v(i) + 1$, where v_{max} is the maximum possible speed.
- Slowing down due to other person :
  If $v(i) > g(i) - 1$ then $v(i) = g(i)$
- Movement : person is advanced $v(i)$ cells.

As a method of discrete simulation, CA is used by some evacuation softwares including EGRESS and FlightSim.

EGRESS is a C++ program developed for modelling the behaviour of evacuees in emergency situations, especially in offshore environments. The physical structure of the offshore installation is represented by using hexagonal cellular grids (see, e.g., Doheny and Fraser [20]). It models evacuation using cellular automata where the movements and interactions of the automata on the cellular grid simulate the movements and physical interactions of evacuees on the platform. EGRESS integrates MOBEDIC (Modelling Behavioural Decisions In Computer) for evacuees's decision making and a movement model

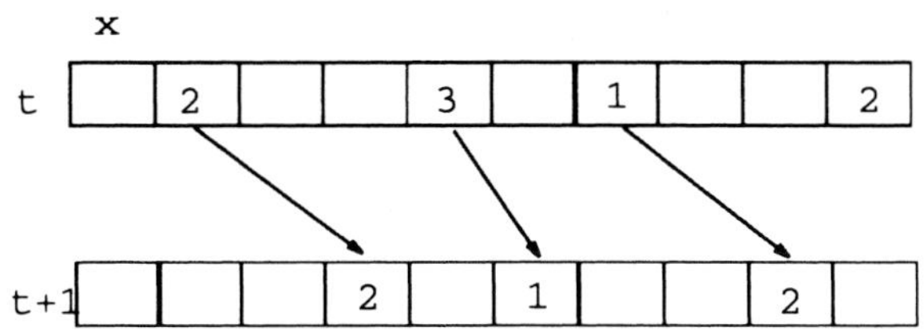

Figure 12. CA Model of Evacuees Moving in a Corridor with $v_{max} = 3$ cells/time-step.

which models the subsequent movement of evacuees throughout the platform. MOBEDIC represents the *brain* of the automata in the movement model. It is responsible for making decisions and for instructing an automaton to carry out some action. The Movement Model distributes informations such as the presence of smoke or fire, the occurrence of alarms and the current position of automaton. These informations are combined with some properties of evacuees, such as reactions to alarm and hazards, familiarity with the building structure, knowledge and experiences about the emergency situation. The combined informations determine the movement of evacuees. In the movement model evacuees are represented using cellular automata which can move about the plan from cell to cell. Movement algorithms determine which cell each automaton should occupy at any given time and move the automata accordingly.

FlightSim was developed at the University of Duisburg, Germany (see Klüpfel et al. [42]). It was originally developed for analyzing the evacuation process of a passenger ship. Some adjusted parameters are included in the model, for example the time that every evacuee needs to enter the saving boats, the distribution of the maximum speed among the evacuee, and the distribution of the number of cells a person looks forward for orientation. Instead of using *parallel updating* when updating the evacuees' positions, FlightSim uses *sequential updating* ([11]). In parallel updating, variables of all evacuees (e.g., position, speed) are changed at the same time. In contrast, the sequential updating selects the evacuees one after the other and updates his/her attributes. The selection on who will be the first is done randomly in order not to give advantage to any specific person. Figures 13 and 14 show the difference of these two updating systems ([11]). Looking at Figure 13, during the first step under parallel updating, evacuee a must wait until the next update, since he is blocked by evacuee b. The left column of Figure 14 describes the sequential updating when evacuee a is always chosen first. The movement pattern is the same as one with parallel update. If evacuee b is chosen first, the situation changes, as is shown in the right column of Figure 14.

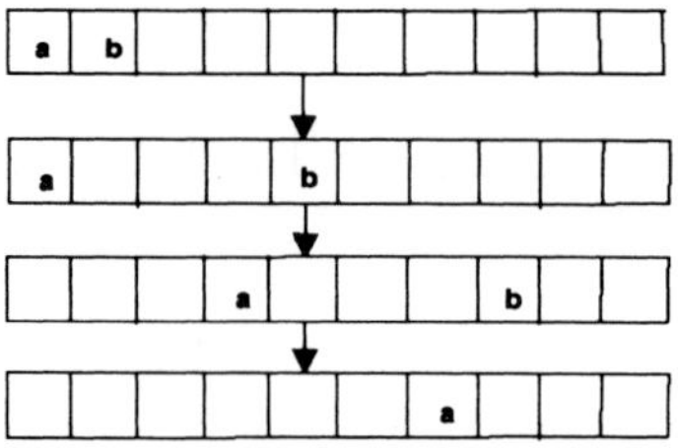

Figure 13. Parallel Update.

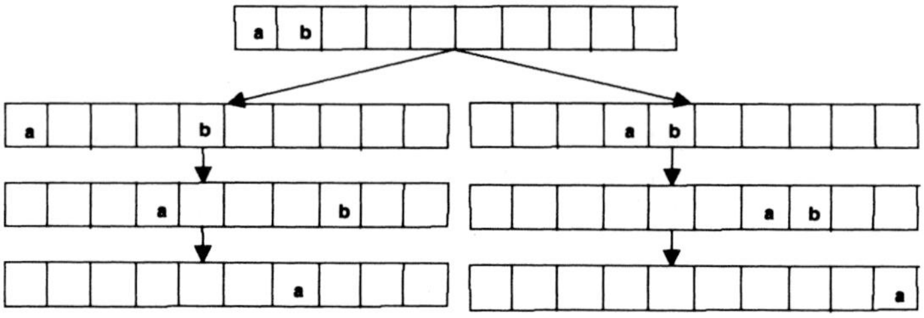

Figure 14. Sequential Update.

Another simulation based evacuation software, EXODUS, was developed by the Fire Safety Engineering group at the university of Greenwich, United of Kingdom (see [29], [32]) and designed to simulate a large number of occupants in a closed environment, for instance building and airplane. It is an expert system-based software which has a set of heuristics or rules to determine the progressive motion and behaviour of each individual. It tracks each individual either making his/her way out of the evacuation area or is being overcome by fire hazards. A more complete survey on evacuation softwares using either macro approaches or micro approaches is discussed by Gwynne, et al. [32].

12 Summary and Conclusion

In this paper, a review of models and algorithms for evacuation planning has been presented. The review covered macroscopic models quite extensively and sketched microscopic models. Both approaches are able to mirror the flows of evacuations over time. The former has its strength in its possibility to optimize the system (while neglecting individuals' behaviour), while the latter is able to capture and utilize properties of each of the evacuees.

Under the macroscopic approach, minimum turnstile cost dynamic network flow models can be applied to estimate the average evacuation time per evacuee. Maximum dynamic flows and universal maximum flows can be used to estimate the maximum number of evacuees which can reach safety during any given time horizon for the evacuation. Quickest flow models allow the estimatation of the minimum time required to bring a given number of evacuees to safety. Considering the source and propagation of hazards, availability of emergency service units and better organization, the evacuation region can be divided into some regions with different priority levels. Therefore, multiple objective models are presented to cope with this problem. Constant travel time is mostly assumed in the literature. This time can be obtained by taking the travel time of the average flow or travel time of a specific queuing level. In order to reflect the congestion phenomenon, it was shown how the constant travel time assumption can be strengthened by considering density dependent travel time. This approach will, however, significantly increase the complexity of the model.

In contrast to macroscopic models, the microscopic approach, usually implemented as simulation, offers possibilities to include individual characteristics and interaction among evacuees. This makes up for the lack of optimization in this approach. Succesful implementations of the macro approach are based on cellular automata.

References

1. R.K. Ahuja, T.L. Magnanti, and J.B. Orlin, *Network Flows : Theory, Algorithms, and applications*, (Prentice Hall, Englewood Cliffs, New Jersey, 1993).
2. E.J. Anderson, P. Nash, and A.B. Philpott, *A Class of Continuous Network Flow Problems*, Mathematics of Operation Research, **7**, 501-514 (1982).
3. E.J. Anderson, P. Nash, and A.F. Perold, *Some Properties of a Class of Continous Linear Programs*, SIAM Journal Control and Optimization, **21**(5), 758-765 (1983).
4. E.J. Anderson, *Extreme-points for Continuous Network Programs with Arc Delays*, J. Inform. Optimi. Sci., **10**, 45-52 (1989).
5. E.J. Anderson and A.B. Philpott, *A Continuous-time Network Simplex Algorithm*, Networks, **19**, 395-425 (1989).
6. E.J. Anderson and A.B. Philpott, *Optimization of Flows in Networks Over Time, Probability, Statistics and Optimization*, (F.P. Kelly, ed.) (J. Wiley and Sons, 1994), 369-382.
7. J.E. Aronson, *A Survey of Dynamic Network Flows*, Annals of Operation Research, **20**, 1-66 (1989).
8. S. Benjaafar, K. Dooley, and W. Setyawan, *Cellular Automata for Traffic Flow Modeling*, (University of Minnesota, Mineapolis, 1997).
9. D.P. Bertsekas, *Linear Network Optimization: Algorithms and Codes*, (The MIT Press, Cambridge, Massachusetts, 1991).
10. V.J. Blue and J.L. Adler, *Using Cellular Automata Microsimulation to Model Pedestrian Movements*, In Ceder, A. editor *Proceedings of the 14th International Symposium on Transportation and Traffic Theory* , Jerusalem, Israel, 235-254 (1999).

11. http://traf2.uni-duisburg.de/bypass/
Assesment and Analysis of the Evacuation of Passenger Vessels by means of Microscopic Simulation, Physics of Transport and Traffic, University of Duisburg, Germany.
12. R.E. Burkard, K. Dlaska, and B. Klinz, *The Quickest Flow Problem*, ZOR-Methods and Models of Operations Research, **37**, 31-58 (1993).
13. M. Carey and E. Subrahmanian, *An Approach To Modelling Time-varying Flows On Congested Networks*, Transportation Research B, **34**, 157-183 (2000).
14. L.G. Chalmet, R.L. Francis, and P.B. Saunders, *Network Models for Building Evacuation*, Management science, **28**, 86-105 (1982).
15. Y.L. Chen and Y.H. Chin, *The Quickest Path Problem*, Computers and Operations Research, **17**, 153-161 (1990).
16. G.H. Chen and Y.C. Hung, *On the Quickest Path Problem*, Information Processing Letters, **46**, 125-128 (1993).
17. H.K. Chen and C.F. Hsueh, *A Model and An Algorithm For The Dynamic User-Optimal Route Choice Problem*, Transportation Research B, **32**(3), 219-234 (1998).
18. W. Choi, R.L. Francis, H.W. Hamacher, and S. Tufekci, *Network Models of Building Evacuation Problems With Flow-Dependent Exit Capacities*, Operational Research, 1047-1059 (1984).
19. W. Choi, R.L. Francis, H.W. Hamacher, S. Tufekci, *Modelling of Building Evacuation Problems with Side Constraints*, European Journal of Operations Research, **35**, 98-110 (1988).
20. J.G. Doheny and J.L. Fraser, *MOBEDIC - A Decision Modelling Tool For Emergency Situations*, Expert Systems With Applications, **10.1**, 17-27 (1996).
21. M. Ebihara, A. Ohtsuki, and H. Iwaki, *Model For Simulating Human Behavior During Emergency Evacuation Based On Classificatory Reasoning And Certainty Value Handling*, Shimizu Technical Research Bulletin, **11**, 27-33 (1992).
22. R.F. Fahy, *An Evacuation Model for High Rise Buildings*, Proceedings of the Third International Symposium on Fire Safety Science, (Elsevier, London, 1991), 815-823.
23. L. Fleischer, *Efficient Continuous-Time Dynamic Network Flow Algorithms*, Operation Research Letters, **23**, 71-80 (1998).
24. L. Fleischer, *Faster Algorithms for the Quickest Transshipment Problem*, Proceedings of 9th Annual ACM-SIAM Symposium on Discrete Algorithms, 147-156 (1998).
25. L. Fleischer, *Universally Maximum Flows with Piecewise-Constant Capacities*, In Cornuejols, G., Burkard, R.E. and Woeginger, G.J., editors, *Proceedings of 7th International Interger Programming and Combinatorial Optimization (IPCO) Conference*, Graz, Austria, 151-165 (1999).
26. L.R. Ford and D.R. Fulkerson, *Flows in Network*, (Princeton University Press, Princeton, New Jersey, 1962).
27. M.L. Fredman and R.E. Tarjan, *Fibonacci Heaps and Their Uses in Improved Network Optimization Algorithms*, J. Ass. Comput. Math. , **34**, 596-615 (1987).
28. D. Gale, *Transient Flows in Networks*, The Michigan Mathematical Journal, **6**, 59-63 (1959).
29. E.R. Galea, M. Owen, and P.J. Lawrence, *Computer Modelling of Human Behaviour in Aircraft Fire Accidents*, Toxicology, **115**, 63-78 (1996).

30. E. Graat, C. Midden, and P. Bockholts, *Complex Evacuation; Effects of Motivation Level and Slope of Stairs on Emergency Egress Time in a Sports Stadium*, Safety Science , **31**, 127-141 (1999).
31. R.C. Grinold, *Continuous Programming, part one: linear objectives*, Journal of Mathematical Analysis and Applications, **28**, 32-51 (1969).
32. S. Gwynne, E.R. Galea, M. Owen, P.J. Lawrence, and L. Filippidis, *A Review of the Methodologies used in the Computer Simulation of Evacuation from the Built Environment*, Building and Environment, **34**, 741-749 (1999).
33. H.W. Hamacher and S. Tufekci, *On the Use of Lexicographic Min Cost Flows in Evacuation Modeling*, Naval Research Logistics, **34**, 487-503 (1987).
34. B. Hoppe and E. Tardos, *Polinomial Time Algorithms for Some Evacuation Problems*, Proc. of 5th Ann. ACM-SIAM Symp. on Discrete Algorithms, 433-441 (1994).
35. B.N. Janson, *Dynamic Traffic Assignment For Urban Road Networks*, Transportation Research B, **25**, 143-161 (1991).
36. J.J. Jarvis and H.D. Ratliff, *Some Equivalent Objectives for Dynamic Network Flow Problems*, Management science, **28**, 106-108 (1982).
37. R. Jayakrishnan, W.K. Tsai, and A. Chen, *A Dynamic Traffic Assignment Model With Traffic-Flow Relationships*, Transportation Research C, **3**(1), 51-72 (1995).
38. D. Kagaris, G.E. Pantziou, S. Tragoudas, and C.D. Zaroliagis, *Transmissions in a Network with Capacities and Delays*, Networks, **33**(3), 167-174 (1999).
39. D.E. Kaufman, J. Nonis, and R.L. Smith, *A Mixed Integer Linear Programming Model For Dynamic Route Guidance*, Transportation Research B, **32**(6), 431-440 (1998).
40. J.L. Kennington and R.V. Helgason, *Algorithms For Network Programming*, (Wiley, N.Y., 1980).
41. T.M. Kisko and R.L. Francis, *EVACNET+: A Computer Program to Determine Optimal Evacuation Plans*, Fire Safety Journal, **9**, 211-220 (1985).
42. H. Klüpfel, T. Meyer-König, J. Wahle, and M. Schreckenberg, *Microscopic Simulation of Evacuation Processes on Passenger Ships, Fourth International Conference on Cellular Automata for Research and Industry*, October, Karlsruhe, Germany (2000).
43. M.M. Kostreva and M.M. Wiecek, *Time Dependency In Multiple Objective Dynamic Programming*, Journal of mathematical Analysis and Application, **173**(1), 289-307 (1993).
44. M.M. Kostreva, *Mathematical Modeling of Human Egress from Fires in Residential Buildings*, em Building and Research Laboratory of the National Institute of Standards and Technology, Technical Report, NIST-GCR-94-643 (1994).
45. S.E. Lovetskii and I.I. Melamed, *Dynamic Flows In Networks*, Automation and Remote Control, **48**, 1417-1434 (1987).
46. G.G. Løvås, *Mathematical Modelling of Emergency Evacuations*, Ph.D Thesis, Department of Mathematics, University of Oslo, Norway, 1987).
47. G.G. Løvås, *Models of Wayfinding in Emergency Evacuations*, European Journal of Operation Research, **105**, 371-389 (1998).
48. E. Minieka, *Maximal, Lexicographic, and Dynamic Network Flows*, Operations Research, **21**, 517-527 (1973).

49. E. Minieka, *Dynamic Network Flows with Arc Changes*, Networks, **4**, 255-265 (1974).
50. C. Montes, *Evacuation of Buildings*, M.Sc. Thesis, (Department of Mathematics, Universität Kaiserslautern, Kaiserslautern, Germany, 1994).
51. K. Nagel and M. Schreckenberg, *A Cellular Automaton Model for Freeway Traffic*, J. Phys. I France , **2**, 2221-2229 (1992).
52. R.G. Ogier, *Minimum Delay Routing in Continuous-time Dynamic Networks with Piecewise Constant Capacities*, Networks, **18**, 303-318 (1988).
53. M. Owen, E.R. Galea, and P.J. Lawrence, *The Exodus Evacuation Model Applied to Building Evacuation Scenarios*, Journal of Fire Protection Engineering, **8**(2), 65-86 (1996).
54. A.F. Perold, *Extereme Points and Basic Feasible Solutions in Continuous Time Linear Programming*, SIAM Journal Control and Optimization, **19**(1), 52-63 (1981).
55. A.B. Philpott, *Continuous-Time Flows in Networks*, Mathematics of Operation Research, **15**(4), 640-661 (1990).
56. A.B. Philpott and M. Craddock, *An Adaptive Discretization Algorithm for a Class of Continuous Network Programs*, Networks, **26**, 1-11 (1995).
57. M.C. Pullan, *An Algorithm For a Class of Continuous Linear Programs*, SIAM Journal Control and Optimization, 31(6), 1558-1577 (1993).
58. M.C. Pullan, *A Study of General Dynamic Network Programs With Arc Time-Delays*, SIAM Journal Optimization, **7**(4), 889-912 (1997).
59. B. Ran and D. Boyce, *Modeling Dynamic Transportation Networks*, (Springer, Heidelberg, 1996).
60. J.B. Rosen, S.Z. Sun, and G.L. Xue, *Algorithms for the Quickest Path Problem and The Enumeration of Quickest Paths*, Computers and Operations Research, **18**, 579-584 (1991).
61. Y. Sheffi, H. Mahmassani, and W.B. Powell, *A Transportation Network Evacuation Model*, Transportation Research A, **16**(3), 209-218 (1982).
62. S.A. Tjandra, *Simulation of Building Evacuation using Simple++*, Software, Institut Techno- und Wirtschaftsmathematik, Kaiserslautern, Germany (1999).
63. S.A. Tjandra, *Dynamic Network Flow Models for Evacuation Problems*, Ph.D Thesis (to appear), (Department of Mathematics, Universität Kaiserslautern, Kaiserslautern, Germany, 2001).
64. W.F. Tyndall, *A Duality Theorems For A Class of Continuous Linear Programming Problems*, SIAM Journal of Applied Math. , **13**(3), 644-666 (1965).
65. W.F. Tyndall, *An Extended Duality Theorem For Continuous Lienar Programming Problems*, SIAM Journal of Applied Math., **15**(5), 1294-1298 (1967).
66. W.L. Wilkinson, *An Algorithm for Universal Maximal Dynamic Flows in A Network*, Operation Research, **19**, 1602-1612 (1971).
67. T. Yamada, *A Network Approach To A City Emergency Evacuation Planning*, International Journal of Systems Science, **27**(10), 931-936 (1996).
68. D.J. Zawack and G.L. Thompson, *A Dynamic Space-Time Network Flow Model for City Traffic Congestion*, Transportation Science, **21**, 153-162 (1987).

Earliest Arrival Flow Model with Time Dependent Capacity for Solving Evacuation Problems

H.W. Hamacher[1,2] and S.A. Tjandra[1]

[1]Fraunhofer Institute für Techno-und Wirtschaftsmathematik (ITWM), 67663 Kaiserslautern, Germany

[2]Fachbereich Mathematik, Universität Kaiserslautern, 67663 Kaiserslautern, Germany

Evacuation problems can be modeled as flow problems in dynamic networks. A dynamic network is defined by a directed graph $G = (N, A)$ with sources, sinks and non-negative integral travel times and capacities for every arc $(i, j) \in A$. The earliest arrival flow problem is to send a maximum amount of dynamic flow reaching the sink not only for the given time horizon T, but also for any time $T' < T$. This problem mimics the evacuation problem of public buildings where occupancies may not known. For the buildings where the number of occupancies is known and concentrated only in one source, the quickest flow model is used to find the minimum egress time. We propose in this paper a solution procedure for evacuation problems with a single source of the building where the occupancy number is either known or unknown. The possibility that the flow capacity may change due to the increasing of smoke density or fire obstructions can be mirrored in our model. The solution procedure looks iteratively for the shortest conditional augmenting path (SCAP) from source to sink and compute the time intervals in which flow reaches the sink via this path.

1 Introduction

In the macro approach, evacuation problems can be modeled as flow problems on the network. Since the time is a decisive parameter in evacuation problems, dynamic network flow models are more suitable than the static ones (see e.g., [4-6,8,10]). A dynamic network is defined by a directed graph $G = (N, A)$ where N is the set of nodes and $A \subset N \times N$ is the set of arcs which have non-negative flow capacities and non-negative integral travel times as their attributes (see for instance [1,2]). In the network presentation of a building, nodes may represent rooms, lobbies or intersection points. While arcs can be used to model corridors, hallways, or stairways. Some locations in the building that house a significant number of evacuees are considered to be source nodes. The building exits or safety locations that might be considered as the final destination of the evacuees' movement, are considered as sink nodes. The supply of source nodes equals to the estimation of the number of evacuees in the corresponding locations.

Dynamic network flow models have been applied to solve evacuation problems under some assumptions or objectives different from ours in this paper, e.g., constant travel time and capacity [3,4,8,10,14], flow dependent exit capacities [5,6], or time dependent vectors of arc costs as part of a multicriteria problem [12]. Our results relate to evacuation problems with a single source where we consider two different building environments, namely: buildings where the number of evacuees is difficult to estimate (public buildings e.g. shopping mals, theaters) and ones with known number of evacuees (e.g. office or residential buildings). In public buildings, we are interested to find the maximum number of evacuees which can be sent out within a time *T*. To handle this problem we work with a so-called Earliest Arrival Flow (EAF) model. EAF is the problem of maximizing flows reaching the sink not only for the alloted time *T*, but also for any smaller time horizon (see e.g. [7,10,11,13]).

The main contribution of this paper is to incorporate the changing of arc capacities over time, e.g. due to smoke or fire, into this EAF problem. We assume that evacuees may wait only in the source node. The solution procedure is applied to solve the evacuation problem with known occupancy number and with objective to minimize the egress time. The latter problem is known as the quickest flow problem (see [3,7,11]).

In the next section we will formally introduce the continuous time dynamic network formulation of EAF and detail the procedure to solve this problem by finding iterative shortest conditional augmenting paths from the source to the sink. Section 3 will contain some interrelations between EAF and the quickest flow problem which is used to find the minimum egress time of evacuation problem with known occupancies. The last section summarizes the results in a numerical example.

2 Solving Evacuation Problems as EAF Problem

Given the network $G=(N,A)$ where node s and d is the source and the sink node, respectively. With each arc (i,j) we associate a nonnegative integer number λ_{ij} which gives the time to traverse the arc, and a nonnegative bounded measurable function b_{ij} on $[0,T]$ which gives its flow capacity. *T*, the time horizon of interest is determined a priori. It is assumed that no node except the source has storage capacity, i.e., no evacuee may wait in any node but the source node. The value $x_{ij}(t)$ defines the nonnegative rate of flow leaving node i at time t, consequently arriving at node j at time $t+\lambda_{ij}$. The continous-time EAF problem is then as follows.

$$\max \; V_t(x) = \int_0^t \sum_{i \in N} [x_{id}(\tau - \lambda_{id}) - x_{di}(\tau)] d\tau, \qquad t \in [0,T], \tag{1}$$

$$\sum_{i \in N} [x_{ij}(t - \lambda_{ij}) - x_{ji}(t)] = 0, \; t \in [0,T], \; j \in N - \{s,d\}, \tag{2}$$

$$0 \le x_{ij}(t) \le b_{ij}(t), \; \forall (i,j) \in A, \; t \in [0,T] \tag{3}$$

A solution procedure for this problem is a dynamic version of the well-known shortest augmenting path algorithm of maximum static network flows (see, for instance [1]). In the classical residual network $G_x := (N, A_x)$, we look for a path from s to d (s-d path) which has positive excess capacity. In the case of the dynamic version we also have to consider the availability of such a path over time. The admissible time S_{ij} for arc $(i,j) \in A_x$ is defined as the set of time intervals when the arc $(i,j) \in A_x$ can carry residual by increasing or decreasing the current flow. Since the flow augmentation on arc (i,j) is possible only when $S_{ij} \neq \varnothing$, the residual network is conditioned by the set S_{ij}. Therefore we introduce the *conditional augmenting network* (similar terminology is used in parametric flows [9]) with respect to feasible dynamic flow x, $G_x := (N, A_x^+ \cup A_x^-)$ with arc set $A_x^+ := \{(i,j) : (i,j) \in A \;\&\; S_{ij} \neq \varnothing\}$ and $A_x^- := \{(i,j) : (j,i) \in A \;\&\; S_{ij} \neq \varnothing\}$. The residual travel times are $\lambda_{ij}^x := \begin{cases} \lambda_{ij} & ,(i,j) \in A_x^+ \\ -\lambda_{ji} & ,(i,j) \in A_x^- \end{cases}$.

Suppose $P = \{s = v_1, v_2, \ldots, v_k = d\}$ is a s-d path in the residual network G_x. Let $\pi_{v_l}^{v_{l+1}}$ be the label measuring the distance of v_l to d along P with respect to travel time $\lambda_{v_l v_{l+1}}^x$. Then $S_{v_l}^{v_{l+1}} := \{t + \pi_{v_l}^{v_{l+1}} : t \in S_{v_l, v_{l+1}}\} \cap [0,T]$ is the set of arrival times of flows at d within the time T that leave v_l at time $t \in S_{v_l, v_{l+1}}$ and move along P, under the assumption that increasing the flow is always possible. Increasing flow along P is possible only when the arrival time at the next nodes v_{l+q}, $q = 1, \ldots, k-l-1$ of flows which leave v_l at time $t \in S_{v_l, v_{l+1}}$, $\forall l = 1, \ldots, k-2$ is also in $S_{v_{l+q}, v_{l+q+1}}$. We define recursively for $l = 1, \ldots, k-1$; $S_{v_l}^{*v_{l+1}} := S_{v_l}^{v_{l+1}} \cap S_{v_{l+1}}^{*v_{l+2}}$ with $S_{v_k}^{*v_{k+1}} = [0,T]$. $S_{v_l}^{*v_{l+1}}$ represents the set of arrival times at the sink node when flows from node v_l reach d with considering the avalaibility of the next arcs along P in the residual network. We say that the sink is *reachable* from node v_l in G_x when $S_{v_l}^{*v_{l+1}} \neq \varnothing$. Path P is a conditional (flow) augmenting path (CAP) in G_x if $S_P := S_{v_1}^{*v_2} \neq \varnothing$. Moreover S_{ij}^P is the set of departure times of flows on arc (i,j)

sent along P for which the arrival time at d is in S_P , i.e., $S_{ij}^P := \{t - \pi_i^j : t \in S_P\}$. Next we compute the maximum increment functions $\varepsilon_P(t)$ on S_P and translated back to S_{ij}^P in order to accommodate the increment on each $(i, j) \in P$. The revised flows for each time $t \in [0,T]$ can be obtained as follows.

$$x_{ij}(t) := \begin{cases} x_{ij}(t) + \varepsilon_{ij}(t) & ,if\ (i,j) \in P^+ \cup \{(d,s)\} \\ x_{ij}(t) - \varepsilon_{ji}(t + \lambda_{ij}) & ,if\ (j,i) \in P^- \\ x_{ij}(t) & ,otherwise \end{cases} \tag{4}$$

In order to find the available path from node i to d, we work with the network $G_x^R := (N, A_x^R)$. In G_x^R, the arc set A_x^R is obtained by reversing the direction of all arcs in A_x. For time dependent capacities the sub-path of the shortest conditional augmenting path (SCAP) might not be a SCAP as well. Therefore we can not update the large distance by a smaller one as in the label correcting algorithm for the static network. The shorter one may reach node j at time $t' \notin S_{jk}, \forall k \neq j \in N$, i.e., the corresponding path can reach node j but may not be able to reach s. Since the longer path may reach node j at time $t \in S_{jk}$ for a node $k \in N$ such that it may become the SCAP, in the case of time dependent capacities, we update the large distance of node j from node i, π_j^i, with the new distance from node $i^\bullet$ only when $\pi_j^i > \pi_j^{i^\bullet}$ and $S_j^{\bullet i} \subset S_j^{\bullet i^\bullet}$ as explained by Lemma 2.1.

Lemma 2.1: *Suppose there are two paths from j to d, P_{jd} and $P_{jd}^\bullet$ with $(j,i) \in P_{jd}$ and $(j,i^\bullet) \in P_{jd}^\bullet$. Let π_j^i and $\pi_j^{i^\bullet}$ be the distance of j to d on path P_{jd} and $P_{jd}^\bullet$, respectively, with $S_j^{\bullet i}$ and $S_j^{\bullet i^\bullet}$ the corresponding arrival times at d. Moreover, assume that $\pi_j^i > \pi_j^{i^\bullet}$ and $S_j^{\bullet i} \subset S_j^{\bullet i^\bullet}$. If there is a s-d conditional augmenting path $P := P_{sj} \cup P_{jd}$, then there is also a path $P^\bullet := P_{sj} \cup P_{jd}^\bullet$ with shorter time distance than that of P.*

Proof: Since P is a s-d CAP, $S_P \neq \emptyset$. By definition, $S_P \subset S_l^j \cap S_j^{\bullet i}$ for any $l \in N$ with $(l, j) \in P_{sj}$. Since $S_j^{\bullet i} \subset S_j^{\bullet i^\bullet}$, we obtain $S_P \subset S_l^j \cap S_j^{\bullet i^\bullet}$. Therefore, we may compose an s-d path $P^\bullet$ with $S_{P^\bullet} \supset S_P$ from path P as described in the lemma. Since $\pi_j^i > \pi_j^{i^\bullet}$ the time distance of $P^\bullet$ is shorter than that of P. □

The SCAP procedure maintains a set of distance labels π, and the time set $S^\bullet$ which are updated iteratively. The distance label π_j^i is either ∞ and $S_j^{\bullet i} = \emptyset$, indicating that we have yet to discover an augmenting path from the node j to the sink node, or it is equal to the length of some augmenting path from the node j to the sink with $S_j^{*i} \neq \emptyset$. For each node j, we maintain a list of predecessor indices Z_j which records some nodes prior to node j via some augmenting paths. We denote S_j^{*i} as $S_j^{*p,q}$ if we want to emphasize that node i is the q-th predecessor of node j during iteration p of the SCAP procedure with $S_j^{*i} \neq \emptyset$. Similar denotation we do for π_j^i. Moreover we may write the q-th predecessor of node j at iteration p as $Z_j^p(q) = (i, q')$ where q' is the predecessor index of node i in the previous iteration. At termination, the predecessor indices allow us to trace the SCAP.

Lemma 2.2: *Consider iteration (p+1) of the algorithm and any* $(i,j) \in A_x^R$. *Suppose that* $\exists q' \in \{1,\dots,|Z_i^p|\}$ *with* $B := \{t + \pi_i^{p,q'} + \lambda_{ij}^x : t \in S_{ij}\} \cap [0,T] \cap S_i^{*p,q'} \neq \emptyset$. *If* $\exists q \in \{1,\dots,|Z_j^p|\}$ *with* $\pi_j^{p,q} > \pi_i^{p,q'} + \lambda_{ij}$ *and* $S_j^{*p,q} \subset B$ *then we get a shorter CAP by updating the value of* $\pi_j^{p,q}$ *with* $\pi_i^{p,q'} + \lambda_{ij}^x$ *, i.e.,* $\pi_j^{p+1,q} := \pi_i^{p,q'} + \lambda_{ij}^x$.

Proof: Follows immediately from Lemma 2.1. □

The iterative processes of SCAP procedure stop when they satisfy the following optimality condition.

Theorem 2.3 (SCAP Optimaliy Condition): *For every node* $j \in N$, *the number* π_j^l *represents the SCAP distance if and only if they satisfy the SCAP condition* $\pi_j^l \le \pi_i^k + \lambda_{ij}^x$, $\forall (i,j) \in A_x^R$ *with* $S_j^{*l} \neq \emptyset$.

We see that the solution procedure consists of two main parts. The first part is to find a shortest conditional augmenting path P and its availability time S_P. The second one is to find the maximum flow increment along P and to revise the current flows. These two parts are repeated until no conditional augmenting path is available. At this point the current flow solves the EAF due to the following lemmas.

Lemma 2.4: *If* $S_P \neq \emptyset$ *then there is a feasible flow with value strictly greater than that of the current flow.*

Proof: Suppose $\{x_{ij} : (i,j) \in A\}$ is the current feasible flow and there is a s-d path P with $S_P \neq \emptyset$. Since $S_P = S_{s=v_1}^{*v_2}$ and $S_s^{*v_2} \subset S_{v_l}^{*v_{l+1}}, \forall (v_l, v_{l+1}) \in P$ we have $S_{v_l}^{*v_{l+1}} \neq \emptyset$. This condition implies $S_{v_l, v_{l+1}} \neq \emptyset$ and moreover $\varepsilon_P(t) > 0,\ t \in S_P$. Hence x_{ds} is not optimal. □

Lemma 2.5: *If* $S_P = \emptyset$ *then the current flow solves EAF.*

Proof: Suppose x does not solve maximum dynamic flow for some $T' < T$. Then there is a s-d path P with $S_P \subset [0, T']$ which is not empty, contradicting the assumption of S_P. □

Solution Procedure for Finding Maximum Earliest Arrival Flows

Initialization

Determine the distance λ of the static shortest s-d path. If $\lambda > T$ then the algorithm terminates with $x = 0$. Set all flows $x_{ij} = 0, \forall (i, j) \in A$.

SCAP procedure: work with G_x^R

Step 0 : Set $p = 1$; $k_i = 1, \forall i \in N$; $\pi_d^{1,1} = 0$; $\pi_i^{1,1} := \begin{cases} \lambda_{di}^x & ,(d,i) \in A_x^R \\ \infty & ,otherwise \end{cases}$ $,i \in N - \{d\}$

$S_i^{1,1} := \{t + \pi_i^{1,1} : t \in S_{di}\} \cap [0, T]$; $S_i^{*1,1} = S_i^{1,1} \cap [0, T]$; $Z_i^1 = \{(d,1)\}, \forall i \in N$

Step 1 : Set $k_j := |Z_j^p|$; $\pi_j^{p+1,q} := \pi_j^{p,q}$; $S_j^{*p+1,q} = S_j^{p,q}$, $q = 1, \cdots, k_j$; $\bar{k}_j := k_j$, $\forall j \in N$

For $l = 1$ to $|A_x^R|$ do

If $a_l = (i, j) \in A_x^R$ then

For $q' = 1, \cdots, k_i$ do

$B := \{t + \pi_i^{p,q'} + \lambda_{ij}^x : t \in S_{ij}\} \cap [0, T] \cap S_i^{*p,q'}$

If $B \neq \varnothing$ then

If $\exists q \in \{1, \cdots, k_j\}$ s.t. $\pi_j^{p+1,q} > \pi_i^{p,q'} + \lambda_{ij}^x$ and $S_j^{*p+1,q} \subset B$ then

Set $\pi_j^{p+1,q} = \pi_i^{p,q'} + \lambda_{ij}^x$; $S_j^{*p+1,q} = B$; $Z_j^{p+1}(q) = (i, q')$

Else if $\nexists q \in \{1, \cdots, k_j\}$ s.t. $\pi_j^{p+1,q} \leq \pi_i^{p,q'} + \lambda_{ij}^x$ and $B \subset S_j^{*p+1,q}$ then

Set $\pi_j^{p+1,\bar{k}_j+1} = \pi_i^{p,q'} + \lambda_{ij}^x$; $S_j^{*p+1,\bar{k}_j+1} = B$;
$Z_j^{p+1}(\bar{k}_j + 1) = (i, q')$; $\bar{k}_j := \bar{k}_j + 1$

$k_j := \bar{k}_j$, $\forall j \in N$

Step 2 : For all $i \in N$ do

If $\exists q \neq q' \in \{1, \cdots, k_j\}$ with $\pi_i^{p+1,q} \geq \pi_i^{p+1,q'}$ and $S_i^{*p+1,q} \subset S_i^{*p+1,q'}$ then

Reduce Z_i^{p+1} by deleting $Z_i^{p+1}(q)$.

Step 3 : If $\pi_i^{p+1,q} = \pi_i^{p,q}$, $\forall q = 1, \cdots, k_i$; $\forall i \in N$ then

If $\{\pi_s^{p,q} : S_s^{*p,q} \neq \varnothing\} \neq \varnothing$ then $q^* := \arg\min_{1 \leq q \leq k_i} \{\pi_s^{p,q} : S_s^{*p,q} \neq \varnothing\}$

Find SCAP by using backtracking procedure on Z_i^{p+1} with $S_p := S_s^{*p,q^*}$

Run maximum flow increment procedure.

Else No *s-d* CAP. Algorithm STOP and the current flow is EAF.

Else Set $p := p+1$ and go to Step 1.

Maximum Flow Increment: work with *G*

Calculate S_{ij}^P, $\forall (i,j) \in P$, $\varepsilon_P(t)$ and get new flow using Eq. (4). Revise S_{ij} using the new flow and return to SCAP procedure.

3 Solving Evacuation Problems with Known Occupancy

For given *T* the building is cleared within the time horizon if and only if the corresponding dynamic network allows a maximal *s-d* flow with flow value at least as large as the initial occupancy *v*. In this section we show how to minimize *T*. The interrelation between EAF introduced in the previous section and the quickest flow problem defined in the introduction is explained by the following theorem.

Theorem 3.1: *Let $V_T(x)$ be the total value of flow x entering the sink within time horizon T, i.e.,* $V_T(x) = \sum_{(i,d) \in A} \int_{t=0}^{T} x_{id}(t)\, dt$. *Suppose x is the earliest arrival flow. If* $T^* := \min\{T' : V_{T'}(x) \geq v,\ T' \leq T\}$ *exists then T^* solves the quickest flow problem with initial occupancy v.*

Proof: If T^* does not solve the quickest flow problem then there is $T'' < T^*$ and flow x'' with value $V_{T''}(x'') = v$. Hence $V_{T''}(x) < V_{T''}(x'')$, contradicting that *x* is also maximum at T'' . □

The theorem suggests a simple algorithm to solve the evacuation problem with known occupancy under the assumption of a single source using the EAF algorithm.

Step 0 : Set the time T large enough such that $V(T) \geq v$. Connect source node s to super source node s' with capacity $b_{s's}(t) = v,\ t \in [0,T]$.

Step 1 : Solve the EAF problem. Find $T^* := \min\{T' : V(T') \geq v,\ T' \leq T\}$ by tracing the x_{ds} value and the corresponding S_p . T^* is the minimum egress time.

4 Numerical Example

Fig. 1 shows the static network structure of a simple building with node 5 represents the safety area. The fire starts burning room 0 near the exit doors to lobby 2 and room 3. The normal capacity 3 persons/unit time of arcs (0,2) and (0,3) is decreased linearly over time due to the increasing of fire. Finally no one can pass those exit doors after three time units. Additional emergency exit is built as arc (0,5) and usable starting at time 6 with capacity only 1 person/unit time. Those arc conditions define the flow rate capacities b_{02}, b_{03}, and b_{05}. The distribution of maximum EAF rates within time $T = 10$ is shown below with maximum flow value is 30. Suppose initially there are 20 evacuees in room 0 and they all are trying to go out to the safety area as quickly as possible. Using solution procedure as explained in Section 3, the minimum time needed to clear the building is 8 time units.

$$x_{01}(t) = \begin{cases} 3 & ,t \in [0,6] \\ 0 & ,otherwise \end{cases} ; \qquad x_{02}(t) = x_{03}(t) = \begin{cases} 3-t & ,t \in [0,3] \\ 0 & ,otherwise \end{cases} ;$$

$$x_{05}(t) = \begin{cases} 1 & ,t \in [6,9] \\ 0 & ,otherwise \end{cases} ; \quad x_{14}(t) = \begin{cases} 3 & ,t \in [2,8] \\ 0 & ,otherwise \end{cases} ; \quad x_{25}(t) = \begin{cases} 4-t & ,t \in [1,4[\\ 7-t & ,t \in [4,7] \\ 0 & ,otherwise \end{cases} ;$$

$$x_{32}(t) = \begin{cases} 5-t & ,t \in [2,5] \\ 0 & ,otherwise \end{cases} ; \qquad x_{45}(t) = \begin{cases} 3 & ,t \in [3,9] \\ 0 & ,otherwise \end{cases} ;$$

$$x_{12}(t) = x_{24}(t) = 0,\ t \in [0,10]$$

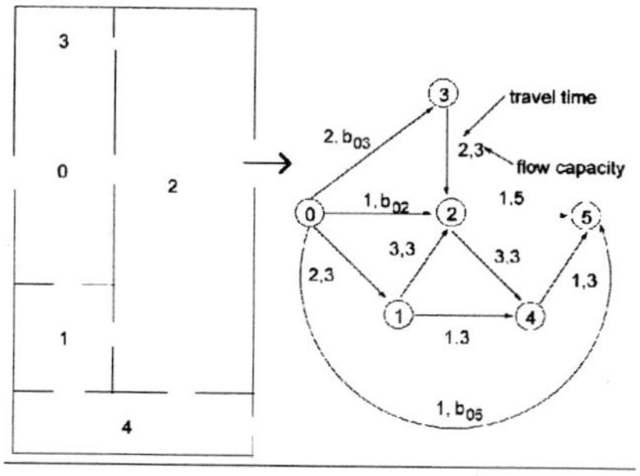

Figure 1: Example Building and the Corresponding Network

References

1. R.K. Ahuja, T.L. Magnanti, and J.B. Orlin, *Network Flows: Theory, Algorithms, and Applications*, (Prentice Hall, Englewood Cliffs, New Jersey, 1993).

2. J.E. Aronson, *A Survey of Dynamic Network Flows,* Annals of Operation Research **20**, 1-66 (1989).

3. R.E. Burkard, K. Dlaska, and B. Klinz, *The Quikest Flow Problem,* ZOR-Methods and Models of Operations Research **37**, 31-58 (1993).

4. L.G. Chalmet, R.L. Francis, and P.B. Saunders, *Network Models for Building Evacuation*, Management Science **28**, 86-105 (1982).

5. W. Choi, R.L. Francis, H.W. Hamacher, and S. Tufekci, *Network Models of Building Evacuation Problem With Flow-Dependent Exit Capacities*, Operational Research, 1047-1059 (1984).

6. W. Choi, R.L. Francis, H.W. Hamacher, and S. Tufekci, *Modeling of Building Evacuation Problems With Side Constraints*, European Journal of Operation Research **35**, 98-110 (1988).

7. L. Fleischer, *Faster Algorithms for the Quickest Transshipment Problem*, in: *Proceedings of 9th Annual ACM-SIAM Symposium on Discrete Algorithms*, pp. 147-156 (1998).

8. H.W. Hamacher and S. Tufekci, *On the Use of Lexicographic Min Cost Flows in Evacuation Modeling*, Naval Research Logistics **34**, 487-503 (1987).

9. H.W. Hamacher and L.R. Foulds, *Algorithms for Flows with Parametric Capacities,* ZOR - Methods and Models of Operations Research **33**, 21-37 (1989).

10. B. Hoppe and E. Tardos, *Polinomial Time Algorithms for Some Evacuation Problems*, in*: Proc. of 5th Annual ACM-SIAM Symposium on Discrete Algorithms*, pp. 433-441 (1994).

11. B. Hoppe and E. Tardos, *The Quickest Transshipment Problem*, in: *Proc. of 6th Annual ACM-SIAM Symposium on Discrete Algorithms*, pp. 512-521 (1995).

12. M.M. Kostreva and M.M. Wiecek, *Time Dependency In Multiple Objective Dynamic Programming*, Journal of Mathematical Analysis and Application **173** (1), 289-307 (1993).

13. E. Minieka, *Maximal, Lexicographic, and Dynamic Network Flows*, Operations Research **21**, 517-527 (1973).

14. C. Montes, *Evacuation of Buildings*, M.Sc. Thesis, (Department of Mathematics, Universität Kaiserslautern, Kaiserslautern, Germany, 1994).

Life Saving Applications of Directional Sound

D. Withington

School of Biomedical Sciences, Worsley Medical and Dental Building, University of Leeds, Leeds, LS2 9NQ, U.K. and Sound Alert Technology plc, 175 Woodhouse Lane, Leeds, LS2 3AR, U.K.

Imagine the scenario of a smoke-filled building (or ferry). The fire alarm is ringing telling you to leave immediately, the most important question facing you is where do I go to find a safe exit? At present exit routes are signed by visual means. Visual signage is immediately obscured with even low levels of smoke, so it is vital that we develop an alternative means of identifying exit routes that does not rely solely on vision. An obvious solution is the use of sound since we hear perfectly well even in smoke. If sound is utilised to identify exits then it is vital that we can instantly pinpoint where the sound is coming from, in other words directional sound. To comply with being directional the sound must be broadband in composition containing the majority of frequencies from the human hearing range. The use of simple sounds, as commonly found in fire alarms, would cause major problems in evacuation beacons confusing the listener and hindering their escape.

1 Introduction

One of the most frightening experiences that we ever face is to be lost and disoriented. Under such circumstances, our ability to process and store environmental information deserts us and, because most of us are dependent upon others knowing their way, or 'information' such as signs pointing the way, our very survival may be threatened. Being truly disoriented may be a relatively rare phenomenon, but given the right series of cataclysmic conditions, it may happen to any of us at any time.

For example, as a newcomer to an *unfamiliar* building such as a department store, hotel complex, or university campus, you will most likely experience some degree of stress or anxiety that progressively worsens as your disorientation increases. As time passes however, by looking at signs, asking others for directions and exploring your environment, you begin to establish a series of inherent spatial relationships that were not apparent when you first entered the building. The more you actively explore, go the wrong way, ask directions etc., the more environmental information you collect, this information beginning to organise itself into a mental representation of the environment. At a certain stage in this 'mental' development, you may class yourself as being a familiar user of the building, or in other words, you are able to move through it efficiently, going from one place to another without too many problems.

1.1 The Emergency Situation

None of us, however, is *totally* familiar with all environments we enter, and this is most evident when an emergency situation arises. Many behavioural studies have repeatedly shown that one of the most natural instincts in the event of a fire is that people evacuate a building by the route through which they entered. More often than not, this is rarely the quickest or most appropriate way. Many people fail to spot nearby exits, and in some cases walk straight past visible fire exits!

The repercussions of such actions have, in several cases, been severe. Certain circulation routes (generally those used for normal, everyday movement) encounter a higher population flow than they were designed for, leading to overcrowding and a slowing down of the evacuation process. As a consequence, some building occupants are exposed to deadly smoke, fumes and flames.

1.2 Current Stategies Employed in Aiding Emergency Evacuation

Given that vision is our primary mode of perceiving our environment, it is not surprising to find that the majority of emergency egress aids, such as emergency lighting, signage, colour coding and photoluminescent guidance strips are solely visual based. How effective are such aids when the part of the building you are in is completely occluded by smoke, or indeed if you are among the one million people in the UK who are registered visually impaired (with a further estimated 750,000 unregistered)? Additionally, there are many thousands of people in the UK who have moderate to severe learning difficulties. Such learning difficulties are particularly manifested in their limited ability to understand the spatial layout of the environment. It is clear that existing emergency way-finding provision offers very limited means by which to escape! This is not to say that 'disabled' provision in the UK is completely inadequate. There are many buildings which are better provided for, such as through the implementation of tactile way-finding aids on handrails, but these methods assume primarily that they are cool enough to touch, and accessible to all occupants, irrespective of age, height or disability. It is a safe assumption that, ***unless*** the victim of a fire situation knows exactly where the exits are, he or she will waste valuable seconds or minutes searching by touch alone. Given the rapid rate with which fires can develop, time becomes the overriding critical factor in emergency egress.

It is clear that this reliance upon visual means just isn't good enough in modern evacuation practice, and it is imperative that another sensory modality is activated, the use of sound being the obvious solution. At Leeds University, such a way-finding aid has been developed, with extensive field trials showing it to offer fast, efficient evacuation for sighted, visually and learning impaired users.

1.3 The Present Role of Sound in Emergency Evacuation

Generally, all uses of sound in emergency evacuation are provided in the form of an "alarm" which merely alerts people to the presence of imminent danger. Irrespective of whether this information is provided by conventional alarm tones or through more sophisticated speech based alert mechanisms, alarms give absolutely no information concerning the direction to, or location of, the nearest exits. Even if such alarms were placed over exit doors, acting as directional beacons, they would still be impossible to locate. To understand why these devices would not suffice as exit locators, it is necessary to describe, in some detail, how we manage to locate sounds in space, including the type of signal necessary for accurate localization.

2 Sound Localization

The ability to localize a sound source is an evolutionary prerequisite for survival. For example, when hearing the crack of a twig as a predator approaches, there is simply not time to wait and look around to check where the sound is coming from. To survive we must react instantly, as soon as the audible signal is received. Similarly, for predators, a rustle of leaves may indicate where their potential prey is hiding, and locating that position will determine whether or not they eat on that occasion. It is, therefore, safe to conclude that pinpointing sound is something we do well. In reality we can localize a sound to an accuracy of about five degrees, given the right type of sound [1]. This level of accuracy is less than that for visual spatial acuity, but more than adequate for survival purposes.

2.1 Brain Site

There is one particular part of our central nervous system that plays a vital role in the detecting of and, equally importantly, the response to a sound source. This area is part of the mid-brain and is called the superior colliculus (SC) [2]. Neurophysiologists studying the properties of neurones in the SC together with psychoacousticians studying human responses to sound have enabled us to understand how the brain processes information relating to a sound source and, importantly, what type of sound is needed for a degree of accuracy to be achieved. It has long been recognised that localizing a sound source requires a vast amount of neural processing [3]. Only certain types of sounds are inherently localizable and what is crucial is that they contain a large spectrum of frequencies, that is broadband noise. Pure tones, simple tone combinations or narrowband noise cannot be localized. To understand why this is the case, the cues given by sound, recognised by the brain, must be considered.

2.2 Localization Cues

We can hear a vast range of frequencies, from approximately 20Hz to 20kHz, although this range diminishes as we age. There are three main types of information that allow the brain to localize sound. The first two are known as binaural cues because they make use of the fact that we have two ears, separated by the width of our head. A sound that emanates from either side of the mid-line will arrive first at the ear closest to it and will also be loudest at the ear closest to it. At low frequencies the brain recognises differences in the time of arrival of the sound between the ears (ITD), and at higher frequencies the salient cue is the loudness/intensity difference between the sound at each ear (IID). The use of these two types of cue is known as the 'duplex' theory and was proposed by Lord Rayleigh as long ago as 1877 (Fig. 1).

For single frequencies these cues are, however, spatially ambiguous. The inherent ambiguity has been described as the 'cone of confusion'. This arises from the fact that for any given frequency there are numerous spatial positions that generate identical timing/intensity differences, these can be graphically represented in the form of a cone, the apex of which is at the level of the external ear. The cone of confusion is the main reason for our not being able to localize pure tones [4,5].

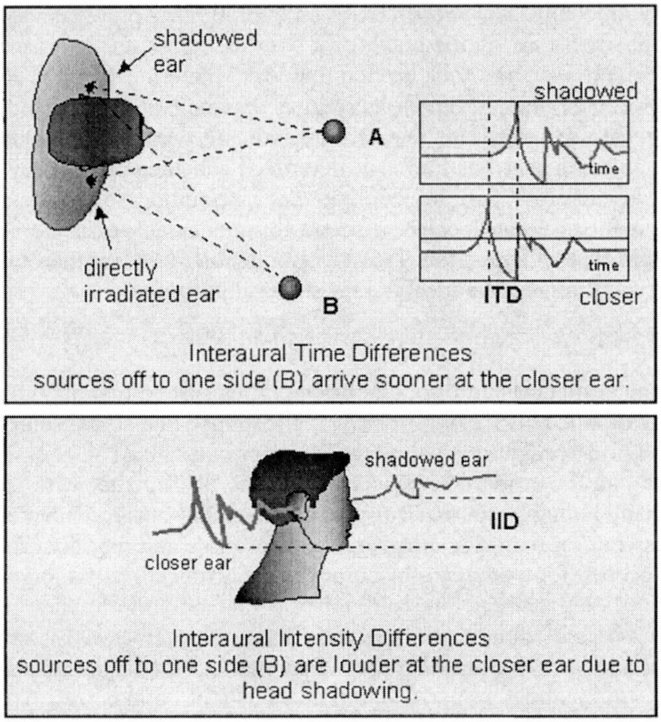

Figure 1: The localization cues of interaural time and intensity differences.

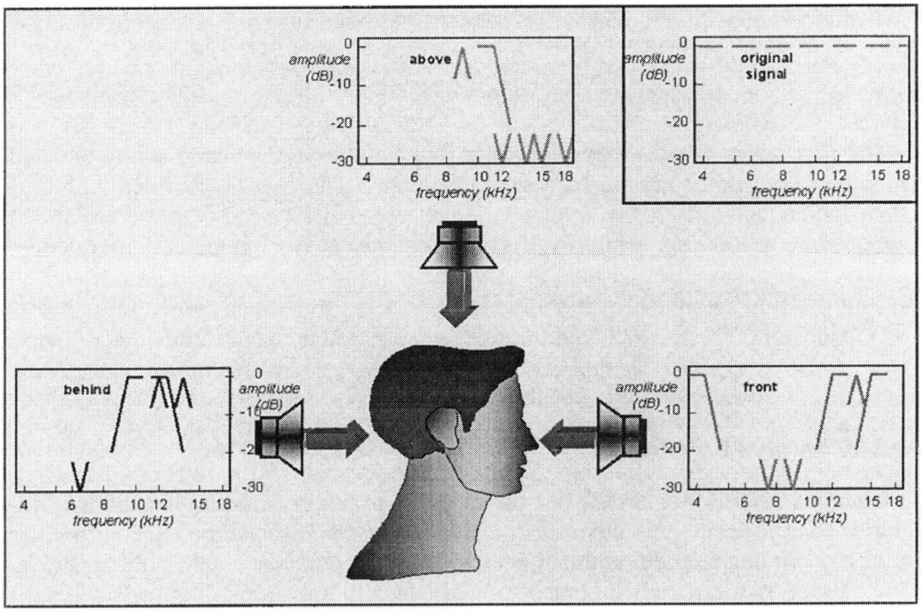

Figure 2: Examples of frequency dependent attenuation for sources in front, above and behind a listener.

The final main piece of information processed by the brain regarding sound localization is called the head-related transfer function (HRTF) [6]. The HRTF refers to the effect the external ear has on sound. As a result of passing over the bumps or convolutions of the pinna, the sound is modified so that some frequencies are attenuated and others are amplified (Fig. 2). Although there are certain generalities in the way the sound is modified by the pinnae, the HRTF of any one person is unique to that individual. The role of the HRTF is particularly important when we are trying to determine whether a sound is immediately in front of, or directly behind, us. In this instance the timing and intensity differences are negligible and there is consequently very little information available to the central nervous system on which to base a decision of 'in front' or 'behind'. So, to locate the direction of a sound source, the larger the frequency content, to overcome the ambiguities inherent to single tones, the better the accuracy.

3 Directional Sound and Evacuation

There are many examples of applications to which the addition of directional sound dramatically improves the safety of the device. One of the most important is the use of directional sound in evacuation guidance. Evacuation beacons, using directional sound, have applications in buildings, ferries, ships, aircraft and many other situations where emergency lighting is currently positioned. The beacons can be used in two ways. Firstly, as perimeter marking. Secondly, for complex routing in which the public must be guided from the centre of a labyrinthine structure. The use of directional sound evacuation beacons has been tested in a variety of situations.

3.1 Simple Scenarios

The first series of tests took place in a relatively large television studio situated in the confines of the campus of Leeds University. The studio was filled with artificial smoke and subjects were placed in the studio and filmed with a thermal imaging camera. Relying, primarily, on their memory of the immediate environment and on touch, it was found that an individual would take some 3 minutes and 50 seconds to find a conventional emergency exit sign. In contrast, when rapid bursts of broadband noise were played through an evacuation beacon immediately adjacent to the exit (which in this case was acting as a perimeter marker), the same individual traversed the wide open space taking 15 seconds to find the way out.

3.2 Complex Routing

Complex routing was tested in a different environment, a deserted school building. A complex route was devised which would test the evacuation beacons to their fullest, including many directional decision making points and also staircases (Fig. 3). Subjects in these trials included sighted, visually impaired individuals and children. Having completely filled the school with artificial smoke, each subject was taken to a starting point on a first floor location. They were taken to the start point via an external emergency escape staircase. By doing this the subjects had absolutely no idea of the route they were about to go through. In addition, they had no prior knowledge of the sound of the evacuation beacons. Once ready all beacons and the building's fire alarm were activated and each subject, or group of subjects, entered into the smoke. Essentially, the whole route was marked by only 4 egress beacons placed at strategic points (mainly above fire doors) on the way. At one point on the route, there was a small flight of stairs which led upwards to a mid-level in the building, and a beacon was designed that, as well as having rapidly pulsing broadband noise, also included an upwardly sweeping "melodic" complex which denoted to the subject "go up the stairs". At another point en route, there was the main staircase that descended to the final intended exit. Similar to the "up sweep" a "down sweep" was designed into this beacon giving the impression of "going down the stairs". As beacons progressed from the starting point of the experiment to the final exit, their pulse rate increased. This concept relies on human intuition with regards to faster events signaling nearing a final goal; the same concept that is used on rumble strips when approaching a roundabout.

Once again, the effectiveness of the beacons was unquestionable. None of the subjects in any of the trials took a wrong turning or ended up in any room that they were not supposed to enter. All subjects reported that the implementation of the melodic complexes denoting "up" or "down" information, informed them not only of the presence of a staircase but also of the intended direction of travel. As previously mentioned, they were not briefed as to the meaning of such tonal sweeps but intuitively understood the "associative meaning" within the sound. Finally, evacuation times were basically reduced close to total travel times that would have been expected under ideal visual conditions with prior knowledge of the building. Indeed, it was interesting to note subjects' responses when the building was eventually cleared of smoke and they were asked to progress through it once again. Without the aid of the navigation beacons and with full visibility, several subjects got lost en route. Even though they had been through it only a few minutes earlier.

From these studies, it is clear that the beacons proved themselves to be a crucial aid for all users under such visually impaired conditions. By providing directional information the beacons remove the need for having prior experience with the environment, reducing hesitancy and totally eliminating way-finding errors. Overall evacuation time was reduced substantially (by more than two thirds in many cases).

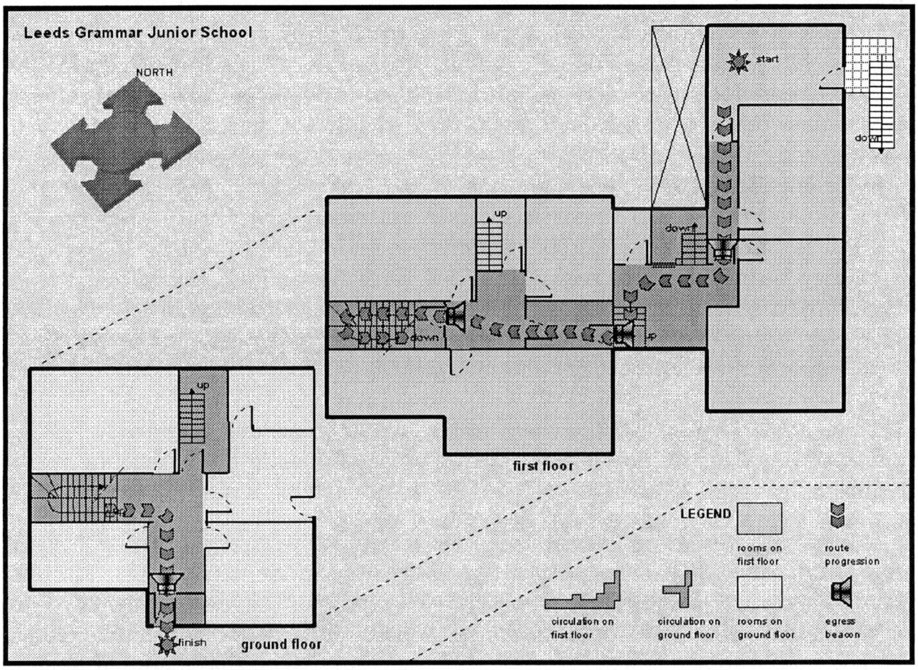

Figure 3: Plan of the school route.

3.3 Ferry Evacuation

Fire is one of nature's most uncontrollable forces, and to be caught in a fire situation is one of the most frightening experiences a person can face. However, although the fire itself is deadly, most casualties die from smoke inhalation which first blinds those trapped in its path, making it impossible to see the emergency exits, before being choked and suffocated. Another complex environment is that of a passenger ferry. We have further tested the evacuation beacons in just such a situation.

Modern ships and ferries are complex structures whose very design and layout can hinder speedy evacuation. Familiar thoroughfares may be hard enough to navigate in the unusual conditions of an emergency evacuation but add an unfamiliar emergency route to the situation and the difficulties are compounded. Even worse add smoke to the passageway, making it difficult or impossible to see, and the situation can rapidly escalate. Behavioural studies have shown repeatedly that one of the most natural instincts in the event of a fire is for people to evacuate a structure by the way in which it was entered. This is rarely by the quickest or most appropriate route, with many people failing to spot nearby exits and in some cases walking past visible fire exits. Ultimately, circulation routes that are used for normal, everyday movement become overcrowded, slowing down the evacuation process.

In April 1990 a fire raged through the ferry Scandinavian Star as it coursed through Norwegian waters on an overnight voyage from Oslo to Copenhagen. Of the 500 passengers and crew on board, 158 died including 29 children. Bodies were found below deck piled up in the corridors close to emergency exits, in cabins and even in the showers. Part of the investigation concentrated on the issue of why people died even though some were so close to emergency exits. Many survivors claimed that it was impossible to see the emergency exits in the corridors when these were full of smoke. In an attempt to understand what went so tragically wrong on board the Scandinavian Star, the Norwegian Fire Research Laboratory performed a series of evacuation trials on a reconstructed section of the Scandinavian Star. Using existing emergency signage provision, it was found that 40% (of their test subjects) could not find the emergency exit. They either passed it, or tried to get out through the wrong door, and some turned round on the way out. In their summary the Norwegian researchers stated "We do know that emergency lighting and marking signs do not help to distribute people among the evacuation routes available."

In January 2000 we were able to put the beacons to a rigorous test on board a ferry in dry dock [7]. Twenty volunteers were used in the experiments. The volunteers were led, by different routes, to two cabins in the centre corridor of an accommodation section comprising three parallel corridors. Each cabin was filled with people who had entered it using different routes. The ship's normal lighting remained on throughout the experiment. Both groups were told that there was going to be one safe route that they could follow to escape from the smoke that would shortly fill the corridors. Some of the potential exit routes would be "non-viable" (i.e. blocked) due to the location of the simulated fire. After filling the corridors with smoke the first group was let out of their cabin in pairs, at five seconds intervals. As you would expect, about half went each way - left and right. Those who turned left found a blind alley and a blocked exit and had to retrace their steps, potentially fatal in a real fire. Those that went right reached a T-junction and had to chose, left or right. Again, about half turned each way. The people turning right found a blocked exit and had to retrace their route. Eventually, due to the fact that it was theatrical, as opposed to real, smoke, all got out.

More smoke was laid down in the corridor before the second part of the experiment began. The evacuation beacons were then activated. The remaining group in the cabin were briefed that the beacons were located on the exit route and then released in the same way as before, i.e. pairs at five second intervals. All turned immediately right and then immediately left at the T-junction and were out in about one third of the time it had taken the first group. No one went the wrong way. There was considerable excitement amongst the volunteers afterwards with volunteers who did not have the beacons to guide them expressing how disorientating and confusing their experience had been. Conversely, those who had used the evacuation beacons made comments like "I couldn't believe how clear and obvious the sound beacons were - I had no doubt which way to go."

4 Behavioural Database

Before the directional sound beacons are used extensively throughout the world in different applications, it is important to be able to state how they function in different scenarios (such as unusual geometric configurations) and how they interact with existing emergency signage and evacuation procedures (eg fire marshals). Data of this nature will enable us to specify optimal configuration of the beacons and refine installation guidelines. A study is currently underway, in collaboration with Professor Galea (Greenwich University, U.K.), to obtain these data. The beacons were supplied by Sound Alert Technology plc. Data from pilot studies are given below.

4.1 Left or Right Test

Participants were led through a door into a room and then given the choice of turning to an exit located at the participant's right or left-hand side. Each exit was equipped with a Sound Alert beacon and a conventional emergency exit visual sign. The beacon and sign units were 9.2 meters apart and the participant entered the room through a door so that they were at equal distance to the exits i.e. 4.6 meters from each. The aim of the study was to direct the participants to a safe exit. The sound beacons were set at 93dB(A) at 1m.

A participant was invited to take part in 1 of 8 situations:

The first scenario was looking at how people respond with just visual exit signs as a guide.

1. No beacon, just lights – no information. In this situation they were told to evacuate as quickly as possible.

The next four scenarios were looking at the degree of training needed to follow the sound system. In the following trials there are two sound beacons and lights, which direct the participant to one of the safe exits, either left or right.

2. No information – In this situation they are told to evacuate as quickly as possible.

3. Information on the spot – In this situation they were told to evacuate as quickly as possible and were informed about the sound.

4. Trained – In this situation the participants were shown the layout of the room, where the sound beacons were placed, they heard the sound the beacons make and were told to find one of the safe exits.

5. Highly Trained – As number 4 with the addition of them already having been through the smoke before.

The following tests had one specific target exit. The participants were told to evacuate as quickly as possible.

6. Conflict – In this situation there was one visual exit sign at one side and one sound beacon at the other side. The participant's chosen exit was recorded.

7. No conflict – In this situation there was one target exit, i.e. beacon and sign unit at one of the exits.

8. Marshals – In this situation there was one target exit, the participant was directed by speech to one of the exits. They were told to 'come this way' to the target exit and told 'this way is dangerous' at the other exit. Both sound beacons were on.

The participants were randomly selected and were made up of staff and students from the University of Leeds. The majority of participants were aged between 18-27 years of age (although the age range was from <18 – 67 years of age, see Table 1). There were 40 participants in all, 19 were female and 21 were male.

Age	Frequency	Percentage
Under 18	2	5.0
18-27	19	47.5
28-37	10	25.0
38-47	4	10.0
48-57	4	10.0
58-67	1	2.5
Total	40	100.0

Table 1: Age range of participants in left/right test.

Those participants, who were told to find the exit by following the visual exit signs only, when in a smoke filled room, took as long as 105 seconds (average 67.8) to reach a safe exit. Whereas, participants who were given sound beacons and visual exit signs to follow and were trained on the sound of the beacons took as little as 7 seconds (average 8) to reach a safe exit.

Interestingly, participants who were just given visual exit signs to follow when in a room <u>without smoke</u> took as long as 12 seconds (average 8.8) to reach a safe exit. Those given sound beacons and visual exit signs to follow when in a room without smoke, and were also trained on the sound of the beacons, took as little as 6 seconds (average 7.2) to reach a safe exit.

4.2 Multiple Choice in Open Space Test

In this set of experiments the participants were led through a door into the centre of a room and then given the choice of turning to one of three exits which were located around the room (see Fig. 4 for details). Sound Alert beacon and visual sign units were placed at exits A & B, only an visual exit sign was placed at exit C. The participants were led into the centre of the room and were requested to find an exit from the choice of A, B or C. The beacon and sign units were around 4 meters from the participant. The aim of the study was to direct the participants to a safe exit and the sound beacons were set at 93 dB(A) at 1m.

A participant was invited to take part in 1 of 7 situations. The first four scenarios were looking at the degree of training needed to follow the sound system. In the following trials there are two sound beacon and lights situated at points A & B and at point C there is a visual exit sign. Initially the target exit was A or B then exit C became the target exit.

1. No information – In this situation they are told to evacuate as quickly as possible.

2. Information on the spot – In this situation they were told to evacuate as quickly as possible and were informed about the sound.

3. Trained – In this situation the participants were shown the layout of the room, where the sound beacons were placed, they heard the sound the beacons make and were told to find one of the safe exits.

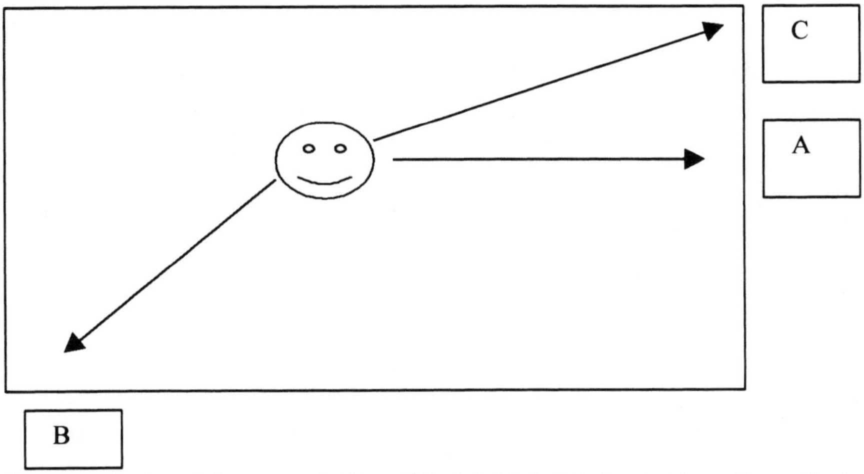

Figure 4: A pictorial representation of the Multiple Choice in Open Space Test.

4. Highly Trained – As number 4 with the addition of them already having been through the smoke before.

5. The following tests had one specific target exit. The participants were told to evacuate as quickly as possible.

6. Conflict – In this situation there was one visual exit sign at point A, and one sound beacon at point B. The participant's chosen exit was recorded.

7. No conflict – In this situation there was one target exit, i.e. beacon and sign unit at one of the exits.

8. Marshals – In this situation there was one target exit, the participant was directed by speech to one of the exits. They were told to 'come this way' to the target exit and told 'this way is dangerous' at the other exit. Both sound beacons were on.

Thirty participants were randomly selected from the University's staff and students. The majority of participants were aged between 18 and 27, 15 were male and 15 were female.

Participants, who were just told to find the exit, when in a smoke filled room, took as long as 22 seconds (average 14.5) to reach a safe exit. Whereas, participants who were given sound beacons and visual exit signs to follow, and were trained on the sound of the beacons, took as little as 7.3 seconds (average 11.3) to reach a safe exit. Participants, who were just told to find the exit, when in a room without smoke, took as long as 5 seconds (average 5.5) to reach a safe exit. Participants who were given sound beacons and visual exit signs to follow when in a room without smoke and were also trained on the sound of the beacons took as little as 4. 2 seconds (average 4.9) to reach a safe exit.

4.3 Complex Maze Test

In these experiments participants were led through a series of rooms to find a safe exit. The maze was equipped with Sound Alert beacon and visual signs to guide the participant to the end of the maze, i.e. the target exit. The aim of the study was to direct the participants to a safe exit. The sound beacons were set at 93 dB(A) at 1m. The participants were randomly selected and were made up of staff and students from the University of Leeds. The majority of participants were aged between 18-27 years of age. There were 21 participants in all, 9 were female and 12 were male.

In the complex maze those participants, who were only told to find the exit, when in a smoke filled room, took as long as 124 seconds (average 97.8) to reach a safe exit. In contrast, participants who were given sound beacons and visual exit signs to follow and were trained on the sound of the beacons took as little as 13.3 seconds (average 51.3) to reach a safe exit.

In the maze participants who were only told to find the exit, without smoke, took as long as 14 seconds (average 9.7) to reach a safe exit. Whereas, participants who were given sound beacons and visual exit signs to follow, took as little as 7 seconds (average 10.86) to reach a safe exit.

One of the interesting facts that stems from the pilot trials is that using sound and lights together, when there is no smoke, actually results in faster evacuation than just using lights alone. There is a neurophysiological explanation for this. Research has shown that the area of the brain that responds to spatial sensory information (the superior colliculus), which also initiates the response to the sensory stimulus, contains cells that are tuned to more than one sensory modality. These neurones respond to light by itself and also to sound alone. However, when light and sound are presented together the response of these cells is far greater than the summation of the response to either modality alone [2]. Some cells respond over 1,000% better when light and sound are used together than the response to either modality alone. In other words, to activate an individual to 'flight', a stimulus containing both light and sound will be far more effective than light or sound alone.

5 Aircraft Evacuation

The ability to unload passengers is an essential requirement in the event of an emergency. It is surprising that this vital act relies solely on one sensory modality, namely vision. If the evacuation is required as a result of a fire, which is likely to generate smoke, the sense that is first affected is sight, rendering the exit lights above the doors and the aisle strip lighting invisible. The layout of aircraft, with fixed forward facing seats, results in the last visual memory a passenger has is of the exit sign in front of him/her. They may have an exit one or two rows behind and much nearer than the one visually cued ahead of them. The use of combined light/sound exit markers would ensure that the passengers' attention is drawn to the nearest exit (as it would sound louder) allowing more structured evacuation using all viable exit routes. Sounders could be inactivated on non-usable exit routes.

Recent tests on aircraft have shown using directional sound at exit locations enhances passenger evacuation. The tests have also answered some important questions in the use of sound in this unique environment. Exits from aircraft are relatively close together compared, for example, with a building and one question that had been asked was "would the use of sound beacons at each exit cause confusion for the passenger?" The tests showed this not to be the case. The listener can be aware of up to four beacons but still able to identify which is the loudest and, therefore, that denoting their nearest exit. Another important question was "would the cabin crew, giving instructions, still be heard when the beacons were utilised?" Again there was a positive answer with instructions from the cabin crew clearly audible when the beacons were in use.

There is always the question of how will people know what the sound means? Aircraft with the pre-flight safety briefing provide an ideal environment in which the public can be informed of the role of the sound. The great advantage of the broadband sound is that it is language independent and will be equally effective for British, Japanese, Russian or any nationality on board the aircraft.

6 Characteristics of the Sound Beacons

It must be made clear that the sound beacons are NOT intended to replace traditional fire alarm sounders. In the case of a fire on board a ferry, for example, the fire alarm will still sound. Indeed, in all the building and ferry testing we have undertaken the fire alarm has been used . The fire alarm plays an important role in alerting people to the potential threat, whether it is a fire or other problem. Once someone has made the decision to evacuate, the fire alarm ceases to fulfil a function, the next most important decision for anyone leaving a building is where to find an exit. The fire alarm sound itself would be entirely unsuitable for use over an exit. It is a narrowband sound and thus not possible to localize. We are capable of listening to many different sounds simultaneously and make the decision which one to attend to based on an analysis of their importance.

The sound beacons are currently available from a licensee in two forms. Either as a stand-alone beacon or as a combined emergency exit light/ sounder unit. The beacons have 2 settings; 93 & 99 dBA at 1m and come with a range of pulse pattern options (including "up" and "down" sounds). Installation of the beacons is based on a process similar to that involved in identifying the location for exit signs. The same guidelines as for the installation of fire alarms can be applied to the broadband sound from the directional evacuation beacons. That is, there is a 20 dB attenuation through normal doors, 30 dB attenuation through fire doors and the sound reduces by 3 dB for every doubling of distance. Interestingly, our studies on the perception of broadband noise show that it is heard at far lower levels than a pure tone (up to 15 dBA difference), thus enhancing its efficacy.

7 Other Applications of Directional Sound

It is worth taking this opportunity to mention life saving applications of directional sound other than in evacuation beacons. Any application in which sound is utilised, and would be enhanced by a knowledge of its location, is a potential candidate for the use of directional sound. There are, however, two very obvious examples; distress signal units (or man down alarms) for firefighters and emergency vehicle sirens.

7.1 Firefighter Distress Signal Units

For some time, CACFOA's (Chief and Assistant Chief Fire Officers Association) National Breathing Apparatus Group had, along with colleagues throughout the Fire Service both in the U.K. and abroad, been concerned at the effectiveness of the existing Automatic Distress Signal Units (ADSU). The existing ADSUs, which have been changed very little over the last 20 years in terms of their function, have a crucial deficiency for such a potentially life-saving piece of equipment. Current models emit a single tone, relatively high frequency signal of approximately 3kHz. These sounds are by no means ideal in terms of perceived urgency and, more importantly, are impossible to localize. In other words, if a firefighter can hear a DSU sounding at an incident then he/she will find it very difficult indeed to identify from which location the sound is coming. This problem could mean the loss of vital seconds for firefighters seeking to locate a colleague under emergency conditions.

In response to this concern, CACFOA and the Home Office initiated an investigation into the potential for an increase in performance that would enhance the localization performance of ADSUs. The investigation was carried out by a team of researchers from the University of Leeds in conjunction with Sound Alert Ltd.

The aim of the field trials was to investigate the increase in localization performance that an ADSU with broadband noise plus the traditional alerting signal would have over the existing units that are currently in use by Fire Services in the United Kingdom. Initial data had proven that, within laboratory-based experiments, an ADSU with broadband noise offers more localization (and detection) acuity, however, it was not known whether such a performance enhancement would be noticeable in the field. Therefore, field trials of both a modified and an existing ADSU were carried out at the specialised facilities of the West Yorkshire Fire Service Training Centre.

In order to evaluate the two different ADSUs many different search scenarios were enacted using various types of protective headgear, flame hoods and breathing apparatus. Additionally, in order to overcome any experimental bias, three subjects were used who had different levels of familiarity with the building used for the search trials.

Within the grounds of the Training Centre there is a multi-purpose Fire Training Building comprising three storeys (ground, first and second floors). A survey representation of this building would depict a spatial layout that is highly labyrinthine. Additionally, the external elevation projections of this building give no idea as to what the internal layout may be.

For the trials, three subjects were used; two experienced fire fighters and one non-operational subject. Subject 1 had some spatial knowledge of the building but his knowledge of the internal layout was more relational as opposed to detailed. Subject 2 on the other hand, had complete spatial knowledge of the building. Subject 3 had no prior knowledge of the interior layout of the building.

In brief, the subject was set the task of finding each ADSU, which was placed on any one of the three floors of the building. To simulate "real" conditions, the building was plunged into total darkness, all subjects having to rely on their spatial knowledge of the building, the torch beams and most importantly the sound patterns emitted from each ADSU. At no time did any of the subjects know where the ADSUs were located. The experiment proceeded as follows. Using the building plans, the location of the Sound Alert Localizer ADSU (SAL) and a traditional ADSU were decided using two separate criteria. Firstly, each ADSU was placed roughly equidistant from the entrance; i.e. it would not take longer to get to one than the other. Secondly, the route progression from the entrance to the ADSUs activated had to be similar, i.e. the same number of doors to get to them, similar number of turns. Having established the locations of the ADSUs, one researcher (the timekeeper) assumed his location within the building where the ADSU would be activated. Carrying a stopwatch, he signaled to another researcher placed outside the building (via radio communication) who would set each subject off on his search and subsequently record the time it took to find the researcher. Having found, for example, the old ADSU, the subject would return to the outside of the building whilst the researcher inside the building would assume the location of the Sound Alert Localizer ADSU. The trial was repeated several times so that an accurate picture could be gained as to the performance of the Sound Alert Localizer over the old ADSU.

To test each ADSU fully, a series of criteria were established and subsequently evaluated.

- Seek times for ADSUs. This was the primary aim of these field trials, to find out how long it would take to find each ADSU.
- Seek times for ADSUs with different headgear. It was thought that the many different types of headgear that can be used by fire fighters would influence adversely the localization ability of both ADSU types.
- Seek times for ADSUs with full breathing apparatus. Given the amount of noise associated with breathing apparatus (with the additionally limited field of view), it was thought that there would be a comparable difference in seek times with and without its use.
- Seek times for multiple, simultaneously activated ADSUs. In this series of trials, two distress signal units were activated simultaneously. There always exists the possibility that two, or more firefighters may "go down" in a fire situation and that the detectability of both ADSUs would be fundamental for their survival.

In all trials, the SAL unit was located significantly faster than the old ADSU.

The SAL unit was located up to **4.625 times** faster than the traditional sounding ADSU.

When SAL and old ADSUs were activated simultaneously, the Sound Alert unit was always found first. Additionally, these trials showed that neither unit interfered with the other.

Familiarity with the building layout had little overall causative effect on determining the location of either sounding distress signal units. Although subjects with a lesser spatial knowledge of the building did take longer to find an ADSU, their performance in finding the Sound Alert ADSU was still significantly faster than the old ADSU.

Different types of helmet made no difference to the overall seek times.

The presence of breathing apparatus did make a significant difference to the overall detection times of both the SAL and old units (taking up to 1.44x longer with BA), however, the SAL units could be found up to 2.44x faster than the old units.

A senior Fire Officer commented on the trials that "the whole of the fire community will, I am sure, welcome this vital work as a major step forward in firefighter safety. I am convinced that this research proves beyond doubt that, by modifying the sound emitted from ADSUs, it will be possible to achieve a significant reduction in the time taken to find a stricken firefighter. The retention of a traditional alert sound, overlaid with bursts of broadband noise, maintains the crucial element of familiarity, so adjustment to the new sound should not be a major obstacle for fire personnel" [8].

7.2 Emergency Vehicle Sirens

An everyday occurrence for the majority of drivers is the sound of an emergency vehicle siren, whether from an ambulance, police car or fire appliance. When the emergency siren is heard, drivers look all around, trying to determine from which direction the sounds are coming. The visual cue is required because the sound alone gives no clue as to which direction the vehicle is coming from. The driver is not able to take appropriate avoiding action until the emergency vehicle is seen, often too late to allow a clear path to be created for the emergency vehicle. The uncertainty regarding the direction of approach costs lives. Any improvement in the sound quality of the siren, which enables road users to take earlier evading action, would both reduce the journey time and enhance the safety for emergency vehicles attending emergencies and thus strengthen the service provided. It would also be safer for road users, pedestrians and drivers alike.

Difficulties in determining the direction from which emergency vehicle sirens are approaching are widely acknowledged. In fact the emergency vehicle siren has been described as 'an extremely limited audible warning device' [9]. A recent study in the Annals of Emergency Medicine has shown that an ambulance is most susceptible to collisions with other vehicles when crossing road junctions. This happens, primarily, because the drivers of the cars or trucks are unable to determine accurately the direction of the approaching ambulance. In one year, in the USA alone, 537 injuries and 62 deaths arose from accidents involving ambulances [10].

So why do they not work? Simply, because the frequency content of the siren sounds is so poor. Typically emergency vehicle siren sounds are emitted over the frequency range 500Hz-1.8kHz, far too narrow a frequency range for localization. Although the anecdotal evidence is plentiful, there has not been, until recently, a thorough study into the localizability of emergency vehicle sirens. The lack of research has been due, primarily, to the difficult nature of tackling such a concept in a meaningful way. It would be of little value, for example, to test siren sounds by asking people seated in a room, or in the open, where they thought the sound originated. The 'realism element' was overcome in a recent study at Leeds University in which participants 'drove' in a simulator [11,12].

Approximately 200 participants (age range 19 to 57 years) were tested. They all had previous driving experience. A hearing test, in the form of an audiogram, which tested the ability to hear a range of frequencies, was performed on most participants. Eight loudspeakers were positioned at 45 degree intervals around the car's azimuthal plane. The speakers were not visible to the participants, and the sound from each speaker was intensity and spectrally matched at a position that marked the driver's head within the car. Matching each speaker in the way described resulted in the only variable each time the sound was played being that of direction. A response panel was mounted by the steering wheel in the car. There were 16 positions marked on the panel that represented the horizontal plane around the car. This allowed greater accuracy in determining the errors made by the participants than just having eight response buttons that corresponded directly to the number of speakers. On hearing a siren sound noise, participant drivers pressed the response button that they judged as equating to the location from which the sound originated. The sound source location was varied randomly, with the constraint that in each session all eight speaker positions were activated twice. The participants began trials after a period of familiarization with the car (a 10-minute 'drive' along a preset rural route). During the trials the participants were asked to maintain a speed of 40 mph which the experimenter monitored continuously.

Four existing sirens were tested. These were the 'hilo' siren characterized by a two-tone sound (670-1100Hz, 55 cycles/min); the 'Pulsar', a pulsating sound (500-1800Hz, 70 cycles/min); the 'wail', a continuous sound rising and falling (500-1800Hz, 11 cycles/min) and the 'yelp', a continuous and fast warbling sound (500-1800Hz, 55 cycles/min). In each trial the same siren sound was used. The siren used was delivered at or around road junctions on the test track.

Data were gathered showing conclusively the poor localization characteristics of existing sirens, though some were better than others; for example, the traditional 'hilo' siren was significantly worse for localization than the 'yelp'. Nevertheless even the best of the current sirens was associated with exceedingly poor front/back accuracy: participants had difficulty identifying whether the sound emanated from the front or rear of the car, to the extent that they were wrong more often than not. It is worth noting that, if participants just guessed front or back, in theory they would have been correct 50 per cent of the time.

In the aforementioned study, in addition to testing the existing siren sounds, a range of new sounds optimised for both alerting [13,14] and localization were also tested. The data obtained from the simulator studies using the new siren sounds, characterized by pulses of rapidly rising frequency sweeps followed by a burst of broadband noise, indicated a much better localization than with the old sirens. With the front/back decision, participants got it right 82 per cent of the time (whereas, with the old siren, front/back was incorrectly judged 56 per cent of the time), and left/right accuracy was increased to 97 per cent (from 79 per cent).

Following the simulator trials it was necessary to test the new siren sounds in real emergency situations. The new sounds were used in trials with ambulances in

West Yorkshire and London, fire appliances in Leicestershire and a police pursuit car in West Yorkshire.

The road trials produced data that complemented the simulator trials [15]. Using an on-board video camera, which recorded the view of the road ahead of the emergency vehicle, road user responses to both the old siren and the new sirens were compared. As a result of discussions with the drivers of the emergency vehicles, various road user behaviours were searched for, such as other road users failing to indicate their proposed direction of manoeuvre. The data showed that with the new siren ambulance lane changes were reduced threefold, so making journeys smoother and faster. Instant recognition of the direction of the ambulance's approach increased by nearly 25 per cent, which meant that other drivers cooperated more quickly and more effectively with the ambulance's passage. Clearer, well-signaled manoeuvres by road users were more achievable and increased by 17 per cent. Finally journey times were reduced by up to 10 per cent. This reduction in journey time was not due to the fact that the emergency vehicle was travelling faster and, often, owing to the natural caution of driving with something new, their speed was actually less. It was due, rather, to the fact that the journey was made easier by road users reacting more quickly, and more appropriately, to the new siren, allowing a clearer path through the traffic for the emergency vehicle.

It seems, therefore, that the two crucial requirements of a siren sound, to alert road users to the presence of an emergency vehicle and to inform the listeners of the direction of approach, can be met by utilising our scientific knowledge of the best types of sounds to fulfil these criteria, resulting in a more effective sound signal.

8 Conclusions

There is a huge versatility in the number of applications that would benefit from the use of directional sound. Undoubtedly the most striking of these is the use in evacuation beacons. As researchers and practitioners in the fire protection industry, it is our duty to provide as many options as possible to all potential occupants of a building, ferry, aircraft etc., using either active or passive means. Although provision for emergency evacuation is getting better, there is still a problem with solely addressing the 'visual' needs of occupants when it is only too likely that these visual means will be made redundant in the event of a fire. Only through a combination of both visual and auditory way-finding aids will optimal safety be achieved.

References

1. J.C. Makous and J.C. Middlebrooks, *Two-dimensional sound localization by human listeners*, Journal of the Acoustical Society of America **87**, 2188-200 (1990).

2. B.E. Stein and M.A. Meredith, *The Merging of the Senses*, (Cambridge, MA: MIT Press, 1993).

3. E.I. Knudsen, S. du Lac, and S.D. Esterley, *Computational maps in the brain*, Annual Review of Neuroscience **10**, 41-65 (1987).

4. J. Blauert, *Spatial Hearing*, (Cambridge, MA: MIT Press, 1997).

5. F.L. Wightman and D.J. Kistler, *Sound Localization*, in: *Human Psychophysics*, W.A. Yost, A.N. Popper, and R.R. Fay, (Eds.), (New York, Springer-Verlag, 1993).

6. S. Carlile and A.J. King, *Auditory Neuroscience: from outer ear to virtual space*, Current Biology **3**, 446-8 (1993).

7. D.J. Withington, *Faster evacuation from ferries with sound beacons*, Fire **30,** (2000).

8. D.J. Withington and S. Smith, *'White noise' can help save stricken firefighters*, Fire **27,** (2000).

9. R.A. De Lorenzo and M.A. Eilers, *Lights and Siren: A review of emergency vehicle warning systems*, Annals of Emergency Medicine **20**, 1331-5 (1991).

10. C.R. Hunt, L.H. Brown, E.S. Cabinum, T.W. Whitely, N.H. Prasad, C.F. Owens, and C.E. Mayo, *Is ambulance transport time with lights and siren faster than without?,* Annals of Emergency Medicine **20**, 507-11(1995).

11. D.J. Withington, The quest for better ambulance sirens, Ambulance UK **11**, 20-21 (1996).

12. D.J. Withington and A.C. Chapman, *Where's that siren?*, Science and Public Affairs, **Summer**, 59-61 (1996).

13. R.D. Patterson, *Guidelines for Auditory Warning Systems on Civil Aircraft*, CAA paper 82017, London: Civil Aviation Authority, (1982).

14. E.C. Haas and J. Edworthy, *Designing urgency into auditory warnings using pitch, speed and loudness'*, Computing and Control Engineering Journal, **August**, 193-8 (1996).

15. D.J. Withington and S.E. Paterson, *Safer Sirens*, Fire Engineers Journal **48**, 6-10 (1998).

Assessment and Analysis of Evacuation Processes on Passenger Ships by Microscopic Simulation

T. Meyer-König[1], H. Klüpfel[2], and M. Schreckenberg[2]

[1]Institut für Schiffstechnik, Gerhard-Mercator-Universität, Duisburg, Germany

[2]Physik von Transport und Verkehr, Gerhard-Mercator-Universität, Duisburg, Germany

The analysis of evacuation processes onboard passenger ships has attracted increasing interest over the last years. Cellular automaton models, widely used for traffic simulations, provide a natural approach towards pedestrian dynamics. Two major difficulties are intrinsical to the problem: two-dimensional movement of pedestrians and the complexity of psychological and social influences. An improved CA-model which allows complex routechoice-behaviour for the description of crowd motion is presented and its implementation in a simulation software outlined.

1 Introduction

Tragic ferry disasters in the past have shown the need for improved safety-standards in the shipping industry. Because of the huge numbers of passengers carried on ships at the same time and the unforgivable environment they mostly sail in, small accidents can quickly result in big casualty numbers. An effective evacuation-procedure should therefore be mandatory. But what will happen, if two thousand or even more people move through a ship, try to gather their belongings or related persons? Where do the bottlenecks occur? How can congestions be prevented effectively? There are many questions that arise, when the evacuation-concept of a ship has to be evaluated and descriptive calculation methods are not able to handle the complexity of the problem in a holistic approach.

In this article, an overview of the work that is done within the *BYPASS*-project is given. It was founded in October 1999 and is executed by the research group *Physics of Transport and Traffic* of Prof. Michael Schreckenberg at the Gerhard Mercator University in Duisburg/Germany. The aim of *BYPASS* is to further develop the Cellular Automaton model formerly developed in Tim Meyer-Koenig's student research project and to support the IMO (International Maritime Organization) in establishing guidelines for the use of microscopic simulations for the safety-assessment of passenger ships [1]. Partners are the German ministry of transport (*BMVBW*), the *Germanischer Lloyd* and the shipyards *Meyer-Werft* and *Flensburger Schiffbau Gesellschaft*.

2 Different Ships Require Different Evacuation Procedures

The variety of vessels seems to be nearly endless. For classification they are therefore divided into the following three groups: High Speed Passenger Craft (HSC), RoPax- and Cruiseships. The exact definition for each type can be found in the appropriate IMO regulations [2].

Figure 1: Pictures of a typical HSC, a RoPax-Ferry and a cruiseliner (pictures: Tim Meyer-Koenig, Jos. L. Meyer shipyard).

Depending on the group that a ship belongs to, the type of the evacuation procedure varies. Because the general layout of a HSC is fairly simple and the passengers are seated in rows (just like in an airplane), it is unlikely that they will be spread all over the vessel in an emergency situation. The evacuation strategy is therefore very straightforward: After the liferafts have been deployed, the passengers will take their lifevests from beneath their seats and leave the ship via the slides.

The procedure is totally different with big RoRo- and cruiseships, because people can have cabins and are able to linger on many public places all over the ship. In case of an evacuation, they have to retrieve their vests in the cabins or at central distribution areas, gather at muster stations and walk on to the embarkation stations were they finally leave the ship. Everybody can imagine what will happen on a cruiseliner where one half of the passengers is seated in the restaurant and the other half sits in the theatre and the evacuation signal is given. Unfortunately imagination is not sufficient for prediction of the possible result.

3 Model Assumptions

The goal was to develop a model that is capable of simulating the various evacuation procedures that can occur in buildings and onboard ships, aircraft, or other kinds of public transportation systems. The basic idea was to simulate the behaviour and movement of every single person involved, which is called a microscopic approach. To increase simulation speeds, a Cellular Automata model was chosen with a minimum set of parameters per person to characterise the behaviour:

- the maximum walking speed,
- the patience before she chooses an alternative goal, if the way is blocked by congestions,
- a factor that characterizes the visual perception of the environment,

- the delay time, during which she will wait before starting to evacuate,
- a dawdle probability with which she will stop for one timestep, to simulate breaks for regeneration or orientation, and
- a factor that characterises the inertia of the person's movement.

The reduction of the parameters to this list is possible, because these are the factors that generally characterize the movement from a physical point of view. Attributes like age, gender and stamina eventually take effect on e.g. the walking speed. As a result of these simplifications, the simulations can be done on standard computers with very high calculation speeds (10.000 persons in real-time on a 500 MHz PC with 128 MB RAM). It is therefore possible to predict evacuations in a very short time which will allow online-simulations in near future. By varying the parameters, the simulation can be adapted to any demographic values.

The floorplan of the structure to be assessed is transformed into a grid of quadratic cells with an edge-length of 0.4 meters. One person thereby occupies one cell. This derives from the maximum density in crowds, when the movement reaches a stop [3].

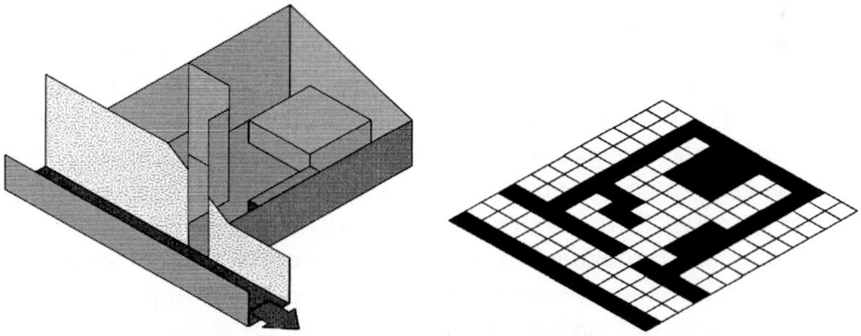

Figure 2: Example for the discretization of the floorplan.

Each cell contains different informations, depending on the way it influences the person standing on it. If it is not accessible, it represents an obstacle like a wall or furniture. Others influence the speed of the persons walking over them, so stairs and doors can be considered. To change the decks (e.g. at stairs), cells are implemented, that cause the person standing on them, to jump to the next upper or lower level. In this manner, the floorplan of complex structures can be modelled in a very simple way.

Depending on her maximum walking speed and the surrounding obstacles, a person moves from one cell to the next adjacent (via the corners or the edges). She thereby tries to avoid others and to maintain her maximum walking speed. Because the persons can move through more than one cell per time-step (one second), a parallel update was rejected, and an improved random-sequential-update was chosen. However for higher speeds the dynamics of the pedestrians converge towards a parallel update. Jamwaves can clearly bee seen in corridors with periodic boundary conditions. The basic procedure is shown in Figure 3:

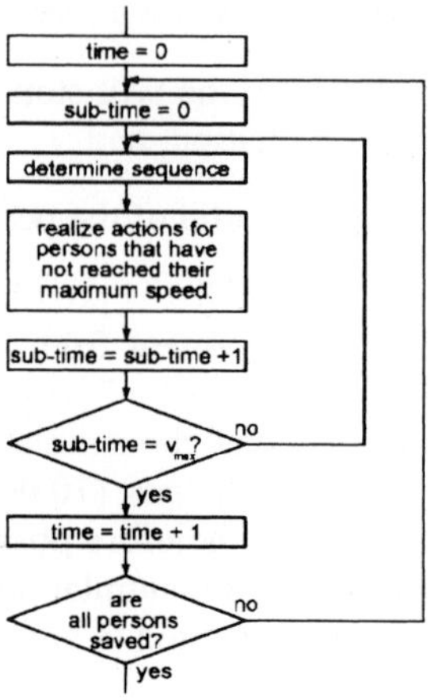

Figure 3: The basic update-algorithm.

4 Orientation

The problem concerning the orientation of a person is the global route-choice. It is easily taken into account that a person evades obstacles and other pedestrians, but the way she chooses where to go and how to get there is the decisive problem. In a first step, the information of which way a person had to walk to reach the exit was implemented by so to speak arrows in the cells that pointed into the direction a person had to walk. In the currently developed new approach, the route-information is given by a potential. This means, that only the goal-cells have to be marked. The potential then automatically spreads from one accessible cell to the following and by this throughout the whole structure. By a smoothening algorithm it adapts to the architectural circumstances, so persons following it will not stick to one side of a corridor, just because they want to bend off at the end of it (Figure 4).

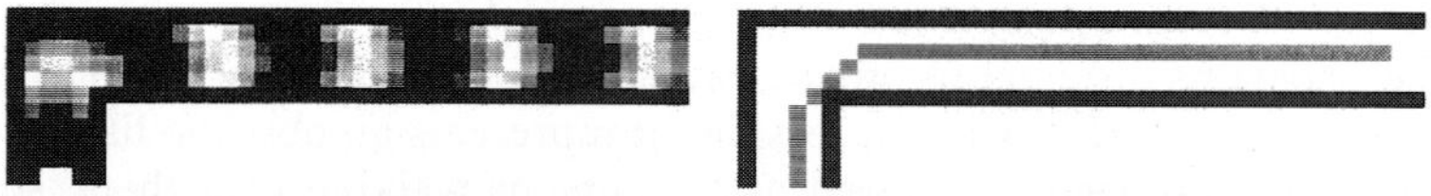

Figure 4: Iso-potential-lines in a hooked corridor, spreading from the bottom-left (left) and a resulting possible path (right).

Because the evacuation-route of a person in a RoPax- or cruiseship can contain many different waypoints (e.g. cabins, musterstations, lifeboats), three different kinds of potentials where developed. The only difference between them lies in the behaviour of the person who reaches the source-cells. Every potential has its own list of possible following goals, out of which a person reaching it can choose.

Cabin-potentials are used for marking corridors. A person reaching these kind of cells will stay on them for a given individual time, before choosing a new goal out of the list. This resembles the behaviour of passengers reaching their cabin and getting their lifevest. On the other hand, cabin-potentials can be used as simple waypoints, when the duration of stay is set to zero, so evacuation-routes can be marked by them.

Reaching a mustering potential, the behaviour of a person will change into a random-walker. This resembles passengers, who are gathering at the assemblystation. However to allow the following people to step onto the musterstation, they do not just stop walking, but randomly move on, staying on the appropriate cells. After a given blocking-time has passed, the people on the musterstation-cells are forwarded to the following goals (mostly the exits), while the new arrivals have to start waiting. This process is repeated in given interval-times, after the first blocking-time has passed. The result of this behaviour is equal to the mustering process: As long as the lifeboats are being prepared, the passengers will be assembled and afterwards led on towards the embarkation stations.

The procedure of embarking into the lifeboats is simulated by the rescue-potentials. They are blocked for a given time (preparing the lifeboats) and after that, the people reaching them have to wait for a certain egress-time before they get rescued. In this egress-time, the delay for entering a lifeboat or sliding on a marine evacuation system (MES) are taken into account. Throughout the simulation, the people which have been rescued per exit are counted and if the maximum capacity is reached, the belonging potential is switched of. Through this, the remaining passengers have to divided towards the other exits.

By dividing passengers into groups (e.g. having cabins onto one corridor) and using various of the mentioned potentials to mark their possible goals, various routechoices become possible, and thus the situation of a big ropax- or cruiseship getting evacuated can be matched (Figure 5).

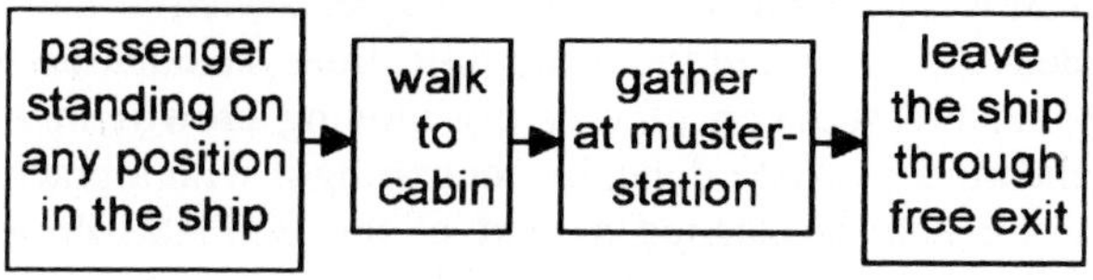

Figure 5: The possible actions of one arbitrary passenger.

5 Characteristics of the Model

Decisive for the closeness to reality of a model for simulating pedestrians are the flow-density-relations it provides. Once again, a simple method was wanted, for simulating the growth of needed space of a person moving at higher speeds. The simulations where done in a corridor (200x20 cells) with different densities of people walking in the same direction and periodic boundary conditions. Three cases where investigated: a person occupies all the cells she passes in one timestep, half the cells and finally only the cell she currently stands on. In the last case, the occurring jam-waves where most remarkable, moving against the flow and growing more and more stable with the density growing (Figure 6).

Figure 6: Corridor with periodic boundary conditions and a density of 4,4 P/m². People are moving from the left to the right and occupying only one cell. Jam-waves are clearly visible (dark areas).

The resulting flow-density-relations were compared with empirical data (4) and it showed, that the results met reality very well, if one person occupied all the cells she passed in one timestep (Figure 7).

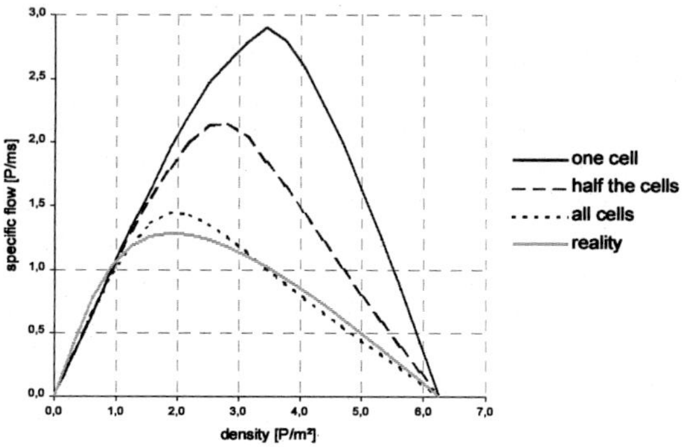

Figure 7: Flow-density-relations of the model.

6 Conclusion

It has shown that it is possible, to simulate complex scenarios and behaviour with elementary models. As a result of these simplifications, calculation-speeds are very high and allow the investigation of very big moving crowds in a short time. A verification of the model will be done, by comparing it with a practical evacuation trial on a cruiseship that is planned as part of the BYPASS-Project in the near future.

References

1. International Maritime Organization, *FP44/19, Report to the Maritime Safety Committee*, Sub-Committee on Fire Protection.

2. International Maritime Organization, *SOLAS (Safety of Life at Sea)*, (Bath Press, UK).

3. D. Helbing, *Verkehrsdynamik, Neue physikalische Modellierungskonzepte*, (Springer Verlag, Berlin 1996).

4. U. Weidmann, *Transporttechnik der Fußgänger*, Transporttechnische Eigenschaften des Fußgängerverkehrs Literaturauswertung, Schriftenreihe des IVT **90**, (ETH Zürich, January 1992).

Simulating Evacuation Processes with ASERI

V. Schneider and R. Könnecke

I.S.T. Integrierte Sicherheits-Technik GmbH, 60325 Frankfurt / Main, Feuerbachstr. 19, Germany

In recent years, the individual-based evacuation model ASERI is increasingly used to perform numerical simulations of egress movement in complex buildings like railway and underground stations, airports, theatres, sport arenas or trade fairs. Observations of passenger movement in sport arenas and large underground railway stations as well as monitored evacuation drills are used to validate the computer program. The basic features of the ASERI computer program and the underlying assumptions with respect to human behaviour modelling are presented, followed by a discussion of case studies, including remarks on code validation, the reconstruction of actual hazards and examples of the use of ASERI to simulate the individual movement of passengers in complex environment. The importance of a careful choice of individual parameters describing the mobility of the occupants is demonstrated.

1 Introduction

Due to the increasing significance of performance-based concepts in fire protection engineering, numerical simulation tools predicting the spread of fire and smoke and the egress movement of occupants are developed and used in the design of fire safety concepts. Furthermore, there is a tendency to replace hydraulic flow type calculation methods by evacuation models simulating the egress movement of individuals, specially if large and architectural complex buildings and facilities and the impact of fire effluents are to be examined. The computer simulation tool ASERI is one of those individual based evacuation models, developed about 10 years ago and now used in fire safety engineering on a regular basis. After describing the basic structure and features of ASERI, case studies are presented. Criteria to be used in the evaluation of design concepts are discussed with special emphasis on the interaction of computational fire modelling and evacuation simulation.

2 The ASERI Simulation Model

ASERI is a computer model designed to simulate the egress movement of people through geometrically complex spaces of arbitrary size. Each occupant is treated as an individual person, moving inside a building, a confined space or any other type of enclosure (e.g. mass transport vehicles). The individual egress movement is governed by certain behavioural aspects that are triggered by external stimuli and restrictions, especially those caused by the presence and simultaneous movement of other occupants. Individual decisions and related actions may contribute to a delay in starting the evacuation or cause interrupts, especially in the initial phase of the evacuation process. Furthermore, egress route choice is strongly influenced by individual aspects like knowledge of the building layout or smoke tolerance. Certain features of behavioural response are modelled in a probabilistic rather than a deterministic way. This probabilistic approach allows for a more profound evaluation of the evacuation process by performing Monte-Carlo simulations. A number of replicate runs with identical boundary conditions are performed and statistically analysed, yielding not only mean values of egress times but also standard deviations and confidence limits. Furthermore, visualisation of the movement of the evacuees and corresponding dynamic graphical information (including the generation of AVI video sequences) on the population and crowd density in sensitive areas contributes to a better understanding of the mechanisms individuals interact with each other and with the physical environment, including fire spread and smoke movement.

The geometrical scenario (building) is defined in a hierarchical way. A building is composed by a number of levels or stories, connected by stairways or ramps. Each level is subdivided into a number of rooms and corridors. Rooms, corridors, stairs and safe areas are the basic geometrical units. These units are defined by the respective ground-plan and by size and position of doors and passages. Inside the units obstacles of arbitrary size and position can be specified. The possible destinations of the occupants are safe areas located either outside the building (thus associated with exits) or inside, in this case representing regions that (temporarily) give shelter. The number of occupants, levels, units, passages and obstacles is only limited by available computer memory. It is therefore possible to model very large and geometrically complex buildings with several thousands of people simultaneously moving.

Body size is defined by shoulder and chest width, according to the well-known concept of body ellipse modelling the horizontal projection of a human body. Geometrically, the body ellipse is represented by three overlapping circles. Furthermore, a minimum inter-person distance and the maintenance of a boundary layer clearance from walls and stationary obstacles is implemented. Shoulder and chest width can be assigned individually, either by explicit input or by specifying a distribution function appropriate for the respective population. By introducing effective size parameters, this concept allows for the definition of persons with increased space requirement, including persons with limited mobility, occupants with luggage or adults accompanied by smaller children.

Occupant movement is defined by the individual walking speed and the orientation of the corresponding velocity vector. The orientation is derived from the local position of the person and the respective individual goal (e.g. nearest exit). Furthermore, obstacles and the presence of other occupants affect the movement.

Route choice is influenced by external impact and the behaviour of the other evacuees. It is thus possible for an individual to alter the original egress route choice in order to avoid smoke or congestion, provided that alternative route choice options are available. Reduced visibility due to smoke spread affects the individual capability of orientation. A reluctance in entering areas obscured by smoke is considered in a probabilistic way, with weight functions depending on visibility.

Time is advanced by discrete time steps of 0.5 seconds. For this short interval of time, the corresponding trajectory of a moving person can be represented by a straight line. The movement algorithm of ASERI ensures that no conflict with walls, obstacles or the movement of other occupants occurs. The concept of sequential updating is applied, that is, persons move one by one according to appropriate priority rules (usually based on the distance with respect to exits or other geometrical boundary conditions that restrict individual movement). Together with the consistency checks that provide valid positions in the progress of individual egress movement, this method is suited to model the effects of anticipation inherent to pedestrian behaviour.

Time-dependent gas temperatures and concentrations of smoke, carbon monoxide, carbon dioxide, oxygen and hydrogen cyanide can be related to each unit. The incapacitating effects of exposure to asphyxiates and heat are calculated using the effective fractional dose model of Purser [Purser 1995]. Based on the movement of the evacuees and the individual dose, exposure is calculated with respect to CO, HCN, CO_2 and O_2 depletion. Synergetic effects, specially hyperventilation caused by the presence of CO_2, are included in Purser's model. In addition, critical concentration thresholds of toxic fire effluents and oxygen are included in ASERI. Thermal stress caused by heat radiation or convection can also be expressed in terms of an effective dose, considering that short exposure to a high temperature is more incapacitating than a longer exposure to a lower temperature. The obscuring effects of smoke are described by assigning a time-dependent visibility to each geometrical unit. Beside the toxic effects of smoke components, the slowing down of walking speed due to reduced visibility and certain aspects of behavioural response to smoke are modelled in ASERI.

Cue perception and first actions prior to egress movement - like information seeking, alarming others or attempts of fire fighting - can either be modelled by assessing individual alarm and reaction times or by introducing intermediate stop positions. These intermediate stops are arbitrary positions inside the building and can be assigned to any occupant, thus forcing him to first find his way to this position and waiting there for a specified time interval before starting his egress movement.

3 Passenger Flow in Underground Railway Stations

Pedestrian flow patterns encountered in subway railway stations are rather complex, including movement to and from incoming trains and the merging and splitting of flow at intermediate levels. Using empirical data on total flow rates of passengers with respect to the incoming and leaving trains and the respective distribution to available exits and wagons as boundary conditions for an ASERI simulation, computational results (AVI video sequences of individual pedestrian movement to analyse dynamic effects like congestion and track maps to analyse route choice and area occupancy) can be compared to available observations and records. It is then possible to make predictions on the effects of increasing the number of passengers, changing the geometrical layout or introducing additional barriers, ticket automata or card readers. Fig. 1 shows the track map of pedestrian movement for the mezzanine level of a subway station, derived for a 15 minute time interval during rush hour.

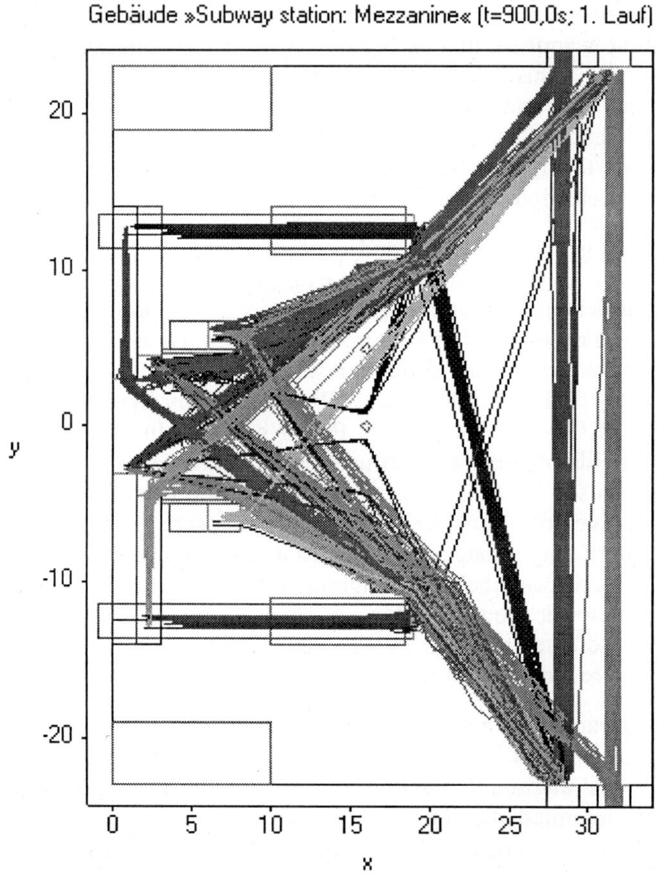

Figure 1: Track map of the individual movement of passengers for the mezzanine level of a subway station. This section of the station connects underground platforms and the upper ground level by stairways and escalators. The distribution is derived for a 15 minute interval during rush hour.

There are six escalators at the upper and lower part that connect to the underground platforms. At the left edge, 4 escalators and two stairways are leading to upper levels. The static information included in the track maps can be used to demonstrate the changes that occur if barriers or card readers will focus and delay the individual movement. Together with the dynamical information derived from the visualisation of the individual movement, the variation of occupation number with time and the individual delay times, it is thus possible to optimise design parameters or establish evacuation concepts on the basis of numerical simulations of passenger flow.

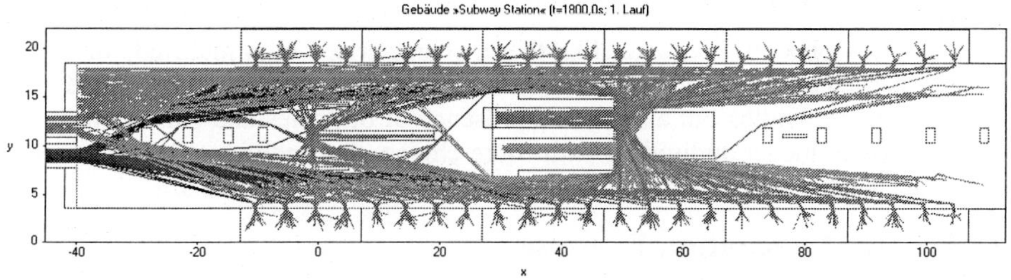

Figure 2: Track map of the individual movement of passengers for an underground platform with trains, stairways and escalators that connect the platform with adjacent platforms and the upper levels of the station. The distribution is derived for a 30 minute interval during rush hour. The dark lines represent passengers leaving the incoming trains, the other lines indicate the paths of passengers to their respective wait positions before entering outgoing trains.

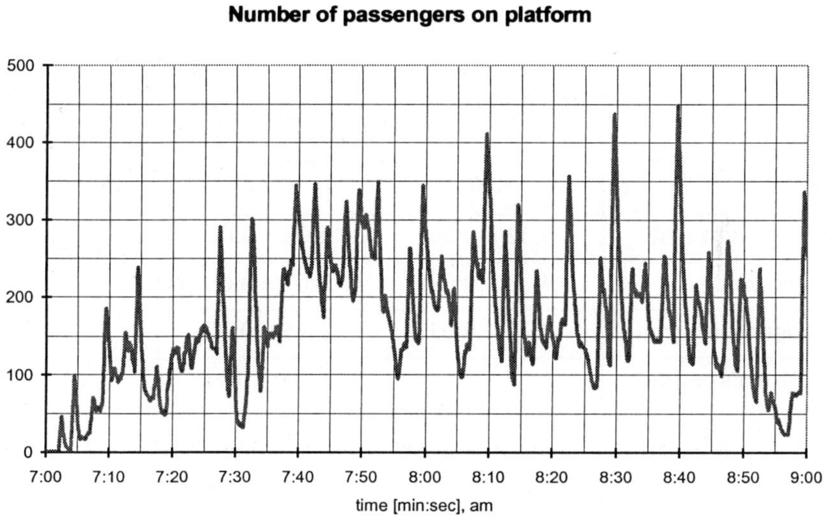

Figure 3: Occupation of a single platform for the time period between 7 am and 9 am.

Figure 2 illustrates the path route choice and occupation density on an underground platform. The position of the trains – six wagons each and every wagon with four available doors – is fixed. Wagons are characterised by a prescribed time-depended passenger flow depending on the arrival time of the trains and the respective occupation. The doors of the wagons are the possible destinations for passengers entering the platform from upper levels or from connections to adjacent platforms. The ASERI simulation also considers persons that do not immediately enter an outgoing train but have to wait for a certain period of time on the platform. Hence at rush hour there is always a mixture of moving pedestrians and waiting passengers. The variation of occupation number for a single platform within a two-hour interval is plotted in Fig. 3.

It is for complex environments like these, characterised by a multitude of possible egress routes, in which flow capacity is not only restricted by well defined exit, stair or corridor widths but also by the presence of obstacles and barriers or other devices that influence individual path route choice, that individual based evacuation models have advantages over hydraulic flow or network models.

4 Theatre Evacuation Exercise

An unannounced evacuation exercise was arranged with over 600 participants in 1995 at a theatre in the City of Tampere. The data of this exercise was used to asses various kinds of evacuation models [Weckman 1999] and to perform parameter variations [Schneider 2000] indicating the sensibility of basic input parameters of individual based evacuation modelling. The auditorium, which is located on the third floor, has 17 seat rows with a total of 750 seats. Two 3 m wide entrances at the left and right side are leading to a foyer. from the foyer at the third floor there is a 6 m wide open staircase to the second floor from which two 3 m wide open staircases lead to the ground floor with exits to the street level.

	total egress time	**egress time auditorium**	**first person at ground floor**
evacuation drill	5:20	3:37	1:10
simulation of drill	5:32 ± 0:06	4:02 ± 0:06	1:15 ± 0:03
danger mode	4:35 ± 0:09	3:07 ± 0:08	1:14 ± 0:03
homogeneous group	3:27 ± 0:03	2:25 ± 0:05	1:17 ± 0:02
100 % occupation	6:17 ± 0:13	4: 48 ± 0:11	1:21 ± 0:06

Table 1: Measured and calculated egress times (building and auditorium) and arrival time of first at ground floor, including a sensitivity analysis.

Due to the symmetry of the floor plan the simulations could be restricted to one half of the building. Fig. 4 shows the right half of the third floor layout (auditorium and part of the foyer). 612 people were present in the auditorium and took part in the exercise. According to the monitored time sequence of events the visitors started evacuation in a time period between 25 s and 47 s after the first fire alarm. Hence a start time of 25 s with an additional random delay time of 0 - 22 s has been used for the individuals in the ASERI simulations. Using a distribution of individual mobility parameters with unrestricted walking speed from 0,7 m/s up to 1,5 m/s and body size from 0,12 m^2 to 0,22 m^2, the results presented in table 1 have been obtained. This distribution is a conservative estimate to characterise the mobility parameters for building occupations including persons with restricted mobility. Such persons indeed were present among the audience, and hence the results should be close to the observed egress time, as it is the case.

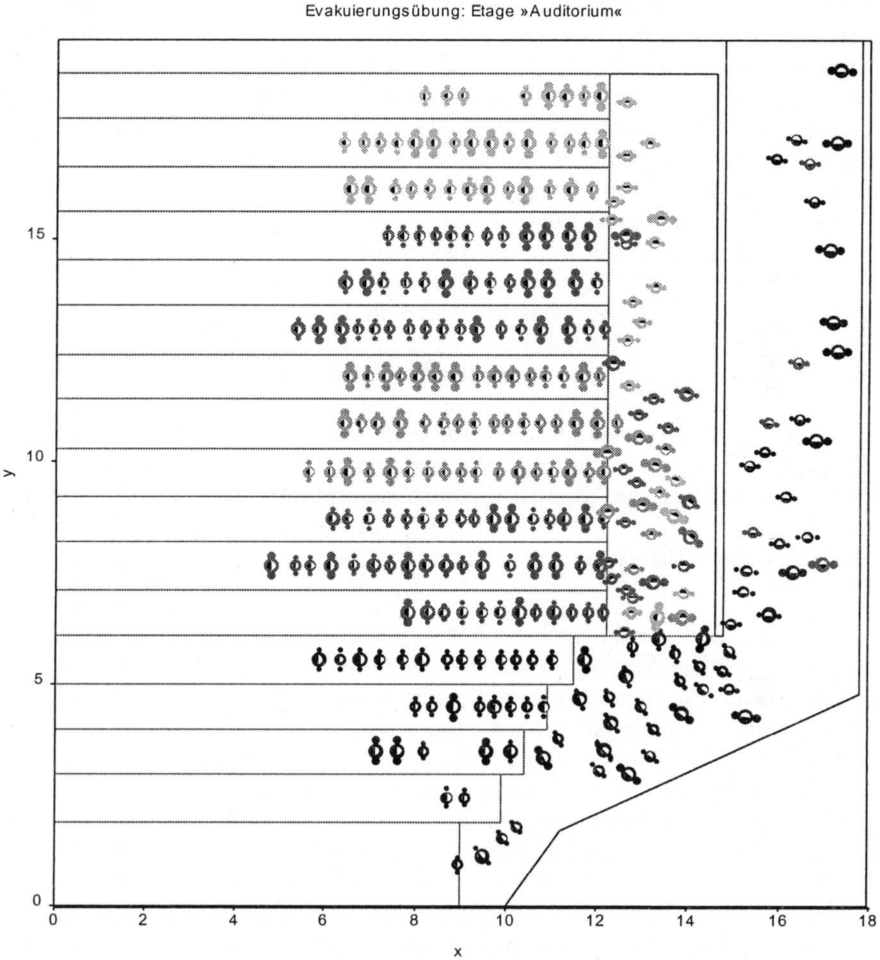

Figure 4: Distribution of evacuees on the third floor one minute after fire alarm.

Table 1 furthermore includes results of parameter variations that demonstrate the effect of different egress behaviour ("normal" versus "danger"), different individual mobility (inhomogeneous group versus homogeneous group with unrestricted mobility) and number of occupants (82% occupation of the drill versus hypothetical 100% occupation). In the danger mode, the clear layer widths with respect to walls, obstacles etc. and the minimum inter-person distances are reduced to model the effect of moderate pushing (not to be confused with the aggressive pushing and struggle usually related to panic). The strongest effects on egress time are clearly produced by the presence of evacuees with restricted mobility.

5 Fire on a Passenger Vessel

One of the test cases of a Norwegian project on the evaluation of evacuation models [Paulsen 1995] was the fire on the passenger vessel "Scandinavian Star", which took place 7. April 1990 at 2 a.m.. The scenario for the numerical simulations was restricted to the part of the passenger deck in stern that had the most severe impact from the fire hazard and where consequently most of the fatalities occurred (Fig. 5). The scenario includes 93 cabins initially occupied by two passengers each. Smoke from the fire in the lower part of the vessel did reach the passenger deck from exits to staircases at both ends of corridor D. Thus, these staircases and the entire corridor D were soon impassable. Passengers in stern could therefore only egress from the passenger deck via the exits to staircases labelled A, B, and C. Passengers in the forward part of the deck could evacuate by passing corridor F. Two fire scenarios assuming uniform smoke concentrations in the corridors had been prepared, based on the results of the official investigation. In the worst case scenario it has been assumed that the concentration of CO increases linearly from 0 to 500 ppm in the first 5 minutes and then from minute 5 to minute 10 linearly from 500 ppm to 20000 ppm, remaining constant later on. In the second scenario called sensitivity case, the maximum value of 20000 ppm has been reduced to 5000 ppm. Similar relations were introduced to estimate the concentration of CO_2 and the visibility in smoke. In the numerical simulations it had to be assumed that the alarm is given after 250 seconds, with a sound level of 55 dB in all cabins.

Figure 5 shows the result of one of the corresponding ASERI simulation. The cabins were free from smoke for the first 15 minutes until the ventilation systems was shut down. The individual behaviour is triggered by the environment, especially the visibility in smoke and the exposure to toxic combustion products. The results of the ASERI Monte-Carlo simulations (10 replicate runs each) for the worst case scenario are 41 successful evacuees and 79 successful escapes for the sensitivity case respectively. These results have to be compared with the fate of the passengers during the actual disaster: 74 successful escapes, 46 fatalities in escapeways and 66 fatalities in cabins. Hence, it seems that the fire scenario of the sensitivity case is more probable than the worst case scenario. However, one has off course to be very cautious in interpreting the results of numerical simulations for such extreme hazardous situations, because there is not enough empirical data available to calibrate and validate the required behavioural models describing deterioration effects and the reluctance to enter areas filled with smoke. In this numerical study, the outcome of the simulation is mainly depending on effects related to the presence of smoke and toxic combustion products:

- the probability assigned to the passengers for leaving their cabins and moving through smoke in order to get to an exit to the upper sections of the vessel,
- the reduction of walking speed in the presence of smoke,
- deterioration effects due to the presence of smoke and
- the physical impact due to exposure to CO and CO_2.

The time of individual incapacitation is determined by applying Purser's fractional effective dose model [Purser 1995]. The reduction of individual walking speed as a function of visibility is based on empirical data presented by Jin [Jin 1978], correlated in a conservative way by a linear decrease of walking speed for visibility in the range between 10 m and 5 m down to a constant minimum value of 0.35 m/s. For the ASERI simulations it has been assumed that the turn-back probability for movement in smoke derived from the studies of Wood and Bryant [Wood] can be used to estimate the preparedness of the passengers to leave (and not to return into) their cabins and move through the corridors of the passenger deck rapidly filling with smoke. Based on these empirical findings a turn-back probability is included in ASERI that has a maximum value of 70 % for a visibility of 1 m and below, decreasing linearly to zero at a visibility of 10 m. For the modelling of deterioration effects, there is an obvious lack of appropriate empirical data. In ASERI, probability weights for path route choice options are introduced that increase linearly with decreasing visibility. Hence it is possible that evacuees depart from the desired or most effective (e.g. shortest) egress route path due to obscuration effects.

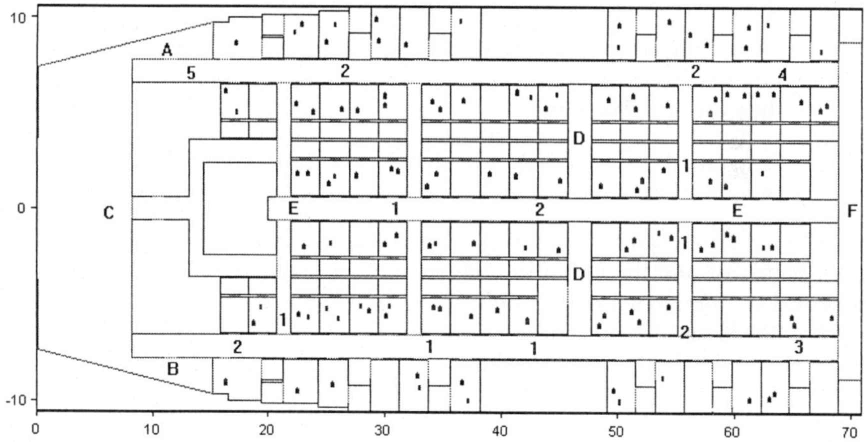

Figure 5: Distribution of passengers in an ASERI simulation of the Scandinavian Star worst case scenario, 458 seconds after the onset of smoke spread in the corridors. The figures in the corridors indicate number and position of persons incapacitated due to the effect of CO and CO_2. There are 38 successful escapes at this time.

6 Conclusion

Evacuation models based on the simulation of individual egress movement play an important role in designing performance-based concepts of occupant safety, specially in the field of modern fire safety engineering. Individual-based models have been proven to be successful in simulating evacuation scenarios of various types, from vehicles like ships and aeroplanes to railway stations, airports or high-rise buildings.

The main advantages of evacuation models based on the simulation of individual occupant movement and human behaviour can be summarised as follows:

- It is possible to model the geometry of the building floor plan resolving any detail that may influence evacuation efficiency.
- The combined effects of unimpeded movement as well as high density crowd movement can be simulated in a consistent way using only fundamental concepts of individual egress behaviour modelling.
- Scenarios including alternative options for egress route choice can be investigated on a more fundamental level based on individual human behaviour. This includes the important cases of merging and splitting crowds and unbalanced egress route choice with the respective implications on evacuation efficiency.
- The algorithms modelling individual egress route choice and movement, the initial distribution of occupants within the building and the corresponding assignment of individual parameters like size, walking speed and reaction times introduce stochastic effects. Hence it is possible to perform replicate runs, yielding a Monte-Carlo type of simulation that gives more comprehensive insight into the processes and parameters effecting the evacuation process.
- If required, it is possible to calculate the impact of fire effluents on the individual capability of the occupants and on certain behavioural aspects.

The fact, that human behaviour in fire, evacuation models, the choice of appropriate scenarios and the definition of criteria to evaluate predictions obtained by numerical simulation are now addressed in fire engineering guidelines all over the world emphasises the importance and acceptance of computational evacuation modelling.

References

1. T. Jin, *Visibility Through Fire Smoke*, Journal of Fire and Flammability **9**, 135 (1978).

2. T. Paulsen, H. Soma, V. Schneider, J. Wiklund, and G. Løvås, *Evaluation of Simulation Models of Evacuation from Complex Spaces (ESECX)*, SINTEF Report STF75 A95020, 1995.

3. D.A. Purser, *Toxicity Assessment of Combustion Products*, in: *SFPE Handbook of Fire Protection Engineering*, (National Fire Protection Association, Quincy, Massachusetts, 2nd Ed., 1995).

4. V. Schneider and H. Weckman, *Rechnerische Simulation des Ablaufs einer Evakuierungsübung*, vfdb-Zeitschrift **4**, 140 (2000).

5. H. Weckman, S. Lehtimäki, and S. Männikkö, *Evacuation of a Theatre : Exercise vs Calculation*, Fire and Materials **23**, 357 (1999).

Models and Algorithms for Evacuation Analysis in Urban Road Transportation Systems

F. Russo and A. Vitetta

University of Reggio Calabria - Department of Computer Science, Mathematics, Electronics and Transportation, Italy

Methodologies for transportation planning in urban area when exogenous events occur and/or in emergency conditions have not received great attention in the literature. In general, there is no systematic analysis of different events connected with transportation system design. Models and algorithms specified and calibrated in ordinary conditions cannot be directly applied in emergency conditions. Users move on the network in different behavioural conditions, they do not know system congestion and reliability in real time, and the user optimum, in general, is different from the system optimum. Different models and algorithms have to be proposed in this scenario. In this paper a general framework is presented and tested with specific methods and models for analysing urban road transportation performance in emergency conditions when exogenous phenomena occur. In the paper a global pseudo-dynamic procedure for analysing supply, demand and demand-supply interaction is proposed.

1 Introduction

Methodologies for transportation planning in urban area when exogenous events occur and/or in emergency conditions have not received great attention in the literature and only emergencies when a nuclear event occurs are treated [11,14]. In general, there is no systematic analysis of different events connected with transportation system design. Models and algorithms specified and calibrated in ordinary conditions [2,4,12,15] cannot be directly applied in emergency conditions. Users move on the network in different behavioural conditions, they do not know system congestion and reliability in real time, and the user optimum, in general, is different from the system optimum. Different models and algorithms have to be proposed in this scenario.

A problem which entails particular attention in terms of coordination and management is that strictly linked to general mobilitation which is created in a densely populated area after a disaster (earthquake, flood, etc.) or in any kind of emergency.

Two different approaches are considered for planning co-ordination: "what if" models (system or user optimum with deterministic user behaviour) and "what to" models (optimal user distribution and departure time, optimal network design system).

In this paper a general framework is reported with specific methods and models (section 2) for analysing road urban transportation performance in emergency conditions when exogenous phenomena occur in the context of a "what if" approach while a specification of a "what to" model is reported in Russo and Vitetta [14]. The procedure was tested in three different real areas (section 3). The obtained results confirm the goodness of procedure and highlight some aspects that need to be developed (section 4).

2 Models and Procedures

The drawing-up of an evacuation plan (*of vehicle types and in all urban areas*) for a city requires a study of the problems connected to landscape characteristics and in relation to a disaster. The complete procedure for developing an evacuation plan can be subdivided into the following variously connected phases: study of the land (not considered in this paper); supply analysis; demand analysis; supply-demand interaction analysis; system design of the "what if" and/or "what to" kind.

In Fig. 1a system of models which may be proposed for elaborating evacuation plans and relative functional connections are represented.

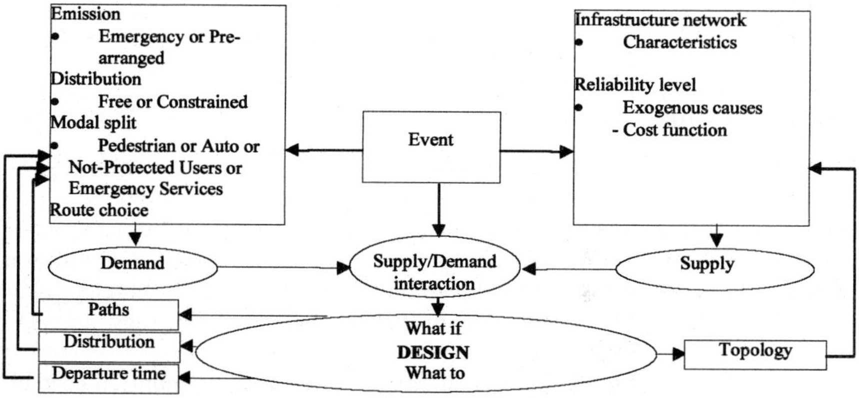

Figure 1: The global procedure for the evacuation plan and relevant links.

2.1.1 Methods for Supply Analysis

The first phase consists in studying the area as follows: the area is delimited, defined as that area comprising the transport system in question and where most of the effects of the planned interventions are believed to be; homogeneous areas are identified; zoning and minimal road network plan.

The second phase involves supply analysis which is conducted by the following: verification of the exogenous disturbances on the transport network; construction of the transport network; graph representation of the network; construction of link cost functions. All the roads are represented by links and intersections between roads with nodes. In each node the control strategy to be used must be defined.

When emergencies occur, the transport system must be optimised, chiefly to allow two operations: evacuation of system users to appropriate assembly centre; access of emergency vehicles and the security forces.

Perturbed conditions on networks may be classified in various ways [13]. As for the cause producing the disturbance, there may be traffic-endogenous and exogenous causes.

Below we will use notation commonly employed in reference works in the transport planning sector [4, 8]. Transport supply is represented by a graph G consisting of an ordinate pair of sets, a set N of elements called nodes and a set L of pairs of nodes belonging to N called arcs or links: G=(N,L).

The link cost vector, **c**, is defined as a vector whose generic component c_{ij} consists of the transport cost (generalised) on link ij. To each link may be associated a link flow f_{ij} (**f** being the link flow vector) which represents the average number of users which, in the time unit, uses the link. The demand flow vector can only assume values belonging to its feasibility set denoted by S_f.

In transport systems, average link cost depends on user flow on the link itself and, in some cases, also on flows on the system's other elements. Cost functions may be either separable or non-separable. In the former case, the cost of a link depends exclusively on the flow relative to the link itself: $c_{ij}(f) = c_{ij}(f_{ij})$.

However, in the latter case, it depends on the flow on more than one link: $c = c(f)$.

In transport networks, various functions may be used to relate travel time c or velocity v on the link to vehicle flow f, capacity C and other characteristics. One of the most commonly used functions is the BPR function. For queuing at junctions Doherty's adaptation of Webster's formula is generally used. In some cases, to simulate the system through a dynamic approach functions linking travel velocity v to the number of vehicles present in a certain instant on the link (spatial density k) are used (Greenshields, Drew, Greenberg, Underwood, Drake functions).

2.1.2 Methods for Demand Analysis

The use of demand models in simulating an evacuation plan involves various problems connected with the many cases that may lead to a state of emergency. Below are considered two decisional units: on the one hand the citizen *user* who has to decide in some cases how to act, and on the other the *system manager* (Prefect, Mayor, etc.) who may impose or advise some user choices. The public decision-maker must predict the effects that implementation that his/her measures have on the general public, and must bear in mind that his/her proposed solutions may interact with the individual reactions of the users served by the transport system in question.

In all choice models, we need to define the set of alternatives considered by the system manager and the user, as well as the type of choice models considered [3]. The sequence of submodels is as follows: emission model supplying trips originating from a certain zone in the study area; departure time model supplying the number of users leaving in each time slice; distribution model supplying the number of users reaching a particular assembly centre from a fixed origin zone; modal split model supplying the number of users using a given transport modality from a certain origin to a certain assembly centre; route choice model supplying the percentage of users who use a certain route between a fixed origin and assembly centre with a transport modality.

The term o is used to indicate a generic trip origin and d an assembly centre, d_{od} being the number of users travelling between the origin and assembly centre pair (o,d). The demand vector **d** (whose generic element is d_{od}) may assume values belonging to its feasibility set denoted by S_d. The choice model from a given choice set, in the context of random utility models, assumes that a generic user, travelling between an origin- assembly centre pair (o, d), associates to each alternative k belonging to the set K_{od} of available alternatives connecting that O/D pair, a perceived utility $U_{k,od}$ which may be expressed as:

$$U_{k,od} = V_{k,od} + \varepsilon_{k,od} \qquad \forall\, k \in K_{od}. \tag{1}$$

The term $V_{k,od}$ denotes the average, or systematic, utility of alternative k while the random residual $\varepsilon_{k,od}$ is usually assumed to include perception errors of the decision maker as well as the modelling approximation of the analyst.

If the $\varepsilon_{k,od}$ are assumed to be independent and identical Gumbel $G(0,\theta)$ variates of zero mean and the parameter θ is more than zero, then the well-known Multinomial Logit model of choice results [2]. The probability of choosing an alternative k between o and d, giving a choice set K_{od}, $p_{od}(k/K_{od})$, is:

$$p_{od}(k/K_{od}) = \exp(V_{k,od}) \,/\, \Sigma_{h\in Kod} \exp(V_{h,od}). \tag{2}$$

If the residuals are assumed to be jointly distributed as a Multivariate Normal of zero mean, the Probit choice model is obtained [9].

Different specifications can be used for term $V_{k,od}$. A possible specification is:

$$V_{k,od} = \Sigma_l\, \alpha_l\, X_l - \Sigma_m\, \beta_m\, Y_m. \tag{3}$$

where α_l and β_m are parameters greater than zero to calibrate; X_l are user utility attributes; Y_m are user disutility attributes. In Tab. 1 an example of attribute specification of the different models is reported.

Choice Dimension	*Utility Attributes X_l*	*Disutility Attributes Y_m*
Emission	Resident, vehicle owner, number or people in family, sick	
Departure time	Accessibility, resident	Congestion
Distribution	Resident, worker, students	Distance, travel time
Modal split	Vehicles owner, modality available	Travel time
Route choice		Travel time

Table 1: Example of attributes specification for choice models.

2.1.3 Methods for Supply-Demand Interaction Analysis

A mathematical model of supply-demand interaction allows us to simulate the way in which the demand uses the supply system, producing flows on network links that in turn generate travel costs.

In the *stationary* flow simulation approach [4]: the stochastic equilibrium approach, i.e. the use of static assignment models with stochastic user behaviour, by which user information is more or less correctly reproduced, non-perfect respect of impositions, alternative choice behaviour; the deterministic equilibrium approach, i.e. the use of static assignment models with deterministic user behaviour, by which complete user knowledge of the system is reproduced; the system optimum approach, i.e. the use of static assignment models with user behaviour imposed by the system manager.

In the Stochastic equilibrium the resulting route flows $\mathbf{F}^*$ may be obtained by the following type of model:

$$\mathbf{F}^* = \mathbf{P}(\mathbf{A}^T \mathbf{c}(\mathbf{A}\,\mathbf{F}^*))\,\mathbf{d} \qquad \text{where } \mathbf{F}^* \in S_F \qquad (4)$$

where: $\mathbf{P}$ is the probability matrix whose generic element supplies the probability of a route k belonging to an OD pair; $\mathbf{d}$ is the demand vector; S_F is the feasibility set of route flows.

Deterministic equilibrium assignment supplies the vector of link flows $\mathbf{f}^*$ (or similarly route flows $\mathbf{F}^*$) with the following relation:

$$\mathbf{c}(\mathbf{f}^*)^T(\mathbf{f} - \mathbf{f}^*) \geq 0 \qquad \forall\, \mathbf{f} \in S_f \qquad (5)$$

termed "variational disequality".

System Optimum assignment is defined by an optimisation model, expressed in terms of link flows, the objective function being total user cost:

$$\mathbf{f}^* = \arg\min z(\mathbf{f}) = \mathbf{f}^T \mathbf{c}(\mathbf{f}) \qquad \text{where } \mathbf{f}^* \in S_f. \qquad (6)$$

In *non-stationary* models [1,5,6,7], it is hypothesised that the flow variables may have a variability not only in space but also in time. There may be three aggregation levels in representing the flow: aggregate flow parameters; analysis of homogeneous vehicle groups (vehicle packages); analysis of individual vehicles.

Table 2 summarises the proposed demand models with relative specifications for the choice set generation and for choice form a given choice set, for system manager and for user.

Choice Dimension	*Choice set*		*Choice model/Assignment*	
	System manager	*User*	*System manager*	*User*
Emission	Emergency (Immediately or Not evacuate) Pre-arranged (Delayed or Not evacuate)	Like system manager	System optimum	Stochastic user equilibrium
Departure time	Constrained (Time slice 1, Time ....) Free (Time slice 1, Time ....)	Like system manager	System optimum	Stochastic user equilibrium
Distribution	Constrained (Centre 1, Centre 2, ...) Free (Centre 1, Centre 2, ...)	Like system manager	System optimum	Stochastic user equilibrium
Modal split	Vehicle, Pedestrian, Emergency services	Like system manager	System optimum	Stochastic user equilibrium
Route choice	Exhaustive, Filtered	Behavioural	System optimum	Stochastic u. e.

Table 2: Choice set alternatives and choice models.

3 Applications

The procedure was tested in three real areas (Crotone and Reggio Calabria cities and the Vesuvian area near Naples), hypothesising that an event occurs and there is a temporal gap between cause and effect on the population. In this temporal gap all or some of the population have to be evacuated. In the city analysis two study areas are considered [16]: the city of Crotone (70,000 inhabitants), with evacuation of all the population by car (44,000 vehicles); the city of Reggio Calabria (180,000 inhabitants), with evacuation of all the population by car (100,000 vehicles) or with evacuation of all the population (35,000 inhabitants) by car (20,000 vehicles).

In the third case the analysed area [10] is the Vesuvian zone near Naples (630,000 inhabitants) with evacuation of all population with two modalities (75% of inhabitants -175,324- by car and 25% of inhabitants –3,272- by bus). The main characteristics of the analysed areas and of the used models are reported in Tab. 3. The application was developed with constrained distribution in terms of "user has to reach spatial nearest assembly centre". For the areas of Crotone and Reggio Calabria free distribution was also studied in terms of "user chooses the assembly centre". The results for the two distributions considered are reported in Tab. 4. Free distribution produces better indicator values than constrained distribution. In terms of global indication, evacuation times in the Reggio Calabria areas are more than 5 hours and in the Vesuvian area they are about 33 hours. Note that the constrained distribution is not performed with the system optimum model.

Area of	*Crotone*	*Reggio Calabria*		Vesuvian area
Evacuation	Total	Partial	Total	Total
Vehicles	44,000	20,000	100,000	175,324
Population	70,000	35,000	180,000	630,000
Supply				
- *Nodes*	157	382	382	269
- *Links*	432	910	910	743
- *Assembly centres*	8	8	8	3
Demand				
- *Emission*	Emergency	Emergency	Emergency	Emergency
- *Departure time*	Free	Free	Free	Free
- *Distribution*	Constrained or Free	Constrained or Free	Constrained or Free	Free
- *Modal split*	Vehicle	Vehicle	Vehicle	Vehicle and Bus
Supply-Demand	Pseudo-dynamic	Pseudo-dynamic	Pseudo-dynamic	Mesosimulation

Table 3: Characteristics of study areas.

Area of	*Crotone*		*Reggio Calabria Partial*		*Reggio Calabria Total*		Vesuvian area	
	TTR (1000h)	*TU (h)*	*TTR (1000h)*	*TU (h)*	*TTR (1000h)*	*TU (h)*	*TTR (1000h)*	*TU (h)*
Constrained distribution	11.8	1.65	11.4	2.40	168.9	5.56	420.0	36.28
Free distribution	9.2	1.37	9.4	1.80	156.0	5.10	---	---

TTR = total time on the network; TU = time of last arrival from starting first departure

Table 4: The indicators for the areas studied.

4 Indications for Further Developments

Research for the analysis and the modeling of transportation systems in emergency conditions requires new studies in the supply, demand and supply-demand interaction. In emergency conditions there is the necessity to develop new methodologies and to rearrange standard procedures such as: network vulnerability analysis in terms of “security coefficient” of the supply system in relation to the events with different levels of hazards and different probabilities of fulfilment; specification and calibration of link cost functions to use in system simulation in over-saturation conditions in the analysis periods; specification and calibration of demand models for the different choice levels for users and for the system manager; study of supply-demand interaction with stationary and non-stationary traffic flow.

The system design is relative to different aspects in relation to model formulation, model calibration and solved approach. The main aspects to investigate are as follows: localisation of the assembly centre; departure user time design; spatial user distribution design; emergency intervention design; network design in term of topology and capacity. We can conclude that some aspects must be developed but the proposed global procedure can be a very useful tool to planning evacuation.

References

1. M. Ben Akiva and A. De Palma, *Dynamic models of transportation networks*, in: *Proc. of the 15th PTRC Summer Annual Meeting,* (University of Bath, England, 1987).

2. M. Ben Akiva and S. Lerman, *Discrete choice analysis: theory and application to travel demand,* (MIT Press, Cambridge, Mass, 1985).

3. M. Ben Akiva, M.J. Bergman, A.J. Daly, and R. Ramaswamy, *Modelling inter urban route choice behaviour,* in: *Ninth International Symposium on Transportation and Traffic Theory*, pp. 299-330 (VNU Science Press, 1984).

4. E. Cascetta, *Ingegneria dei Sistemi di Trasporto*, (UTET, 1998).

5. E. Cascetta, *A stochastic process approach to the analysis of temporal dynamics in transportation networks,* Transportation Research **23** B, 1-17 (1989).

6. E. Cascetta and G.E. Cantarella, *A Day to Day and Within Day Dynamic Stochastic Assignment model,* Transportation Research **25** A, 277-291 (1990).

7. E. Cascetta and G.E. Cantarella, *Modelling dynamics in transportation networks*, Transportation Science, (1993).

8. M. de Luca, *Manuale di Pianificazione dei Trasporti*, Franco Angeli, (1999).

9. C.F. Daganzo, *Multinomial Probit: the theory and its applications to demand forecasting*, (Academic Press, New York, 1979).

10. M. Di Gangi, A. Longo, and R. Polidoro, *Una procedura di carico dinamico per la valutazione dei piani di evacuazione,* in: *Proc. of 2° seminario scientifico su Metodi e Tecnologie dell'Ingegneria dei Trasporti*, Università di Reggio Calabria.

11. R. Goldblatt, *Development of Evacuation Time Estimates for the Davis Nuclear Power Station,* T. R., (1993).

12. J.deD. Ortuzar and L.G. Willumsen, *Modelling transport*, (John Wiley and Sons, 2a ed., 1994).

13. F. Russo and A. Vitetta, *The road network design problem to improve the safety during exogenous flow perturbations,* in: *Proc. of the 29th ISATA*, (Florence, Italy, 1996).

14. F. Russo, and A. Vitetta, *Metodi per l'analisi dei sistemi di trasporto urbano in condizioni di emergenza,* QD/SS n. 9, DIMET, (Università di Reggio Calabria, 1999).

15. Y. Sheffi, *Urban transportation networks*, (Prentice Hall, Englewood Cliff, NJ, 1985).

16. P. Velonà, *Metodi di assegnazione per l'analisi dell'evacuazione di una rete: applicazioni a casi reali,* Tesi di Laurea, (Università degli Studi di Reggio Calabria, Facoltà di Ingegneria, 1999).

17. J.G. Wardrop, *Some Theoretical Aspects of Road Traffic Research*, in: *Proc. Of Inst. Civil Engr.* Part II, (1952).

Detection of Pollutant Sources in an Indoor Environmental Monitoring System

S. Bandini, G. Frisoni, and S. Manzoni

Dipartimento di Informatica, Sistemistica e Comunicazione, Univerisità di Milano - Bicocca, Via Bicocca degli Arcimboldi, 8, 20126 Milano, Italy

The main aim of this paper is to present a knowledge-based system (SER) integrated with an indoor environmental monitoring system for historical buildings, and to draw its possible extension for pedestrian evacuation. SER concerns the identification of rooms containing sources of pollutant after that another module of the same system has detected unhealthy conditions or potentially dangerous situations for people. This system is based a general model to represent a building and the set of sensors installed in it. Moreover SER incorporates a representation that helps the domain expert to analyze microclimatic conditions in buildings and, when necessary, to identify sources of detected dangerous pollutants. To this aim the knowledge-based module uses a two-level model of the monitored building and specific knowledge about fluid dynamic. This work has been developed as part of the RAID project, funded by MURST (Italian Ministry for University and Scientific Research). It has already taken two years and will continue for one year more. A prototyping version of the system is now going to be tested on a real building.

1 Introduction

The purpose of the RAID[1] project was to realize a wireless indoor environmental monitoring system for historical buildings. The project involved several multinational companies and two academic departments. Target buildings (include ancient churches, museums, etc.) are such that it is not always possible to freely lay communication cables. For this reason the chosen communication infrastructure had to be wireless. For wireless transmission hasn't been chosen GSM (Global System for Mobile telecommunication [1]), the European standard for mobile telephony, but DECT (Digital European Cordless Telecommunications) [2]. DECT is not a standard for wide-area wireless communication (like GSM), but rather one for *cordless telephony*. The main reasons behind this choice are related to advantages that can be obtained with the use of DECT instead of GSM (related to bandwidth, cost, availability of DECT in many European buildings).

---

[1] The RAID (Environmental Data Detection with DECT Interface) project was funded by MURST, the Italian Ministry for University and Scientific Research.

Apart from pure monitoring, another peculiarity of this project was the development of an intelligent module (SER – RAID Expert System). The SER module is able to identify the source of a specific pollutant within the monitored building. To that aim, a knowledge-based system (rule-based) has been designed and integrated in the RAID system. The expert module is triggered when the concentration of a monitored pollutant exceeds one or more thresholds, and is able to infer where the pollutant comes from (e.g., a defective heater). Therefore, it is equipped with expert knowledge about passive air circulation and other expertise in the field of energy engineering. The description of the SER module is the main aim of this paper and it will be covered further.

UML has been chosen as the analysis and design language. Java has been used as implementation language for a number of reasons, among which portability, wide availability of pre-built, excellent libraries, and ease of use, developers' experience. For the implementation of the rule-based module a Java-based shell (JESS- Java Expert System Shell [3]) has been used. Jess is the Java implementation of the popular CLIPS expert system shell. This choice provided for very easy integration between the expert module and the RAID system.

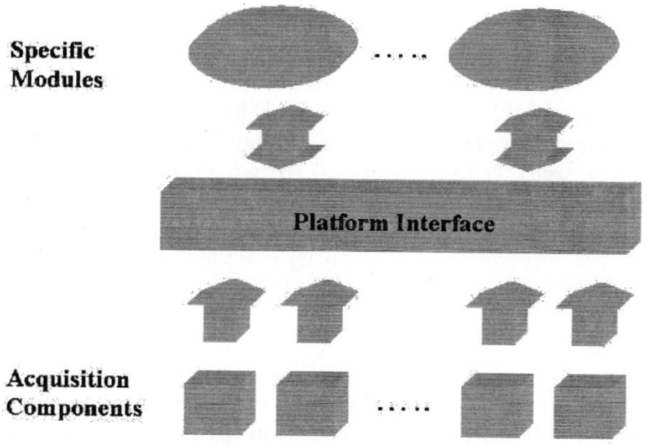

Figure 1: Overall system architecture.

In Fig. 1 the overall RAID system architecture is reported. As the figure shows, the system comprises three classes of components:

- Sensors, the acquisition components which collect microclimatic measurements of physical sensors; there is typically one component for each sensor;
- Central platform, the main system component. It contains processing images for all sensors and all the necessary items, such as domain model for the modules of the system dedicated to specific tasks;

- Specific components, single modules that perform specific tasks. In general an integration interface is dedicated to the communication between each specific component and the central platform. The main subject of this paper is one of these specific components: the knowledge-based system SER and its integration interface endows it with necessary items from the central platform.

2 Domain Representation Model: The Two-Levels Net

The task of the SER module is to elaborate measurements collected by acquisition components (sensors) and to elaborate information belonging to a different conceptual level (the topological one). In fact SER has to identify rooms eventually containing source of the pollutant detected by sensors. SER is endowed with a topological description of the building and the set of measurements collected and elaborated by sensors installed in it. For this reason the domain model constructed for the SER module had to be a two-levels net (topology and sensors levels).

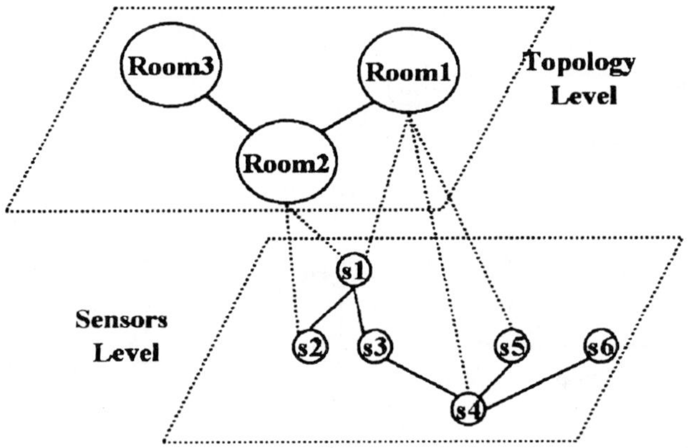

Figure 2: A 2-levels net that represents a portion of a building. Topology level contains three rooms while six acquisition components are in the sensors level.

Figure 2 shows a portion of the net describing a building with three rooms in which six sensors are installed. A particular type of arcs links the two levels. In particular, the sensor s5 (represented in the sensors level) is connected to the room Room1 (represented in the topology level) if the domain expert considers the measurements s5 collects informative for the microclimatic state of the room Room1. This means that each sensor can be connected to more than one room, and vice versa each room can be monitored by more than one sensor. Note that sensors monitoring a room are not necessarily all contained in it. A net of rooms connected by arcs represents the topology level. Each arc does not represent a physical adjacency among rooms but possible air (and then pollutant gases) flows. For this reason if two rooms are physically adjacent but airflow between them has to pass through, for example, a corridor, in the topological net a link is not present between the two rooms but there is one between each room and the node that represents the corridor (see for example in Fig. 3 the case of nodes with identifiers 5 and 6). The domain expert defines through a 'configuration' module which topological elements have to be represented in the topological level of the net. This means that, for example, he has to decide the set of rooms, corridors and lifting vans to be there included. Each node of the topological level contains information to identify the room (the floor it belongs to, its identifier and name), its use (for example if it is a bath, a scale, a simple room, etc), its dimensions (volume and surface), the set of sensors it contains, and the mean value for the concentration level of each pollutant. The SER module uses this information to try to identify pollutant sources.

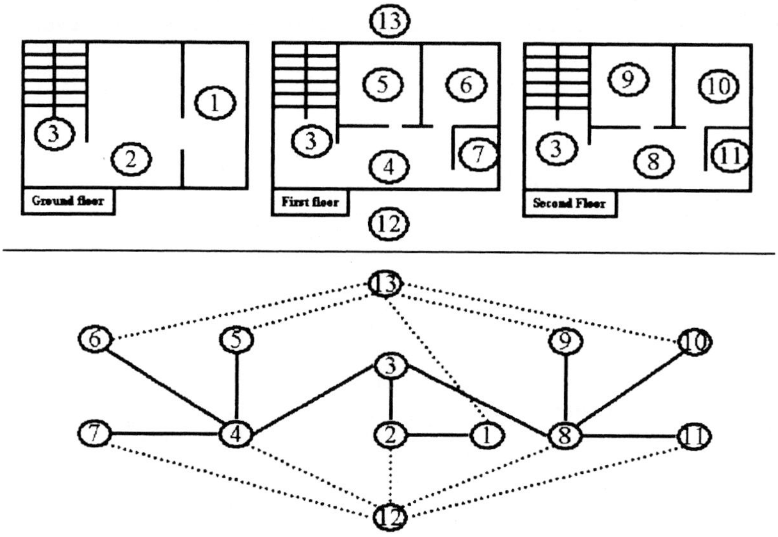

Figure 3: The topological representation of a building.

The sensor level represents in a net-structure the set of sensors installed in the building. What is described at this level as a sensor (a node in the graph), at the physical level is constituted by a set of measurement devices. In particular each sensor can measure data of different nature, that is microclimatic data (temperature, enthalpy, etc) and instantaneous data about pollutants (one value for each gas), and in general there is a device for each different type of data.

Arcs in the sensor level net represent a spatial adjacency relation between sensors. While the domain expert decides positions for installing sensors, he indirectly sets this kind of relation among them. Each node of the sensors level contains information to identify it, its location in the building, the set of rooms that are interested by its measurements (in general a sensor measurement can be informative also for rooms that do not contain it). Moreover each sensor contains the set of values about instantaneous measurements of pollutants and microclimatic data collected and three boolean variables. The latter assumes a non-null value if the instantaneous quantity of pollutant is greater than a given threshold. A specific module of the RAID system computes these thresholds in a dynamical way, as will be outlined in the next section.

3 The SER Module

SER is triggered when potentially dangerous conditions are detected on one or more sensors of a specific room inside the building. By dangerous conditions we mean that a room has become unhealthy for people. RAID takes into account three different thresholds to declare a place unhealthy; they are calculated on the basis of the guidelines of the World Health Organization [4]. As mentioned above, the thresholds are not simple numbers that a quantity must reach, but they are functions depending on some variables such as the concentration of the pollutant, the time of the year, and the temperature. A RAID module makes those computations outside SER, which is triggered only if one or more of the thresholds are reached; in doing so SER knows that there is a potentially dangerous situation inside the monitored building and starts searching the pollution source.

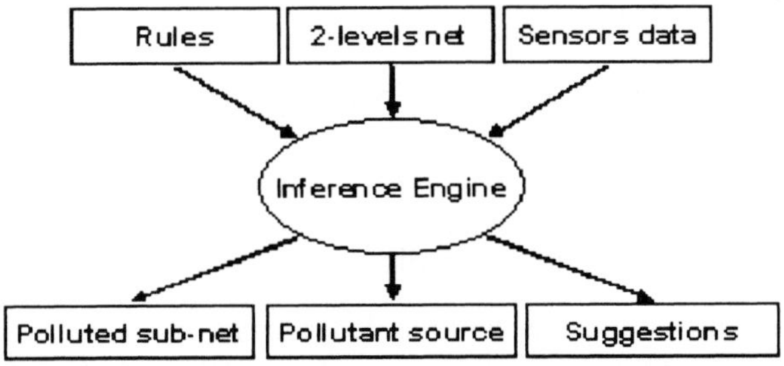

Figure 4: SER Inputs and Outputs.

As Fig. 4 shows, SER receives many different inputs, but they can be divided into two main groups: static inputs or dynamic inputs. Static inputs are loaded only at the system start-up, and represent the knowledge on the topology of the building, on the airflows dynamics and on the relations between pollutant and source of pollution. Dynamic inputs correspond to sensors measurements; they are updated and sent to SER every time there is an alarm in the building. We refer to them as dynamic because, once the system has been installed in a real building and the 2-levels net has been defined, sensors data are the only inputs that change. There are two kinds of dynamic inputs: the values of the instantaneous detections, and the so-called flags, that are variables that assume non-null value if the threshold is reached. The threshold is computed by one of the specific modules of the RAID system but it is given to SER as a 2-state variable.

As usual for knowledge-based systems, SER rules regarding the domain of air-flows dynamics don't want to be omni-comprehensive: with the aid of an expert of the field we have tried to implement his knowledge and experience on the airflow dynamics inside buildings that are supposed to be museums or, in general, great buildings with wide rooms. For example, we forecast that there could be more than one sensor in each room and there could be more than one boiling room, possibly a kitchen, and so on. The knowledge on the relations between pollutant and source of pollution is a typical case of knowledge that could be implicit: in fact, we exploit the information on the destination of a room to infer that that room is one possible source of the pollutant. For example, in some cases it is more relevant to look for a boiler room if there is a specific pollutant, instead of going around blindly.

In a traditional way of viewing things, we can say that the knowledge coded in the rules is the code of the program, the 2-levels net is the configuration data and the sensor data are the inputs of the program. But, as the Fig. 4 shows, the inference engine threats all them as inputs, that is you can use the same rules for different buildings (you can load different 2-levels nets), or you can use different rules depending on the building. The results of the computation of SER are the locations where there could be the source of the detected pollutant. In addition to that, SER gives back the sub-net of the polluted rooms connected to the pollutant source: this is a part of the topological net where each node is a room that belongs to the path followed by the pollutant. SER tries to associate a possible source of pollution to each polluted sub-net, if there is one.

In some cases, when the knowledge about the pollutant is enough, SER gives suggestions on how to stop the pollution source or, at least, what it is best to do in that situation. For example, if traffic pollution is detected, knowing that it comes from outside, the best thing to do probably is to close some windows that are on the main street next to the monitored building.

Three main modules compose the knowledge base: Association Module, Building Module, and Suggestions Module. The *Association Module* is the one dedicated to the identification of probable physical sources of the pollution, that is, for example, traffic pollution, bad boiling pollution, and so on. To do that, as stated above, the Association Module exploits the implicit knowledge about the use of each room to associate a pollutant, or a set of pollutants, to a probable source; this means that the other two modules will look for a specific room whose destination could be compatible with the probable source found by the Association Module. The *Building Module* operates onto two levels of representation of the monitored building: the rooms level and the sub-nets level. When SER is triggered, it tries to build the pollution path; it is the path that the pollutant followed from the source to reach the room where sensors detected the pollution. The pollution path is a net whose nodes are the rooms where the pollutant passed through: it is a sub-net of the net that topologically represents the building. To assemble the sub-net, the Building Module uses the information about sensors measurements and about the topology of the monitored building. For example, the Building Module can add to the polluted sub-net all the polluted rooms connected to the current room with an air passage. But it can add to the sub-net a room even if it is not polluted, for example exploiting information about the volume of the rooms and the concentration of pollutants. When there is no more room to add to the existing sub-nets, the Building Module tries to find the probable source of the pollutant for each sub-net; to do this it exploits the knowledge on the topology of the building and, basing on the results of the Association Module, it tries to find out if there are rooms that are close enough to the considered net and that can be the source of the considered pollutant. If there is a sub-net that doesn't have a proper probable source associated, the Building Module tries, at the sub-net level, to merge it to other sub-nets that have the related probable pollutant source, exploiting the topological knowledge of the building. For example it can infer that if two sub-nets are not connected, but there is a single room between them, it can merge the nets together supposing that the pollutant must have past even in the unpolluted room. At the end of the computation of the two modules described till now there are, in general, one or more polluted sub-nets with the relative pollutant source, if SER found it. At this point the *Suggestion Module* tries to give some suggestion on how to handle the detected problems; it exploits the knowledge on the topology of the building, on the polluted sub-nets and on the destination of rooms; for example, if there is a pollutant coming from a boiling room, it suggest what to do with that boiling, if and how to switch it off; then, if the boiling room is at the ground floor and the sensor that alerted the monitoring system are at the first floor, it could suggest to close the doors that lead from first to second floor, to avoid the pollution of that floor.

4 Further Extensions

The SER architecture is characterized by a modular structure that allows an easy extension to other specific tasks; for instance, the integration of modules for supporting sensors diagnosis or pedestrian evacuation. Obviously in each case specific knowledge should be added. In particular for pedestrian evacuation information could be added about the possibility for people to pass through an existing air passages. This kind of information could be coded in the 2-levels net that describes the monitored building while explicit knowledge to manage it could be added as rules. Moreover the algorithm for the polluted sub-net construction could be exploited for the safe path definition for evacuation.

References

1. M. Rahnema, *Overview of the GSM System and Protocol Architecture*, IEEE Communications Magazine **31**(4), pp. 92-100 (April 1993).

2. ETSI - European Telecommunications Standards Institute, Doc. Nb. ETR 178, Edition 2, Ref. RTR/DECT-000088.

3. http://herzberg.ca.sandia.gov/jess, Java Expert System Shell, Internet site.

4. World Health Organization, *Air Quality Guidelines for Europe*, (1997).

The Precautions in Istanbul in Case of a Possible Earthquake

R. Bozdogan and H.I. Polat

Transportation Department, Metropolitan Municipality of Istanbul, Turkey

The undesired results of the earthquake in Marmara region on 17th Aug. 1999, in Istanbul City evidently showed that, new measures for the natural disasters must be taken. All of the transportation channels and transportation elements like airports, helipads, quays etc. have the vital importance to transport the search and rescue crews first aid crews, the injuries to the secure areas and in transporting the ruins. Istanbul Metropolitan Municipality; gave start to a series of studies related to keep those transportation channels available in accordance with a disaster scenario. The studies are called "Emergency Transportation and Action Plan" and they are grouped under these titles;

i) The main arteries and alternative arteries will be used to keep transportation possible in case of a disaster, were planned and map formatted. Also, studies about the efficient use of the air, waterborne and other transportation elements were complete.

ii) The integration of local government units and secure areas (areas will be constituted of tents for very short term accomodation) were determined and converted to map format.

iii) Helipads were determined which are close to secure areas, health institutes and other logistically important units.

iv) The emergency transportation alternatives among the health institutes, fire brigades, administrative and education institutes, logistic support units, water, natural gas, power supplies and communication establishements. In the scenario, the other aspects of the rescue and evacuation facts are considered and the possible needs of emergency supplies, technical equipments were determined.

1 The Seismological and Geographical Situation of Istanbul

Istanbul is situated on a connection point that intersects Asia and Europe geographically and culturally. The Bosphorus connects the Black Sea and Marmara and divides the city into two sides called as Anatolia and Trace. Transportation between two sides is realised through two bridges and through various waterway modes.

Istanbul is the only city which has been established on two continents. The City has a known history of about 3.000 years and in time, it was ruled by 119 emperors as the capital of three greatest empires of their ages.

The population of the city is over 10 millions (1650 people/km^2). In addition, 25% of total GDP is produced in Istanbul and more than 50% of total GDP is produced in this region.

Turkey is placed on the Alp-Himalayan earthquake zone- one of the biggest three zones in the world. Earthquakes in Turkey occur on the live faults through tectonic processes among Eurasia, Africa and Arabic continents. Anatolian Plate is a smaller part of Asian Plate but a major part of Anatolia and the region that Turkey takes place, includes many smaller plates between big plates. This threatens Turkey to be situated on the earthquake zone.

During the last 2500 years in Istanbul and it's surround, there have been more than 50 big earthquakes with 8.0 Mw and over. The 92% of total area, 95% of total population and 98% of industrial and economic formation of Turkey is under the earthquake threat. In the last century, there have been 130 eartquakes in 5.0 Mw and over. The result of the quakes is approximately 80,000 deaths, 150,000 casualties and 600,000 damaged buildings.

The last one was happened in 17. Aug. 1999 with Mw 7.4. The base was Izmit cty, a 100 km close to Istanbul and resulted with losses of lifes and assets in big numbers. After the earthquake, the ineffectiveness of all the transportation systems obstacled the search and rescue, first aid, transportation, communication and other vital efforts. These undesired situation showed the lack of precautions in case of a natural disaster.

2 The General Analysis of the Problems Occured in Izmit Earthquake

The Izmit Earthquake has happened in 17. Aug. 1999 at 03:02 a.m. and took 50 seconds. The result of this earthquake was 18,000 deaths, 300,000 damages and over 1,000,000 homeless people.

In the region affected from the earthquake, all power and water supplies were demaged, telephone lines were offline, roads and railways were not available to be used, waterborne quays were defected and bridges were destroyed. Rescue teams and machines could not reach to the region at the proper time and failed to give the first aid to the victims.

Damaged Buildings	Residence	Business
Totally Damaged	3052	448
Medium Damaged	15100	2530
Less Damaged	14065	1943
Total	**32217**	**4921**

The area affected from the earthquake was 5,5% of whole Turkey (Fig. 1). However, the population in the given area is 23% of the total country and the number of industrial and economic units is very high. The density of population and establishements evidently shows that, if no precautions are taken there will be big catastrophe and huge loss of humans and materials.

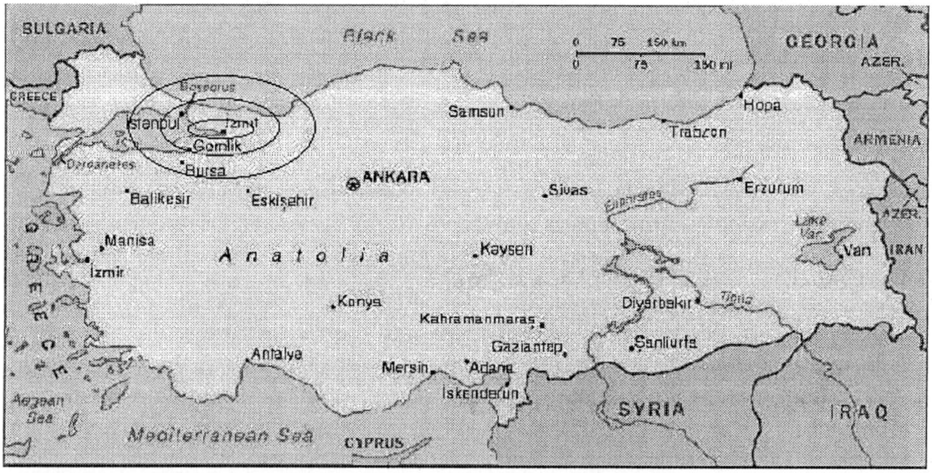

Figure 1: Centre of the earthquake.

3 The Possible Transportation Problem in Istanbul and the Precautions for the Evacuation of the Citizens

Municipality of Metropolitan Istanbul is actually studying on a serial of precautions. One of these studies is about maintaining the transportation possible in such case is called "The Emergency Transportation and Action Plan"[1]. This study is prepared in two sizes; citywide and regional (Fig. 2). All the alternative transportation modes and their alternatives were determined.

3.1 Determination of Secure Areas

In case of a possible earthquake in Istanbul, the appropriate 416 areas were selected in cooperation with the local units. These areas will be used to provide accomodation, food and health care.

The transportation systems and their alternatives for transporting the citizens to secure areas and transporting crews that will organize, train and control the citizens were determined. The crews will also be responsible of arranging accomodation and health services.

[1] "The Emergency Transportation and Action Plan" was prepared by a technical team under the supervision of Dr. Rafet Bozdogan and Simsek Deniz from Metropolitan Municipality of Istanbul, Transportation Department.

The important places like health institutes, fire brigades, schools, security centers, graveyards, water, gas, electricity distribution points, communication centers and helipads are shown on the detailed map (Fig. 3).

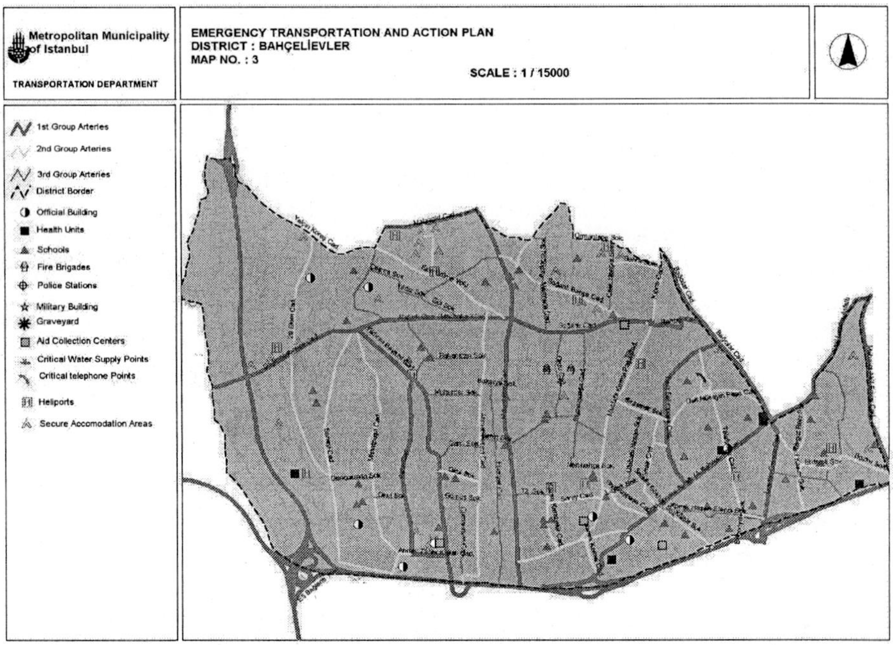

Figure 2

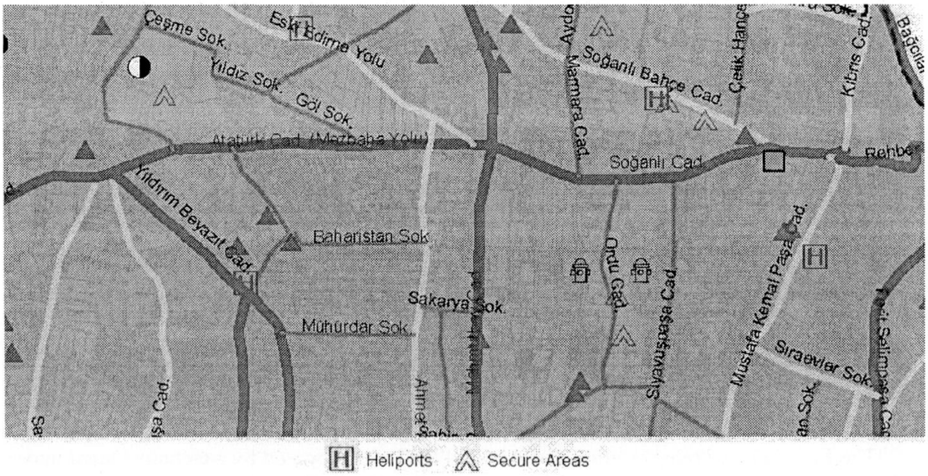

Figure 3: Part of the map showing heliports and secure areas.

3.2 Offers for Transportation Systems

In accordance with a scenario, the transportation systems in Istanbul were grouped under three modes –land based, waterborne and airway and integration of these systems were examined.

Land-Based Transportation System Offers

The traffic circulation is heavily dependent to the main arteries that are in the responsibility of Istanbul Metropolitan Municipality. These arteries were categorised into three groups, according to their importance (Fig. 4). This categorisation can be seen in the Fig. 4. For maintaining these arteries open, the first group arteries were closed to parking and second group arteries were open to parking only under control. All the alternatives of city roads are determined and shown on the maps and main arteries were integrated by two international motorways. The reason to use these arteries is reaching the strategic buildings and crisis coordination centers.

If the magnitude of the earthquake in Istanbul exceeds Mw 7.0, the roads that are 500 meters close to the sea may be flooded. At this point new main arteries will be the alternatives of the ones close to the shore.

In case of destruction of the alternative roads or lack of transportation possibilities, the crews of Istanbul Metropolitan Municipality will intervene and re-open the roads. These crews are located at 6 different points of the city.

There are 160 bridges on the roads in Istanbul. 64 of them are multipurpose structures 19 of these 64 structures are important viaducts and bridges. There is the possibility that these structures are closed by the ruins of the close buildings and this probability is considered and the plans were prepared to re-open the routes.

The bridges on the Bosphorus may be defected and become out of use. At this point, waterborne alternatives will be evaluated. There are mobile quays and they will be used in case of emergency needs. By this way waterborne transportation will be provided.

Waterborne Transportation Offers

It's estimated that, the land-based transportation systems will be seriously damaged in an earthquake so that waterborne transportation will gain more importance when the roads are out of use. The existing quays and ports are examined and improved for such case, and the number of mobile quays were increased.

In addition, the other transportation modes like passenger ships, motorboats, sea buses need to be organised to increase the effectiveness. This subject was presented to the local governors.

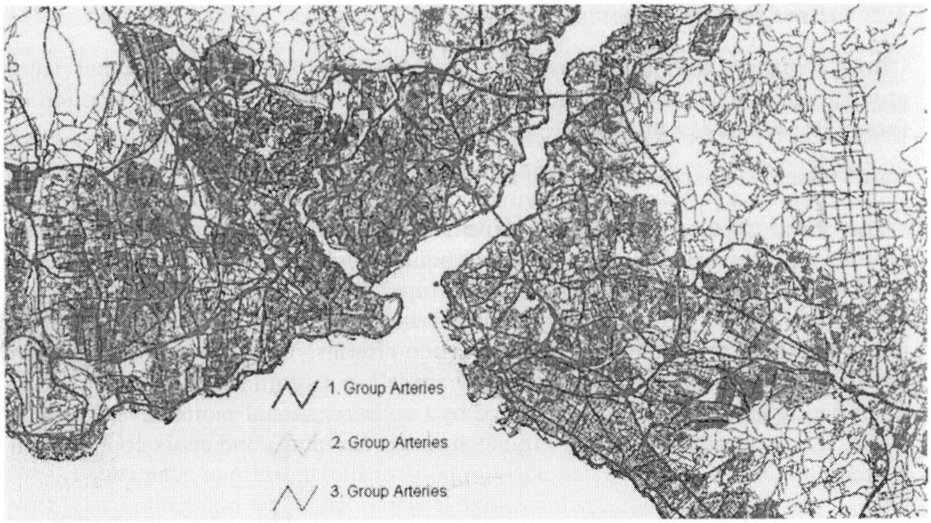

Figure 4: Map showing the categorization of the arteries.

Airway Transportation Offers

Probably, the existing international airport, Atatürk Airport; maybe damaged in an earthquake and stays out of use. At this situation, there are four alternative airports. One of them belongs to private sector, one belongs to the military, third one is a smaller airport 100 km away from Istanbul and last one is a just constructed airport and open for the civilian flights.

Helicopters have a strategic importance for the transportation in the city. Because, all the types of crews, injured people, can be transported more efficiently and more fast by the use of helicopters.

The heliports and their criterias were determined (Fig. 5). There are 250 heliport areas determined in Istanbul, to be used in case of an earthquake. The distance to the secure areas were considered in selecting the heliport areas.

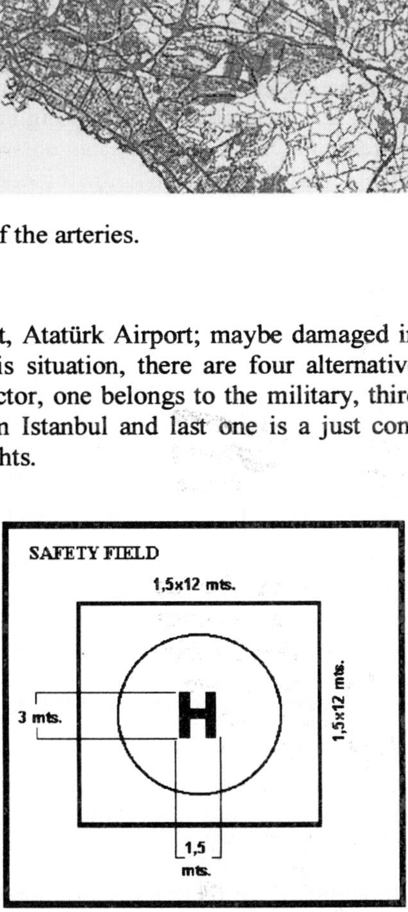

Figure 5: Heliport criteria.

Ship Evacuation

An Overview of Present and Under-Development IMO's Requirements Concerning Evacuation from Ships

M. Dogliani

Head, Innovation, Research & Products Section, RINA SPA, Genova, Italy

The maritime community is painfully aware that, in the event of a major casualty, the ship may have to be abandoned, a process which is neither easy nor necessarily successful. The need to consider this matter in the design was firstly recognised for high speed passenger craft for which an "escape path analysis" is requested by the HSC Code. In the wake of the Estonia tragedy, evacuation was carefully considered and the 1995 SOLAS Conference promulgated a regulation requiring ro-ro passenger ships built after 1st July 1999 to undertake an analysis early in the design stage in order to identify and solve by design, the potential critical points in the escape routes and life saving appliances. In January 1999, the Fire Protection (FP) Sub-Committee of the IMO developed guidelines for the evacuation analysis. However, recognising that a very limited experience and data were available on the matter and knowing that efforts were ongoing world-wide for producing a more scientifically grounded method, the Sub-Committee considered these guidelines as "interim" and allocated 3 years time for their improvements and further development. Aim of this paper is to review the background of the guidelines and to illustrate recent and on-going IMOs' developments on the matter.

1 Introduction

The maritime community is painfully aware that, in the event of a major casualty, the last resort is to abandon the ship, a process which is neither easy nor necessarily successful.

The need to stress this aspect for certain passenger ship types was firstly recognised for high speed passenger craft for which an "escape path analysis" is to be carried out at the design stage according to the HSC Code.

In the wake of the Estonia tragedy, however, the evacuation process was seriously considered and eventually the 1995 SOLAS Conference issued a new regulation, namely SOLAS II-2/28-1.3 [1] which impose to ro-ro passenger ships built on or after 1st July 1999 to undertake an analysis early in the design stage in order to identify, and solve by design, the potential critical points in the escape routes and life saving appliances.

Sometimes, when a completely new requirement enters into force, its potential benefit is initially reduced by the lack of experience which causes problems of understanding to the designers and possible lack of uniform application.

In January 1999, the evacuation analysis for ro-ro passenger ships was likely to be one of these cases: no direct experience and/or feedback was available within the maritime community and, notwithstanding several R&D studies which were initiated world wide shortly after the ESTONIA accident, little input was available for those designers who were already facing the problem of the evacuation analysis.

This situation was an unfortunate one since, in a passenger ship, escape route layout is one of the most important design premises and must be finalised, often, more than six months before starting to build the ship.

Therefore the FP Sub-Committee of the IMO decided in January 1999 [2], to develop interim guidelines for the execution of the evacuation analysis. The method outlined in the guidelines (see section 3 of this paper for further details) is the so-called "macroscopic" approach and is based on a simplified method put forward, for application in civil buildings, by the US National Fire Protection Association (NFPA) [3].

Having so solved the most urgent problem, namely provide designers with interim indications on how to carry out the analysis, the FP Sub-Committee focused their efforts on [2]:

- collecting feed-back of application of MSC/Circ. 909 (Interim Guidelines for a simplified evacuation analysis of ro-ro passenger ships [4]) with a view of their future improvement and of their possible application to other types of passenger ships;
- develop similar guidelines for application to High Speed passenger Craft (HSC);
- monitor R&D developments on the issue of evacuation from ships aiming at adopting a more refined method of evaluation when available and applicable to ship situations.

The activity developed through correspondence and through working groups meetings, as usual for IMO's regulatory developments, since January 1999 and is scheduled to be completed by February 2002. In this paper this activity and current trends are summarised.

More specifically, in chapter 2, presently in force requirements connected with evacuation are outlined and their effects on the method for evacuation analysis shortly discussed; chapter 3 is devoted to a short analysis of the most important aspects of the guidelines as well as on the simplifications, assumptions and data used; in chapter 4 the current lines of development are briefly discussed and the possible future requirements outlined while, finally, comments and conclusions are provided in chapter 5.

2 High Speed Passenger Craft (HSC)

2.1 Present Requirements

When the IMO's Code of Safety for High Speed Passenger Craft (HSC Code) was firstly developed in 1992, the matter of passengers evacuation was considered and explict requirements were included in section 4.8 of the code [5]. These are the following:

a) section 4.8.2: an evacuation procedure should be developed and should cover, among other aspects, the supervision of passengers and their orderly evacuation. The evacuation procedure should include donning of lifejackets and manning of survival craft and emergency stations before the order to evacuate the craft is given;
b) sections 4.8.3 and 4.8.4: a practical demonstration of the evacuation procedure should be conducted under controlled conditions in the presence of the Administration and should be fully documented and verified for high speed passenger craft by the Administration;
c) section 4.8.4: the evacuation time on a category B craft (these are the craft which carry more than 450 passengers) should be the time elapsed from the moment the order to abandon the ship is given until the last person has embarked in a survival craft; passengers and crew may be wearing lifejackets and prepared for evacuation and they may be distributed among assembly stations; for category A craft, passengers should be assumed distributed among passengers spaces in normal voyage conditions;
d) section 4.8.1: the provisions for evacuation should be designed such that the craft can be evacuated under controlled conditions in a time of one third of the structural fire protection time (SFP) after subtracting a period of 7 minutes for initial detection and extinguishing actions;
e) section 8.1.10: embarkation station is the place from which a survival craft is boarded; an embarkation station may also serve as an assembly station, provided there is sufficient room and the assembly station activities can safely take place there.

2.2 Effects on Evacuation Analysis

Looking at the requirements for HSC under the perspective of an evacuation analysis, it turns out that:

an evacuation analysis is to be developed and a practical demonstration is to be carried out before the craft enters in operations and that both must be approved by the flag administration;

the scenario (location of passengers, need or not to wear lifejackets, role of the crew) is specified;

the evacuation time is prescribed and includes 7 minutes time for detection and passengers' awareness;
a safety factor of 3 is considered in the evacuation time.

Moreover an HSC can not have passengers' cabins, lifejackets are typically placed under each seat and passengers spaces are very closely placed around the muster stations (which are normally the public spaces themselves). This implies that the largest portion of the total allocated evacuation time is due to people embarking liferaft through the slides of the MESs rather than the time needed for persons to travel from where they are to the embarkation station.

Having denoted with t_M (mustering time) the time needed to travel up to the embarkation station and t_E (embarkation time) the time needed for all passengers to embark liferafts, in Table 1 the results of full scale evacuation demonstration of 4 typical HSC built under RINA's class are shown. From inspection of these data, it turns out that embarkation is the most contributing portion of the evacuation.

#	pass.	crew	life saving appliances	SFP time	Evacuation time		measured t_M (*)	measured t_E (**)	Notes
					Allowed	Measured			
1	1800	27	4 MES	60'	17' 40''	15' 50''	unknown	unknown	NN
2	608	9	4 MES	60'	17' 40''	11' 11''	2' 37''	7' 39''	Evacuation broken down into groups of max. 100 persons. Attained embarkation time (t_E) for 100 persons using 1 MES: 5' 13''
3	538	6	2 MES	60'	17' 40''	15' 18''	2' 42''	12' 36''	Evacuation broken down into groups of max. 100 persons. Attained embarkation time (t_E) for 100 persons using 1 MES: 5' 15''
4	780	12	4 MES	60'	17' 40''	14' 30''	unknown	unknown	Evacuation broken down into groups of max. 100 persons.

Table 1: Results from evacuation demonstrations of HSC under RINA's class.

2.3 IMO's (Interim) Guidelines for Evacuation of HSC

Although being the first type of ship for which an evacuation analysis was prescribed, no guidelines were developed until January 2001 [6], i.e., 2 years after the guidelines for evacuation of ro-ro passenger ships were issued. The reason for this is that, as already mentioned, the evacuation analysis for an HSC is comparatively simpler than for a ro-ro passenger ship and therefore the priority for development of guidelines was placed on the latter.

In developing the interim guidelines for evacuation analysis of HSC, whose entry into force is July 1st 2002, subject to the final decision of the IMO's MSC (to be taken in May 2001, i.e., after this paper will be published), several simplifications were made within the spirit of "simplified" analysis which characterise the method adopted. The main assumptions are the following:

passengers will evacuate orderly and under the supervision of trained members of the crew; this simplification is based on the fact that an ad hoc evacuation procedure for the craft is requested and is approved by the flag administration;

recommended values of passengers travel speed are given based on data taken from civil building experience (published by the US NFPA); this is due to the fact that no specific and well documented data pertaining to this type of ships was available; it is to be noted that it is the wish of the IMO to replace these data with more appropriate ones when available;

passengers moves in "ideal" conditions, i.e., their disabilities and the effects of, e.g., craft movements in waves, smoke and possible obstructions in the escape ways are accounted for in the (large) safety factor used.

The full text of the guidelines can be downloaded from IMO's web site and is not reproduced here, however, in Appendix A an example of application is provided.

3 Ro-Ro Passenger Ships

3.1 Present Requirements

As mentioned already, ro-ro passengers ships built on or after 1st July 1999 must comply with SOLAS regulation II/2-28-1.3 whose contents are summarised in Fig. 1.

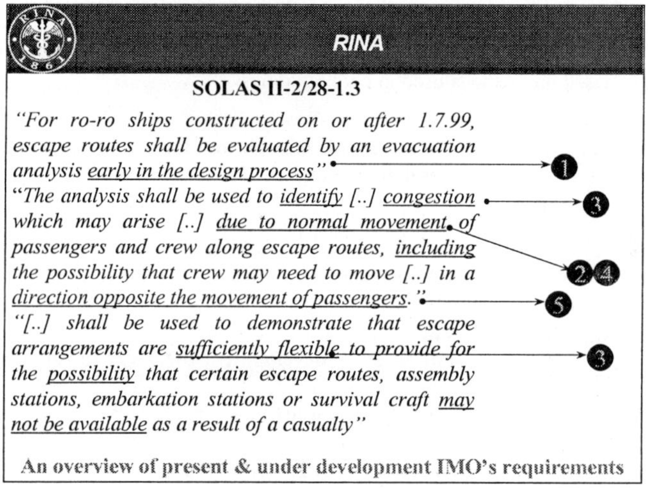

Figure 1: Presently into force requirements for ro-ro passenger ships.

In addition to that, several requirements, already in place, directly affects evacuation on this type of ships, the most important are:

- SOLAS-III-21/1.4: the maximum time allowed for embarkation and launching of survival boats is 30 minutes;

- 1995 SOLAS Conference: the evacuation of ro-ro passenger ships should be completed in 60 minutes;
- IMO's resolution A757(18), prescribing the dimensioning criteria for escape routes, stairs, stairs' landings and corridors.

3.2 Effects on the Design and Verification Processes

The entry into force of the new SOLAS requirement has considerably influenced ro-ro ship design and verification issues as follows (see also Fig. 2): accommodations layout are defined at a very early design stage and influence the structural design and the passive fire protection design; therefore the evacuation analysis need to be quickly carried out and quickly verified by the flag Administration; SOLAS II-2/28-1.3 was added to existing requirements, therefore, in developing the methodology and guidelines for its execution, care is to be taken in: not insert, in way of describing the methodology, additional requirements; considering how escape ways, stairs and doors are dimensioned (Res. A757(18)) so as to be congruent with it.

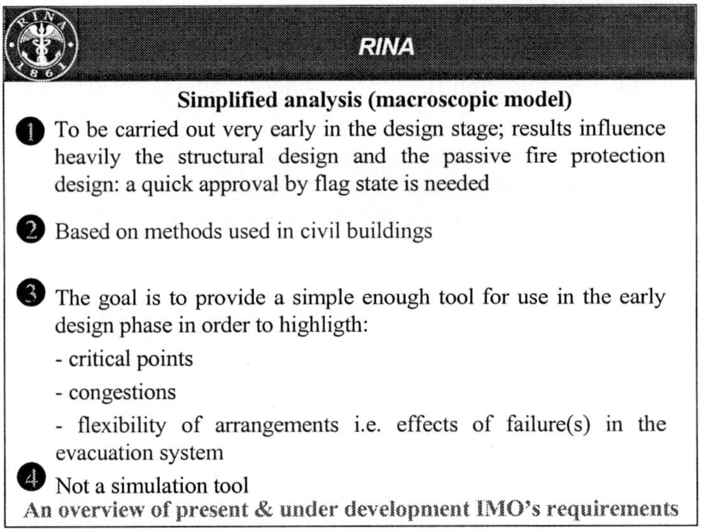

Figure 2: Main characteristics of the evacuation analysis.

3.3 IMO's (Interim) Guidelines for Evacuation of HSC (MSC/Circ.909)

The main principles of the presently into force guidelines (MSC/Circ.909) are illustrated in the following:

a) the total time for evacuation, 60 minutes, is broken down into two main phases, namely mustering (people after becoming aware that the ship is to be evacuated, move to the muster stations) and embarkation (people from muster stations pass to embarkation stations and life boats are launched); it is noted that, by SOLAS III-21/1.4 the latter phase should last no more than 30 minutes;

b) awareness time is assumed to be 5 and 10 minutes respectively in day or nigth scenarios; mustering and embarkation phases are assumed to have a certain degree of superposition, see Fig. 3;

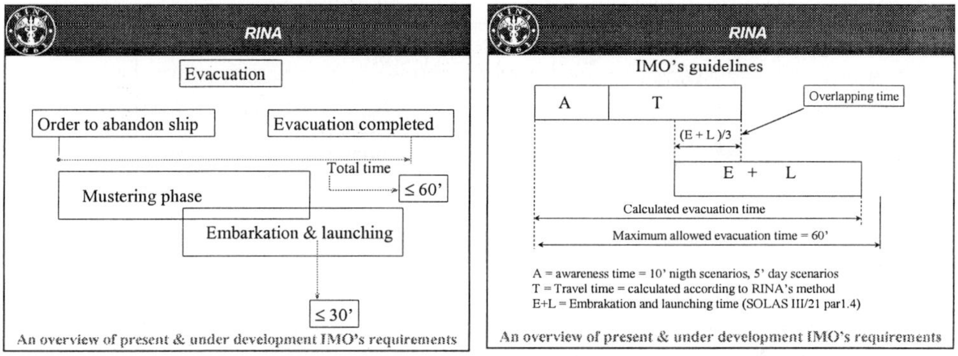

Figure 3: Phases of evacuation as considered in the interim guidelines.

c) the analysis is to be carried out in at least two scenarios; these are the scenarios specified by the IMO for the design of escape ways (IMO's Res. A757(18)), namely night scenario (all passengers in cabin, a fixed % of the crew in cabin) and day scenarios (passengers in public areas, crew distributed in working spaces);

d) the analysis is to be made with 100% of passenger loads i.e. by considering the ship with all the passengers she is certified to carry;

e) the evacuation time is then calculated as the sum of the mustering time and of the embarkation time; the latter is to be based on data provided by the manufacturer of the life boats and launching system or on results from full scale trials, in case no such information are available, the highest allowable value, namely 30 minutes, is taken;

f) in the calculation of the mustering time, besides the awareness time already discussed, the assumption is made that persons move in ideal conditions, therefore an "ideal" mustering time is calculated which is them multiplied by a safety factor 2 and then an additional 30% is added to account for the possibility of counterflow in a real emergency (see Fig. 4);

g) the speed of people in an ideal evacuation is taken from US NFPA (civil buildings) and varies as a function of type of facility (ascending stair, descending stair, corridor) and on the density of people (see Fig. 5).

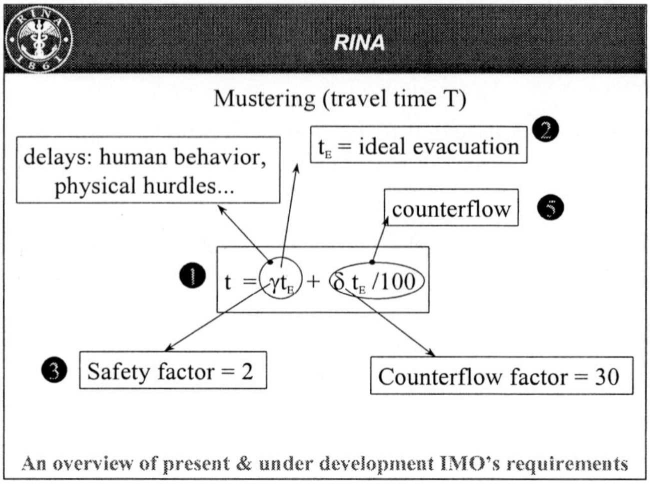

Figure 4: Calculation of the travel time.

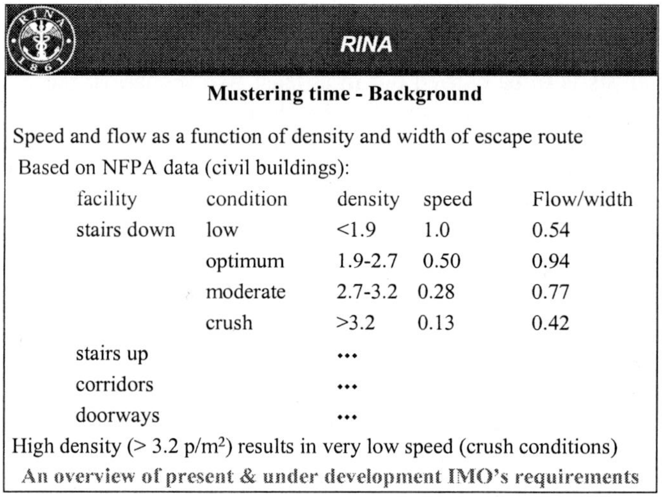

Figure 5: Person's speed in ideal conditions.

4 Current Developments at the IMO

Although guidelines have been already developed for ro-ro passenger ships and HSC, the matter of evacuation is not completely finalised at the IMO, the reason being that the guidelines are, for the time being, interim and the process of their improvement and revision is already underway.

However, this is the least important reason why evacuation analysis is still high on IMO's agenda, more important issues are the following:

1. development of guidelines for the evacuation analysis based on more scientific methodology then the one presently considered;
2. extension of the applicability of the present guidelines (suitably modified) to deal with all passenger ships, i.e. including cruise ships;
3. application of the evacuation analysis to existing passenger ships and development of the relevant acceptance criteria;
4. safety of large passenger ships, the so called jumbo-cruise ships, expected to be requested by the market in the next 5 to 10 years: how to deal with evacuation of 6000 people?

The latter issue is perhaps the most important and evacuation is just one of the items which the MSC group on large passenger ship safety is considering.

5 Conclusions

In December 1995 the decision was taken at the IMO that ro-ro passenger ships should undertake an evacuation analysis early in the design stage. This decision resulted in a new SOLAS regulation which entered into force on 1st July 1999.

Since then, considerable progress was achieved: guidelines, although interim, are available for executing the evacuation analysis on ro-ro passenger ships and on HSC.

As illustrated in this paper, however, significant work is still to be carried out, in particular as far as:

- extension of the applicability of the guidelines, as necessarily modified, to cruise ships; this step should be finalised by January 2002;
- development of guidelines for application of more sophisticated and scientifically grounded evacuation models; this step should be finalised by January 2002;
- assess the applicability to existing ships and determine criteria of acceptance;
- consider the evacuation process in the wider problem of safety of very large cruise ships.

In the nest few years considerable assistance on these aspects is likely to be requested to Universities and Research Centres in order to contribute to solve these issues and, at the end of the day, increase safety of shipping.

References

1. International Maritime Organisation (IMO), *International Convention for the Safety Of Life At Sea (SOLAS)*, Chapter II-2 (Fire protection), Regulation 28-1.3, Consolidated edition, (1998).

2. International Maritime Organisation (IMO), *Draft report of the 43rd Session of the Fire Protection (FP) Sub-committee*, FP43/WP.6, (January 1999).

3. US National Fire Protection Association (NFPA), *The SFPE Handbook of Fire Protection Engineering*, 2nd Edition, (1995).

4. International Maritime Organisation (IMO), MSC Circular n. MSC/Circ.909, *Interim Guidelines for a simplified evacuation analysis of ro-ro passenger ships*, (4 June 1999).

5. International Maritime Organisation (IMO), Code of Safety for High Speed Craft (HSC Code), (1992).

6. International Maritime Organisation (IMO*), Draft report of the 45th session of the Fire Protection (FP) Sub-committee*, document FP45/WP.8, (January 2001)

Appendix A – Example of Application to an HSC

A.1 General

The example provides an illustration on the application of the interim guidelines. Therefore it should not be viewed as a comprehensive and complete analysis nor as an indication of the data to be used. More specifically, the embarkation time and the deployment time used in paragraph 4 below are purely illustrative.

A.2 Craft Characteristics

The HSC considered is a Category B craft with a total capacity of 800 persons (784 passengers and 16 crew members). As shown in Fig. 1, **when the order to abandon the craft is given**, passengers are distributed on **the public spaces on** two decks (210 on the upper deck and 574 on the lower deck), the lower one is equipped with 4 MES. The structural fire protection time (SPF) is 60 minutes.

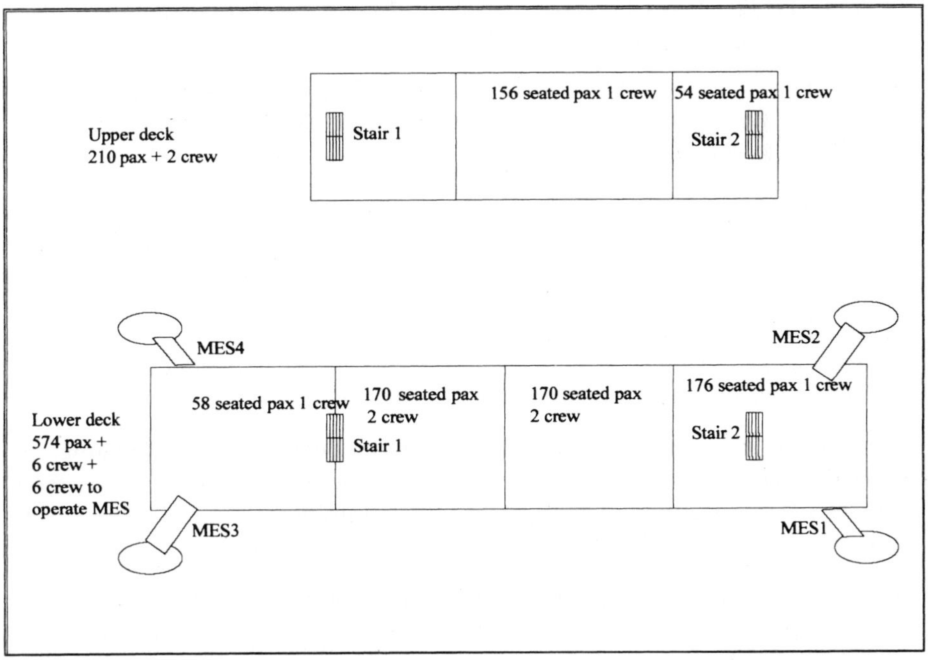

Figure A1: Sketch of the considered HSC.

A.3 Description of the System

1 Identification of Assembly Stations

Assembly stations coincide with the public spaces where passengers are located (seated). Passengers are wearing life jackets.

2 Identification of Embarkation Stations, MES and Liferafts

2.1 Embarkation stations (4, one for each MES) are located at the lower deck.

2.2 Each MES consists of an inflatable slide with an attached platform.

2.3 Liferafts (11), 100 persons capacity each, are stowed in racks on the lower decks, in proximity of the MES. The aggregate capacity of liferafts is therefore 1080 persons, or of 810 persons if one embarkation station is not available in accordance with the 2000 HSC Code.

2.4 Two rescue boats are available for marshalling the liferafts.

3 Description of the Evacuation Procedure

3.1 When the order to abandon the craft is given, crew members start operating the MES (total 6 crew members), the rescue boats (1 crew members per boat) and to direct the passengers (as shown in Fig. 1: 2 crew members on the upper deck and 6 crew members on the lower deck).

3.2 PHASE 1: for each MES, the slide is inflated and the first liferaft launched, inflated and connected to the slide's platform. In the mean time the first 4 groups of persons (each composed of 99 passengers + 1 crew member) move to the embarkation stations.

3.3 PHASE 2: once the first liferaft is ready for boarding, the first group for each MES descends to the liferaft using the slide and platform. When boarding is completed, the liferaft is detached from the slide and the second liferaft is launched, inflated and connected to the platform.
In the mean time the second 4 groups of persons (each composed of 97 passengers + 1 crew members) move to the embarkation stations.

3.4 PHASE 3: once the second liferaft is ready for boarding, the second group for each MES descents to the liferaft through the slide and platform. Finally, the 6 crew members operating the MES board. When boarding is completed, the liferaft is detached from the slide. The evacuation is now completed.

4 Identification of Groups and their Escape Routes.

In total 8 groups, each composed of 100 persons, are considered. Their (primary) escape routes are shown in Fig. 2 for the first 4 groups and in Fig. 3 for the second 4 groups.

5 Calculation of t_E and of t_M

5.1 Embarkation time t_E
According to the evacuation procedure, each MES is used by 200 persons, if all four MES are available. Based on full scale trials on craft having similar arrangements and using the same MES, the total time needed to deploy, inflate and moor-

ing the life raft and to embark 100 persons is 330 s (5 minutes and 30 seconds). Accordingly, the total embarkation time is 660 s (11 minutes).

5.2 Deployment time t_M
Based on full scale trials on craft having similar arrangements and using the same MES, the total time needed to deploy and inflate an MES is 150 s.

6 Calculation of t_I

6.1 For the purposes of this example, it is assumed that calculations have been carried out for all the 8 groups of people into which the evacuation is organized, according to the evacuation procedure (para 3.3 above) and that the highest travel time is that of the first two groups i.e. the groups of people moving from the afterward passenger area in the upper deck down to MES 3 and 4 on the lower deck.

6.2 The schematization of the escape route is shown in Fig. 4. As it may be seen, the elements composing the escape path are 2 doors, 2 corridors and 1 stairway.

6.3 The characteristics of the escape path's elements are given in Table A1.

Element	L (m)	W_c (m)	F_s	S (m/s)	F_c (p/s)	N people
Door 1	N.A.	1.4	1.3	N.A.	1.82	165
Corridor 1	14	4.2	1.3	0.67	5.46	165
Stairway 1	4.7	3.5	1.1	0.55	3.85	165
Corridor 2	14	3.0	1.3	0.67	3.90	200
Door2	N.A.	1.4	1.1	N.A.	1.54	100

Table A1: Characteristics of the escape routes.

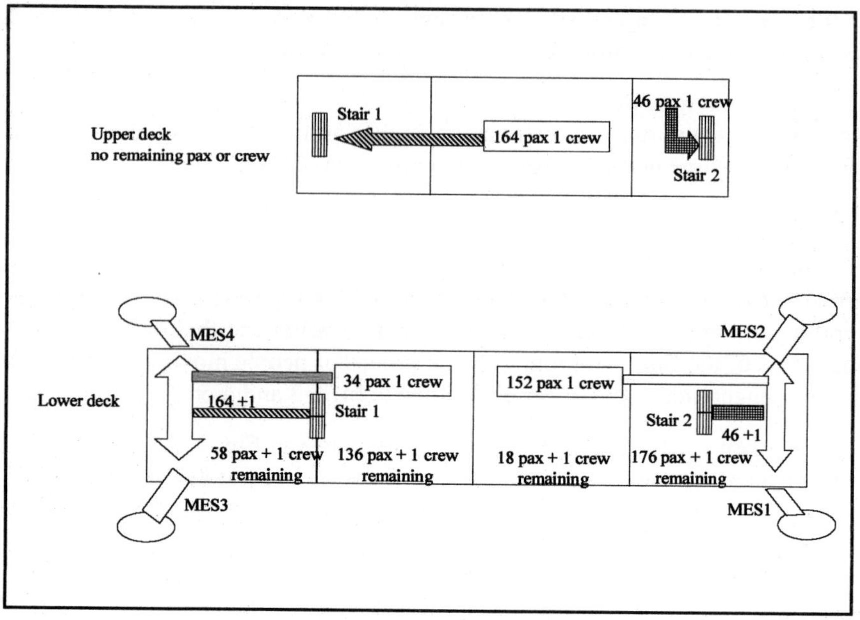

Figure A2: First 4 groups of persons (each 100 people).

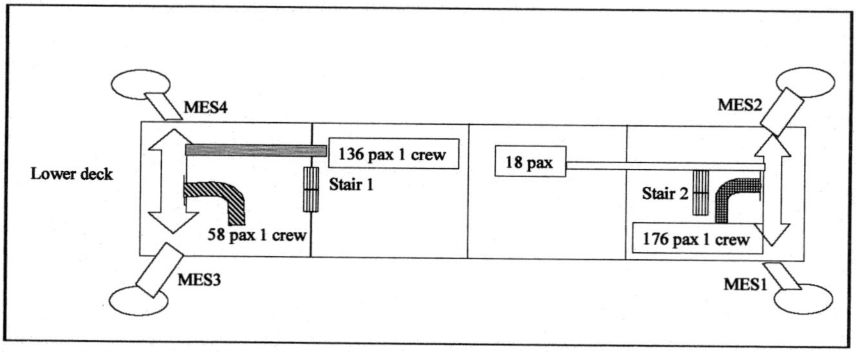

Figure A3: Second 4 groups of persons (each 100 people).

The values of specific flow (F_s) and speed (S) are taken from the guidelines, accordingly, the value of calculated flow (F_c) is obtained as $F_c = F_s W_c$.

The resulting walking time (t_w) and flow time (t_F) are given in Table A2.

Element	L (m)	W_c (m)	N people	t_W (s)	t_F (s)
Door 1	N.A.	1.4	165	N.A.	91
Corridor 1	14	4.2	165	21	30
Stairway 1	4.7	3.5	165	9	43
Corridor 2	14	3.0	200	21	51
Door2	N.A.	1.4	100	N.A.	65

Table A2: Walking and flow time for each element in the escape route.

The resulting total walking time is the sum of the walking time of each element in the escape path and totals 51 s. The flow time is the highest among all the elements in the escape path and corresponds to 91 s.

Accordingly, the ideal travel time, t_I is 142 s.

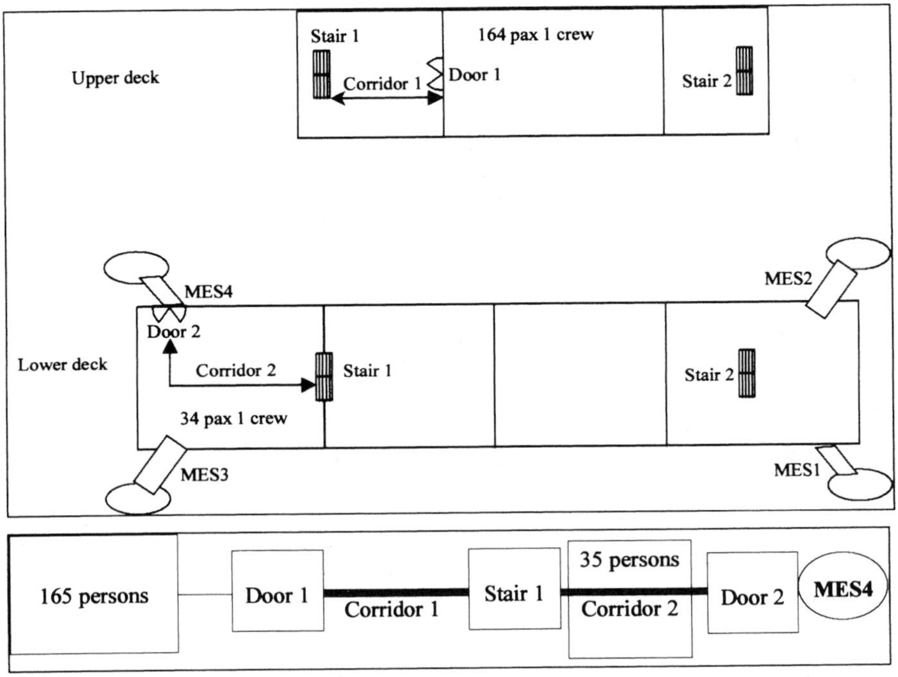

Figure A3: Sketch of the evacuation path and its schematization.

7 Performance Standard

The overall evacuation time: $t_M + t_E = 150 + 660 \leq$ (SFP–7)/3 minutes = 1059 s

$t_I + t_E = 142 + 660 \leq$ (SFP–7)/3 minutes = 1059 s

The requirements are fulfilled.

EVAC and LBL

The Mustering and Evacuation Computer Models Resulting from the BriteEuram Project MEPdesign

K.H. Drager[1], H. Soma[2], and S. Orset[1]

[1]A/S QUASAR Consultants, [2]SAFETEC-Nordic, Norway

In the last years there have been some severe accidents with passenger vessels. The accidents have demonstrated a clear need for improvements, and highlighted deficiencies inherent with the safety regime in force. The basis for this regime is international co-operation through the International Maritime Organisation (IMO) and specific safety requirements on a detailed level, in combination with approvals from national authorities. IMO has recognised that computer simulation of the Evacuation may be required for new ship concepts, and work is ongoing with specification of the requirements for such computer tools. The Norwegian consulting company A/S QUASAR Consultants has through a 15 years period developed computer programs for Evacuation simulations, and performed Evacuation studies of about 30 installations and buildings. The ownership of the programs has now been transferred to the Norwegian consultant company SAFETEC, while QUASAR and SAFETEC co-operates in program development and projects. These programs have been based upon available data from accidents, research projects and empirical observations.

1 Background

The evacuation process of a passenger vessel may suitably be divided into the following phases; Mustering, Embarkation and Evacuation. The paper discusses the problem area by addressing, decision taking, behaviour of passengers, guidance, effect of ship motion and list and effects of mustering group wise.

Seven European companies with different expertise decided to co-operate in resolving the above explained problems related to passenger ship evacuation. They agreed on a project plan in 1997, and applied for support from the EU commission. The support was granted from the *Industrial and Materials Technological Research and Technological Development program.* Hence, in 1997 the EU project *Mustering and Evacuation of Passengers: Scientific Basis for Design (MEP-design)* was launched.

The mustering simulation program EVAC is developed as part of the MEP-Design Project. It utilises data and knowledge created by the MEPDesign project, but the program is also based on experience from earlier programs. The objective of EVAC is to simulate the mustering operation on passenger vessels. The mustering time is the main result from the simulations.

The initial passenger distribution is user defined. Individual passengers are assigned properties based on weighted drawings from data for the population. Hence, individuals, each with his distinct property, constitute the basic entity for the simulations. The motion of each passenger, including his interactions with other evacuees, is simulated. The EVAC model is therefore microscopic, as it is very detailed. (The opposite would be typical flow models, which are macroscopic).

Prior to the full scale Evacuation test with Kronprins Fredrik, the mustering was predicted with EVAC, and documented in a report, based on assumptions with respect to number of passengers and their distribution. These assumptions proved afterwards to be reasonably correct. The predicted mustering time was 20 minutes, while the recorded mustering time at the test was 24 minutes. These figures show an acceptable correspondence, and give an indication that the program may be reasonably correct, even though it also is possible that this correspondence may be coincidental.

The computer program LBL was developed several years ago for simulation of lifeboat launching from offshore installations. In the MEPdesign project, the program has been updated to be more suitable for simulating the launch of a lifeboat from a ship exposed to waves. Further, the simulations are visualised by a powerful 3D graphical package. The objective of LBL is to simulate the launch of lifeboats from a ship exposed to waves. The main results are the collision probability and collision speed distribution. Further, slamming loads and effects of possible release delay are calculated.

Based on the work carried out, as well as on earlier experience with evacuation analyses of ships and offshore installations, the above described computer programs represent a significant potential for safety improvement of passenger vessels, provided that the movement away from detailed requirements towards Goal Setting requirements proceeds.

Ship	Date of Accident	Accident	Fatalities
Herald of Free Enterprise	March 1987	Rapid capsize, high list	193
Scandinavian Star	April 1990	Fire, toxic smoke	158
Moby Prince	April 1991	Fire	140
Estonia	Sept 1994	Rapid capsize, high list	852

Table 1: Accidents with passenger vessels.

In the last years there have been some severe accidents with passenger vessels, as referred in table 1.

The accidents have demonstrated a clear need for improvements, and highlighted deficiencies inherent with the safety regime in force. The basis for this regime is international co-operation through the International Maritime Organisation (IMO) and specific safety requirements on a detailed level, in combination with approvals from national authorities. The regime suffers from similar deficiencies as those strongly criticised in the UK investigation following the severe accident with the oil platform Piper Alpha. The answer to that accident was the Safety Case Regime and Goal Setting Requirements.

2 Legislation Regime

In IMO, it is a major concern that safety legislation shall not impose competitive benefits for any group of ship owners. This concern has gone so far that IMO hardly recognises that the requirements to evacuation means must be far stricter in areas where evacuees may be exposed to a very cold and hostile environment than in warmer areas. With respect to evacuation programs, IMO intends to develop a standard, which shall yield "equal" strict requirements for all ships. This promotion of equal competitive conditions goes on the cost of realism and ability to adopt new knowledge in the analyses. Evacuation and human performance are such complicated matters that it is not possible to develop an objective model, which experts may agree upon. Further, development of new knowledge is an ongoing process, and standards are therefore deemed soon to be out-dated. The oil production industry has faced similar problems in performing risk assessments. They have realised that even though there are different weaknesses with most models, they give a high contribution in the achievement of a high safety standard.

If passenger vessels had been treated in a similar way as offshore installations, the ship owner would be responsible for identifying and evaluating all potential accidents, and demonstrate the efficiency of the safety measures implemented to prevent and mitigate the accidents. Hence, he would be responsible for demonstrating that evacuation was sufficient safe for the accidents in question.

Work is ongoing in IMO, which may imply a step in the above direction. It may be required to demonstrate that evacuation can be performed within a certain time limit (1 hour). However, this limit will be a specific requirement, and not be based upon an analysis of accidents that may occur. Further, reliability requirements with respect to evacuation means will not be stated, despite the fact that potential failure and accident with lifeboats, slides and chutes is a highly recognised problem. Anyhow, the process goes in the right direction, and IMO has recognised that computer simulation of the evacuation may be required for new ship concepts, and work is ongoing with specification of the requirements for such computer tools.

3 Experience with Computer Simulation

The Norwegian Consulting Company QUASAR has through a 15 year period developed computer programs for Evacuation simulations, and performed Evacuation studies of about 30 installations and buildings. The ownership of the programs has now been transferred to the Norwegian consultant company SAFETEC, while QUASAR and SAFETEC co-operates in program development and projects. These programs have been based upon available data from accidents, research projects and empirical observations.

Evacuation of ships involves some particular problems, relating to passenger performance in case of list of the ship and on their moving capability in a ship exposed to waves. Lack of relevant empirical data has represented a severe limitation with respect computer simulation of passenger ship evacuation, but this situation is now greatly improved, as explained in this paper.

4 Evacuation of Passenger Vessels

The evacuation process of a passenger vessel may be divided into the following phases:

* Mustering
* Embarkation
* Evacuation

It should be noticed that the term "evacuation" is ambiguous, as it usually is used both to denote the total evacuation process as well as to denote the particular operation of leaving the ship in lifeboats or other evacuation means. It has therefore to be seen from the context what is meant with the word.

When the captain has decided to muster the passengers, the alarm will be activated and PA announcements will inform passengers and crew about the situation. The passengers are assumed to leave their cabins and walk along the marked escape ways to their muster areas, where the crew will support them in donning their life vests. If the situation deteriorates, embarkation of the lifeboats, which in the meantime have been swung out to a suitable position, will be started. The lifeboats will be sequentially launched. This is an operation involving severe hazards, in particular in adverse weather conditions.

Some of the crew will systematically search the cabins and others will be posted at strategic positions to guide passengers to their muster stations. The crewmembers shall be dressed in their uniforms to be perceived as authority persons by the passengers. This contributes in preventing inappropriate or dysfunctional behaviour among passengers. Otherwise, they may adopt the behaviour of other persons appearing to behave with authority, who in the worst case this may be panicking people.

5 Problem Areas

Decision Taking

In case of a severe accident on a passenger vessel, the captain is responsible for deciding whether the ship has to be evacuated. The history shows several examples that delay to take a decision or incorrect decisions have contributed in creating disasters. On the other hand, mustering and evacuation always involve hazards. The stress involved may cause heart attacks on passengers, and launching of lifeboats is always a risky operation. For this reason it is understandable that captains may hesitate to start evacuation when they are uncertain about the seriousness of the situation. Hence, it is of the utmost importance to install safety systems, which provide the captain with adequate decision support in emergencies.

Behaviour of Passengers

Passengers may show a range of behaviour patterns when the alarm sounds. Hearing disabilities or extremely deep sleep caused by sleeping medicine or drugs may prevent some passengers from hearing the alarm. Some passengers may wait for the crew to rescue them in their cabins, while others immediately may start to evacuate. Different types of disabilities may prevent some passengers from mustering. Most passengers will try to find some persons with authority to follow. In situations with flooding or list of the ship, passengers will try to get up in the ship and out as fast as possible, for fear of being trapped. In severe accidents, dysfunctional behaviour like panic and apatia may dictate the behaviour of some passengers.

Guidance

Guidance to passengers during mustering is given in terms of signposting along the evacuation route and by crewmembers who are posted along the escape route.

The traditional way of signposting is to attach luminescent signs with arrows and symbols at eye height at decision points along the escape route. There are two main deficiencies involved, as highlighted by the Scandinavian Star disaster: In the exposed areas where passengers perished, escape way marking lead towards the scene of the fire. However, due to short visibility in the dense smoke, signposting played a minor role and did neither guide evacuees to a certain death nor to safety.

The Scandinavian Star disaster shows that it would have been very desirable to provide a dynamic signposting system carrying directional information, by which the captain could direct passengers through the safest routes in a fire event. This is possible to achieve with the present technology, i.e., by moving lights (Light Emitting Diodes -LED technology). To utilise the potential of this system, it has to be supported by an efficient decision support system, which gives the emergency management correct information about the accident development. Otherwise accidents could occur, similar to the fire at the King Cross station in London, where evacuees were directed in the wrong direction with several fatalities as the consequence.

The other problem relates to how efficient information is communicated to the evacuees. There are several reasons that evacuees may overlook the information, as follows:

- Competing information and announcements which make it easy to oversee the evacuation signposting.
- The signposting may not be sufficient striking to catch the attention of the evacuees.
- Darkness or bad illumination.
- Reduced visibility due to smoke.

Effect of Ship Motion or List

Ship motion and list may make walking difficult and even impossible. Such events may further create dysfunctional reactions among the passengers. The experience from accidents shows that passengers in a listing ship have a strong desire for walking up in the ship and out, for fear of being trapped in a capsizing or sinking vessel. Until recently there did not exist much empirical data on the effect of list and motion of a ship on the motion capabilities and behaviour of evacuees.

Effects of Mustering Group Wise

The experience from accidents, shows that people who are related to each other are likely to muster as a group. If they are separated when the mustering starts, they may try to find each other rather than to muster individually. During the mustering, they will move as a group and take decisions together. The effects of such behaviour have not been studied until recently.

6 The MEPdesign Project

Seven European companies with different expertise decided to co-operate in improving the evacuation safety by resolving the above problems. They agreed on a project plan in 1997, and applied for support from the EU commission. The support was granted from the *Industrial and Materials Technological Research and Technological Development program.* Hence, in 1997 the EU project *Mustering and Evacuation of Passengers: Scientific Basis for Design (MEPdesign)* was launched.

The project will be finished in March 2001. The main objectives of the project are to improve the evacuation process of passenger vessels and to provide computer models with adequate data bases. The studies, which are most relevant with respect to computer program development, are shortly described in the following.

TNO (Netherlands) is the project leader of MEPDesign. Their main professional contribution relates to empirical studies on behavioural aspects of evacuating passengers. TNO used mock-ups of passenger cabins and a motion simulator to study passenger motion in corridors and stairs, which were listed to represent a damaged ship or moving to represent ship motions in defined sea states. The volunteers, who participated in the empirical study, constituted a representative sample of the population. In the presentation of the results, they were divided into four age categories. TNO also performed an empirical study on *Improved Signposting for the Evacuation of Passenger Ships.* In this study the following variables were addressed:

- Traditional route signs, Low Location Lighting (LLL) as required by IMO, photo luminescent stripes with directional arrows and moving lights (LED).
- Mustering in light or dark.
- Mustering as group or individually.
- Ship list or level.

The performance of the evacuees with respect to way finding errors and mustering time, were recorded and analysed with respect to statistical significance. The main result was insufficient potential of existing way finding systems, despite they were compliant with existing regulations. Measures for improvement were implemented and tested.

DMI and Scandlines (Denmark) were responsible for the practical studies. They collected population and behavioural data for passengers with particular focus on group binding and with respect to their tendency to collect their luggage before mustering.

A full scale mustering exercise was performed with the passenger vessel Kronprins Fredrik. About 600 passengers participated in the exercise. They were informed they were going to participate in some sort of tests, and they were offered meals and "tax-free" for about 25 US dollars. When the alarm sounded, the majority was dining in restaurants. The mustering was performed in an orderly way according to the procedures. The mustering process was monitored and recorded for later references.

KTH (Sweden) performed comprehensive model tests with launching of lifeboats and with a slide Evacuation system. A particular free fall system was also investigated. Kronprins Fredrik was chosen as mother vessel, and a simplified model was developed, with the same motion characteristics.

Launches of a lifeboat, both from the windward and downwind sides, were tested in 1, 2 and 3 m. high regular waves. The ship was oriented broadside the oncoming waves. The impact on the collision probability from varying initial clearances between the lifeboat and the shipside and different launching speeds were investigated. Accelerations in collisions with the shipside was measured, and the likelihood of the "open" lifeboat to be flooded or to capsize when it become sea borne was assessed. The tests showed a high likelihood of failures in 3 metres, and even in 2 metres wave heights.

Tests performed with a slide system showed also this system to be prone to failures.

DNV (Det norske Veritas) and IFN (Institute Francais de Navigation)) are responsible for evaluating the pragmatic value of the project.

QUASAR/SAFETEC develop computer programs.

Note that the views expressed in this paper, represents the view of the author and not of the MEPDesign consortium.

7 The Mustering Simulation Program EVAC

Background

The mustering simulation program EVAC is developed as part of the MEPDesign Project. It utilises data and knowledge created by the MEPDesign project, but the program is also based on experience from earlier programs.

Objective

The objective of EVAC is to simulate the mustering operation on passenger vessels. The mustering time is the main result from the simulations. Bottlenecks in the escape way system are also identified.

Method

The initial passenger distribution is user defined. Individual passengers are assigned properties based on weighted drawings from data for the population. Hence, individuals, each with his distinct property, constitute the basic entity for the simulations. The motion of each passenger, including his interactions with other evacuees, is simulated. The EVAC model is therefore microscopic, as it is very detailed. (The opposite would be a typical flow model, which is macroscopic).

EVAC performs Monte Carlo simulations as several of the variables in the program are assigned values based on weighted drawings from probability distributions. Hence, if the same simulation case is repeated several times, the results will differ depending on the results of the weighted drawings. Therefore, a large number of replications are calculated in a particular case. The number of replications should at least exceed 15-20. The results from the replications are used for predicting evacuation times in statistical terms.

EVAC models the evacuation as a stationary Markov process. There are several states of the process, which may change from one state to another one, depending on the transaction probabilities between states. The process is stationary and memoryless as exemplified in the following. When an evacuee leaves his initial position and start evacuation, the process changes state. During each time step before he starts to evacuate, the probability that he starts walking is the same, implying that the process is stationary. Further, the process is memoryless because the probability that he starts to evacuate does not depend upon how long he has waited or on what has happened while he has been waiting

Scenario Definition

An evacuation scenario with specified environmental conditions, ship heading relative to oncoming waves, potential initial heel, etc., forms the basis for the analysis.

Individual Properties

The user may chose to use representative default distributions with respect to age and social relation among passengers, or he may define his own distribution.

Each individual has his own properties with respect to walking speed in different situations, reaction time, group relation and binding to his luggage. He will be assigned his individual properties based on weighted drawings from statistical data. First, his age category is predicted, based on statistical data on age of passengers. Then his behavioural data, like walking speed, are predicted based on statistical data for the age category in question.

Escape Way System

The escape way system is modelled by marking escape ways and stairs on raster copies of arrangement drawings, which are shown on the computer screen. The program automatically links together stairs on different levels. This way of collecting data for describing the escape way system is rather fast as tedious measurements and input of numbers is avoided. At a later stage, however, it may be relevant to take the data directly from CAD models.

Personnel Distribution

The initial personnel distribution must be defined by the user. He must specify how many persons are present in different areas and rooms.

Initial Delay

The expected initial delay of passengers may be based on a default distribution, or it can be specified by the user. The delay should depend upon the accident scenario. In some scenarios the evacuees are not likely to know whether the evacuation is initiated by a false alarm or by a real accident. In other scenarios, the initiation may be violent, causing either an immediate evacuation or severe delay due to apatia. Hence, the user has to select appropriate data for the scenario in question.

Way Finding

Evacuees are in general assumed to follow the defined and marked escape route. However, they may choose an incorrect route at some instances. In particular, this may happen when they leave their cabins. If they have chosen a bad route, they are assumed to follow this route until they reach the next decision point.

Walking towards Destination

The walking speed of a passenger will vary along his route. He walks with his individual speed when walking in corridors and stairs. If the ship lists, the speed depends on the orientation of the corridor or stair relative to the inclination axis, as they may become more or less steep. EVAC calculates inclination data for all escape ways, and interpolates correct factors for walking speed reduction, based on the empirical data from the TNO tests. Further, there may be obstructions and corners throughout the escape route, which may cause further delay.

Result Presentation

EVAC calculates evacuation times for each replication of a simulated case. Instead of developing a dedicated routine for the statistical analysis, it has been chosen to apply a statistical program package for this purpose. SAFETEC uses SYSDAT for this purpose, but any statistical package can be used. The main results will usually be mustering time for the last evacuee, the time when 95 percent of the evacuees have mustered and average mustering time. These times will be described by result or confidence intervals. It may, e.g., be shown that the maximum evacuation time will be 20 minutes, the most likely 15 minutes and the minimum 12 minutes. With respect to average evacuation time, the confidence interval will be calculated. An increased number of replications will reduce the confidence interval, but not the result interval.

In addition to the above results, a lot of intermediate results from individual replications can be presented. Counters can be defined at selected positions on the escape ways, and the flow of passengers can be calculated at these locations. The evacuation process, showing the motion of evacuees in the escape way system, can be visualised on the PC screen.

Validation

EVAC is based on empirical data with respect to the motion capabilities of persons. The validity of the program in this respect may therefore not be questioned. What is of much concern, is the validity with respect to human behaviour in severe accidents. The experience from accidents shows that people will behave relatively rationally, but this does not imply they will follow the procedures.

They may walk up and out in a listing ship whatever the procedures tell, and they may try to find and support friends and relatives instead of mustering. Some of them will even be so attached to their luggage that they may attempt to pick it up before mustering. Such behaviour is implemented in EVAC, based on experience from passenger vessels and also some judgement. Dysfunctional behaviour has been decisive for the outcome of some accidents, but these events should rather be considered exceptions than the rule. Hence, extreme apatia and panic are not implemented in EVAC.

Prior to the full scale evacuation test with Kronprins Fredrik, the mustering time was predicted with EVAC, based on assumptions with respect to number of passengers and their distribution. These assumptions proved afterwards to be reasonably correct. The study was documented in a report. The predicted mustering time was 20 minutes, while the recorded mustering time at the test was 24 minutes. These figures show an acceptable correspondence, and give an indication that the EVAC predictions may be reasonably correct, even though this correspondence may be coincidental.

8 The Lifeboat Launching (LBL) Program

Background

The computer program LBL was developed several years ago for simulation of lifeboat launching from offshore installations. In the MEPdesign project, the program has been updated to be more suitable for simulating the launch of a lifeboat from a ship exposed to waves. Further, the simulations are visualised by a powerful 3D graphical package.

Objective

The objective of LBL is to simulate the launch of lifeboats from a ship exposed to waves. The main results are the collision probability and collision speed distribution. Further, slamming loads and effects of possible release delay are calculated.

Method

The program is based on theoretical models of the physical phenomena involved. Regular waves as well as stochastic sea states based on Jonswap or Newman energy spectra are included. The wind model is also based upon a stochastic Davenport energy spectrum. The waves are unidirectional, and the ship can be oriented at any heading. The ship response is calculated based on Response Amplitude Operator (RAO) data for the ship.

The main assumption in LBL is that the launching is initiated at random time in relation to waves and ship motion. This assumption is reasonable, and makes the model probabilistic.

The motion of the fallwire suspension points are calculated based on the ship motion. The pendulum motion of the lifeboat during descent is calculated based on the impact of the moving suspension points, wind gusts and collisions with the ship side.

The lifeboat may be assumed to be immediately released from the fallwires when it is freely floating, or alternatively the time required for release of the lifeboat may be assessed based on a probability distribution. In this case it is possible that the boat may be snatched out of the sea again when a wave passes and the boat moves down behind the crest or because the fallwire suspension points moves up. This event may cause breakage of a fallwire.

As the boat is small as compared to the length of the dominating wave components, the motion of it when sea borne is calculated based on drag forces, thrust, surfing force on sloping wave surface, rudder moment and inertia forces. This implies that focus is put on the horizontal motion of the boat and the possibility that it may be swept back by waves and collide with the ship side. In this event, the collision speed is calculated.

Collisions may damage the lifeboat. However, even without severe damages, the acceleration loads to the poorly protected passengers may cause severe injuries and fatalities.

LBL performs Monte Carlo simulations in a similar manner as EVAC. As the simulations are relatively fast, it is possible to perform several hundreds of replications overnight. Cumulative probability distributions with respect to collision speed, slamming speed, etc., are then directly produced by sorting the results as basis for presentation of cumulative probability distributions.

During the development, the different models in LBL have been compared with the results from analytical models on simplified cases. Further, the LBL results have been compared with the results of the model tests at KTH. Due to the probabilistic nature of the launching operation, a stringent comparison was not possible, but it was demonstrated that the results were similar.

The results of the LBL simulations as well as the results from the KTH tests shows that launch of lifeboats from passenger vessels is a high risk operation if the wave height is, say 3 metres or higher.

9 Conclusion

Based on the work carried out, as well as on earlier experience with Evacuation analyses of ships and offshore installations, the above described computer programs represent a significant potential for safety improvement of passenger vessels, provided that the movement in IMO away from detailed requirements towards Goal Setting requirements proceeds.

References

1. Ship motions, ship motions for the passenger ferry m/f Kronprins Frederik. BriteEuram project 97-4229 "MEPdesign" WP 1c.DMI.

2. Way finding Systems, A proposal for new way finding systems for passenger ferries. Brite/EuRam project BE97-4229. MEPdesign- Work Package 1a. DMI.

3. Design Features. A proposal for new design features on board passenger ferries. BriteEuram project 97-4229 "MEPdesign". WP1b. DMI.

4. Group binding: Emergency behaviour of ferry passengers. BriteEuram project 97-4229 "MEPdesign" WP2c. DMI.

5. Influence of Waves and Ship Motions on Safe Evacuation of Passenger Ships. KTH.

6. Experimental Investigation of "Slide" and "Fall" Evacuation Systems. BriteEuram project 97-4229 "MEPdesign" WP2d. KTH. Olle og Elena

7. Experimental Investigation of Lifeboat/Davit Evacuation System. BriteEuram project 97-4229"MEPdesign" WP2d. KTH.

8. Software Quality Assurance Plan WP2a, Mustering. BriteEuram project 97-4229 "MEPdesign" WP3a. QUASAR.

9. Software Quality Assurance Plan WP3a, Mustering. BriteEuram project 97-4229 "MEPdesign" WP3a. QUASAR.

10. Lifeboat launch LBL, Theoretical bases. QUASAR consultants.

11. Design Features:effects of ship listing and ship motion on mustering speed. BriteEuram projects 97-4229 "MEPdesign" WP2b. TNO (Bles and Boer).

12. Design Features: effects of ship listing and shipmotion on mustering speed. BriteEuram project 97-4229 "MEPdesign" WP2b. TNO (Bles, Groen, Boer).

13. Signposting for the Evacuation of Passenger Ships. TNO, Boer.

A Mesoscopic Model for Passenger Evacuation in a Virtual Ship-Sea Environment and Performance-Based Evaluation

D. Vassalos, H. Kim, G. Christiansen, and J. Majumder

The Evacuation Simulation Group of the Ship Stability Research Centre (SSRC), University of Strathclyde, UK

The paper outlines a mesoscopic passenger evacuation simulation model, based on multi-agent techniques and a multi-level planning structure, developed by SSRC in collaboration with Deltamarin Ltd. Code-named *Evi* (Evacuability index), it uses a virtual environment for enhanced effectiveness of evacuation performance evaluation. It represents the state-of-the-art computer simulation-based capability for the prediction of passenger mustering and evacuation involving a number of escape and rescue scenarios (abandon ship, transfer to refuge centres or a combination of these) in a range of incidents (fire, collision, progressive flooding, cargo shift, foundering) whilst accounting for ship motions in a sea environment. Unlike earlier models, *Evi* has been developed from the outset for application to passenger ships in a sea environment, including the largest cruise liners and Ropax. Valuable input and feedback from the owners/operators helped refine and render the model a practical tool which, coupled to modelling of uncertainty in all the parameters that may affect evacuation times and the ability to play back a given scenario as video, provide for wide-ranging capability in modelling realistically the most complex of evacuation scenarios thus allowing for routine application of passenger evacuation simulation analysis. The term *Evacuability* entails a wide range of capabilities including evaluation of evacuation time, potential bottlenecks, assessment of accommodation module layout and sensitivity analyses to assist design for ease of evacuation, passenger familiarisation with a ship's environment, "what if" scenarios for training, devising effective evacuation planning procedures/strategies and decision support to manage a crisis. *Evi* is available in the form of a computer program that can be readily customised to any vessel environment with an efficiently tailored user interface and Run Time Simulator (RTS). Typically, simulation of 5,000 passengers mustering on a 15-deck vessel can be achieved at real time.

1 Introduction

1.1 Background

A number of drivers have brought passenger evacuation to the forefront of priorities of the European shipbuilding industry triggering the need for the development of tools and procedures in support of performance-based design for evacuation to ensure cost-effective treatment of this important issue:

- Ro-Ro ferry accidents have brought about the realisation that "*ship and cargo survival*" might have to be addressed separately from "*passenger survival*" in that these vessels can capsize very rapidly, when damaged, thus not allowing sufficient time for evacuating passengers and crew. Accidents have also shown that life saving appliances (LSA) do not always fulfil the latter objective, particularly so in heavy weather conditions [1].
- An amendment to SOLAS '74 requires "Ro-Ro passenger ships constructed on or after 1 July 1999, to have escape routes evaluated by an evacuation analysis early in the design process". This derives from the recommendation of the POE suggesting that Ro-Ro vessels will carry only as many passenger as can be demonstrated to be evacuated within typically 1 hour.
- After investing for decades in ship's hardware for purposes of increased returns, it has been recognised that marked ship safety improvement necessitates investment in people. This is paramount when addressing the very survival of passenger, in that the consequence of accidents involving large loss of life could drive shippers out of business, as the Estonia tragedy has amply demonstrated. Such consequences are bound to reach intolerable levels when addressing new concepts such as cruise liners carrying well over 5000 passengers. Moreover, the impracticality of evacuating such a large number of passengers exercises the minds of all concerned in that they all appreciate the potential of a disaster waiting to happen.
- There is a clear shift from rules-based and deterministic to performance-based and probabilistic approaches in the shipping industry necessitating a fresh approach to problem solving and state-of-the-art tool development with emphasis clearly on cost-effectiveness as key measures to increasing the competitiveness of the European shipbuilding industry.

Deriving from the above, there is an immediate need to address the capability of the whole passenger evacuation system pertaining to mustering routes and procedures, life-saving appliances, decision support and management. In turn, this leads to the necessity to focus on the development of evacuation analysis and simulation tools for the prediction of evacuation performance, thus allowing for a meaningful evolution of passenger ship designs with enhanced evacuation performance (minimum time for safe evacuation of passengers and crew). Successful mustering and evacuation can avert disaster as last lines of defence even after the safety measures linked to structural reliability and enhanced ship survivability have failed. In this respect, the development of tools in the form of computer simulation models for the prediction of evacuation scenarios, evacuation time and probability of success in different conditions must be addressed as a top priority. The same tools could also be used to aid decision making onboard the ship, thus tackling the same problem as an operational rather than a design issue. Attempts in this direction by the SSRC-Deltamarin team are the subject of this paper.

1.2 Ship-Based Evacuation Problems Explained

Before addressing modelling and implementation issues, it would be worth pondering on the critical differences between passenger evacuation simulation pertaining to passenger/Ro-Ro ships and to problems pertinent to land-based experience or aircraft (or even offshore structures on a number of issues). Much as there are generic elements in the simulation of evacuation, solutions are drastically affected by these differences and ought to be addressed at the outset. These include the following:

Modelling the Ship (Virtual) Environment

Geometry: One does not need to be a ship designer to appreciate that the complexity of ship geometry requires innovation in capturing only its essential features and allow for cost-effective data reading, handling and evolution (removal, decimation and healing) whilst providing models of humans (and indeed real people) sufficient knowledge of the environment for decision making. The difference between innovative and brute-force modelling could be an order of magnitude in the time taken to produce a virtual ship model. The ship geometry is modelled in *Evi* on the basis of CAD drawings of the GA plans, "building" the ship deck by deck, Fire Zone by Fire Zone until the requisite 2½D model is completed (note that a full 3D model may be required for virtual reality applications).

Topology: Closely connected to ship geometry and hence unique to ships are topological issues pertaining to accommodation layout, location of assembly and embarkation stations and schemas of evacuation "flow".

Semantics: Most semantic specific information crucially affects *Evacuability*, mainly because of the geometric complexity but also due to adversity of the sea-ship environment, reinforced by uncertainties in the time available, distance to land, functionality of LSA to mention only a few.

Platform: Unfortunately ships move, on occasions severely whilst other contributing factors exacerbate this situation further (e.g., progressive flooding and cargo shift, which may also curtail evacuation time to the extent that evacuation became impossible). Fire - the most usual reason for evacuation in buildings - is also a high-risk hazard in ships but because of the potentially disastrous consequences, particularly in large passenger/Ro-Ro ships, risk control, evacuation strategies and decision support become considerably more important and again unique to ships.

Sea environment: Evacuation pertaining to land-based scenarios, means escaping to safety. In ships, more often than not, it means escaping to sea, where rescue is far from complete.

Modelling Human Behaviour

Passenger numbers: In the majority of cases, evacuation from other enclosed spaces does not involve very large number of people. This in itself presents modelling problems in terms of macroscopic and microscopic movement of people, processing capacity and information handling. These, in turn, give rise to needs of multi-threaded programming and parallel processing, particularly for use in immersive technology and/or employing the navigation interface of a virtual reality integrated environment.

Way-finding and path selection problems: Deriving from the ships' geometric complexity, these problems are exacerbated by the moving base, often severe time constraints and the anticipation of an awaiting unfriendly environment, thus complicating most aspects of human behaviour.

Uncertainty modelling: Primarily because of the inherent uncertainty deriving from unpredictability of human behaviour but exacerbated in ships (because of the passenger numbers involved, the possible co-existence of multiple scenarios in a crisis and the complexity of information in all respects), the individual results obtained through any one simulation will not be suitable to evaluate evacuation performance or for sensitivity analysis and some aggregate statistical quantities, evaluated over several simulation results, would have to be defined. This fact, coupled with the multi-purpose utilisation of evacuation simulation for training, legislation, design and operational purposes demands a close scrutiny of efficiency in every aspect of modelling and analysis of evacuation simulation.

Passenger Mustering/Evacuation Process Modelling

Procedures: Evacuation strategies, procedures and decision support systems are likely to affect drastically success of passenger evacuation in ships, more so than in other enclosed spaces, again primarily because of the plethora of parameters likely to affect *Evacuability* in such a complex environment with so many people.

Evacuation scenarios: In addition to evacuation strategies that may be considered (abandon ship, transfer to refuge centres or a combination of the two) and the range of possible incidents (fire, collision, progressive flooding, cargo shift, foundering), most of which are mainly ship specific, it is in the range of scenarios that may have to be considered in a passenger ship evacuation situation, that a novice to the ship-sea environment may find himself lacking in knowledge. To this end, passenger profile, passenger distribution in any one time during night or day and crew functionality and responsibilities during a crisis are all paramount to undertaking a meaningful evacuation analysis.

Macroscopic or microscopic modelling: Macroscopic is the flow model advocated by the IMO Interim Guidelines whilst microscopic is associated with human behaviour models. However, neither of the two alone is appropriate as a generic model of evacuation analysis in that a mesoscopic (macroscopic and microscopic combined) model will be needed to account for high level and low level planning, a typical feature of even the simplest of evacuation scenarios. However, such models may be used for specific functions (e.g., a macroscopic model for accommodation layout design purposes through calculation of evacuation times or a microscopic model for investigating the effect of behavioural parameters on *Evacuability*).

Holistic or specific approach: To optimise *Evacuability*, it is obvious that all the elements of evacuation ought to be considered, as clearly many functions will take place in parallel. Hence, it would be sub-optimal to model the various procedures (mustering, embarkation, launching of life boats, etc) separately or indeed sequentially. A holistic approach is therefore key to understanding the evacuation process in ships and to properly model and analyse it for design, operational and regulatory purposes. Having said this, local optimisation may be used for optimising accommodation module layout for easy of evacuation.

Simple or complex models: It is true that the simpler the model the better but this cannot be used as the only argument for adopting unrepresentative and unrealistic models! Once decided to resort to computer simulation models, then the emphasis must be on developing friendly user interfaces for pre-processing (modelling ship information, passenger profile and distribution, crew functions and setting up scenarios) and post-processing (analysis and presentation of results).

Ship abandonment: When transfer to refuge centres is not an option, this most important aspect of passenger evacuation is wholly ship specific (albeit there is strong similarity with offshore platforms) involving aspects of ship dynamics and LSA functionality issues whilst also accounting for human behaviour.

Following this preamble, it should be obvious even to the acclaimed experts in evacuation analysis that direct transfer of knowledge from land-based experience to passenger ship evacuation is highly inappropriate, demanding a wholly different approach. This assertion is further explored in the following.

2 Mathematical Modelling

2.1 Multi-Agent Modelling

The lowest common denominator of the many definitions of 'agent' is an encapsulation of code and data, which has its own thread of control and is capable of executing independently the appropriate piece of code depending on its *own state* (the encapsulated data), the *observables* (the environment) and the *stimuli* (messages from other parts of the system or interactively provided). The agent's action model is essentially a 'sense-decide-act' loop. The *sense* and *decide* steps may be coalesced, as the sensing is nothing more than the interface of the agent with the data structures representing the environment. The decision process requires access to the perceived information, thus perception is not a complex process but rather a simple access interface between the environment and the agents. Notably, the actions of agents may also change the environment, giving rise to what is called *interactive fiction.* Multi-agent is a further generalisation of process-based modelling methods where the environment is very well defined and the agents may communicate in a fairly versatile manner. In natural systems, all component parts "live" in some sort of topological space (predators and prey may live on a two dimensional forest floor, data packages traverse a network graph and the evacuees move around on a 2D deck or offshore installation layout). An *environment* is defined to be an artificial representation of this space. Autonomous agents can perform the activities defined by a computer program in this environment. This strong sense of environment does not exist in a process-based simulation. Processes are only aware of themselves and the resources they wish to acquire. In the implementation the environment will be an appropriate collection of data structures in the computer. Communication in multi-agent simulation describes all interaction between real-life entities. This makes multi-agent simulation an extremely powerful tool but also one, which is hard to verify in the context of known mathematical theory. The essence of using agents requires a rigorous definition and full implementation of the environment and its interfaces with the agents as well as an inter-agent communication protocol, as described next.

2.2 The Environment Model

Modelling of the environment is one of the most important aspects of multi-agent modelling. In the whole, this consists of three aspects - geometry, topology and domain semantics. The perception model for the agents will be able to use the information in these three abstractions at different levels of the decision processes. A multi-deck layout may be modelled as a *manifold* of topological dimension 2. A ship layout can be a pedagogical example of a topological space where one can assume uniform co-ordinate charts for large finite neighbourhoods. Such two dimensional Euclidean co-ordinate charts will hold for large regions, typically a corridor. Therefore, the overall model may consist of many interconnected regions with local co-ordinate systems where the structure of a linear space may be assumed. The ship area manifold is segmented into convex subsets whose mutual intersection can at most be subsets of one-dimensional topological manifolds for which a Euclidean structure holds. Furthermore, this segmentation is done in such a way that three regions may intersect only at points (i.e., zero-dimensional topological sets). These subsets are called *regions* and the one-dimensional submanifolds (curves) along which two of them intersect are termed *gates.* Two regions are called *directly connected* if they have a common gate. Similarly, gates are defined to be directly connected if they lie on the same region. This connectivity, for all computation and analysis purposes can be represented by a *graph.* In ship layout terms regions are defined as cabins, corridors, public areas (or subsets of these), each with its own co-ordinate system and connectivity, defined by gates (these may be actual or artificial doors). Figures 1-3 next illustrate schematically these ideas. The path of the agents leading to the embarkation station is determined by searching the connectivity graph. A *depth first* exhaustive search over the gate graph is used in *Evi* to choose the optimal path to be used for high level planning activities. Currently, the length of the path is taken as the criterion of optimality for network flow.

Finally, the availability of 2½D and 3D models allows for real time visualisation, in which the complete geometric details of the ship and human agents will be utilised to give rise to an extremely realistic representation. As an alternative, the code can also be executed separately, allowing a much faster evaluation of a simulation and leaving visualization as a post-processing alternative.

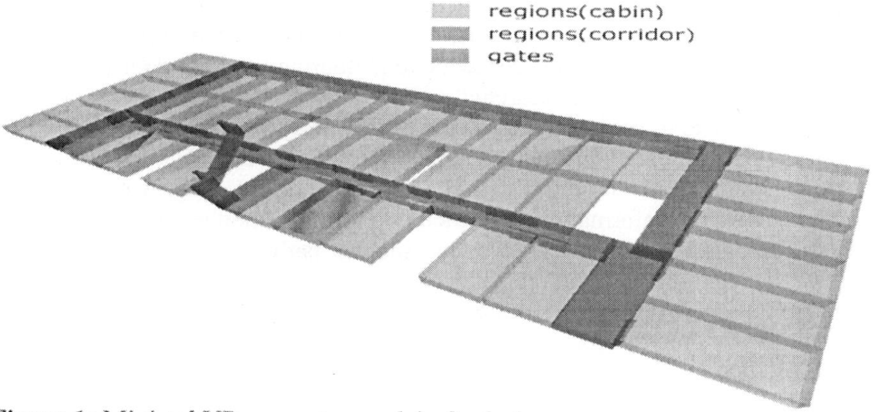

Figure 1: Minimal VR geometry model of a deck.

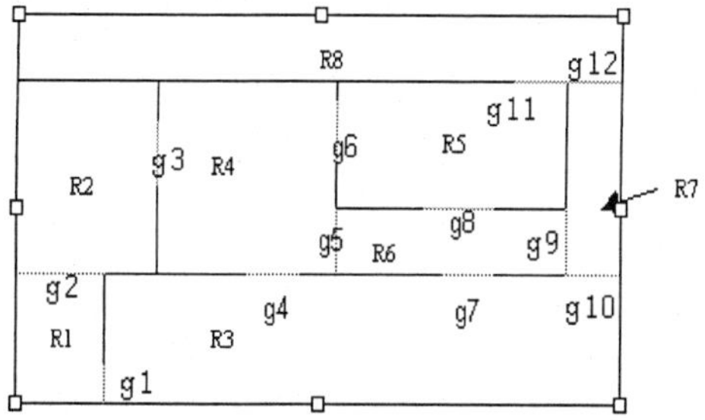

Figure 2: An example layout of regions and gates.

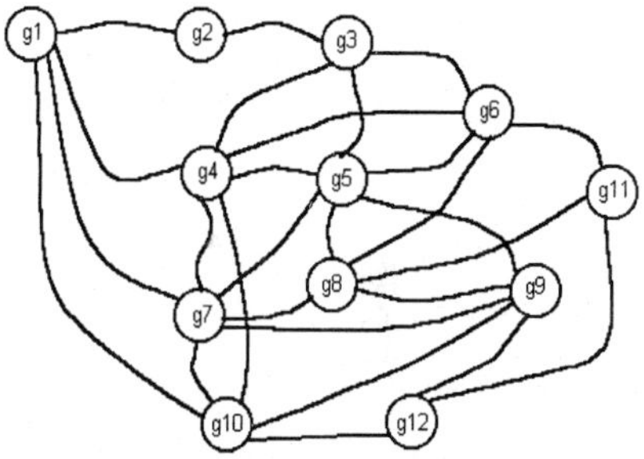

Figure 3: Gates graph corresponding to Fig. 2.

A minimal description of the ship layout will enable designers to modify the layout easily (add a new corridor or a staircase in virtually no time without having to draft the details of it using an elaborate CAD tool), hence obtaining evacuation performance faster and thereby making simulation an ideal design tool. The contrary can be also easily achieved – by simply blocking areas, regions or whole fire zones one can examine the effect of these changes and therefore the sensitivity of each different part of the vessel on evacuation capability. This will make 'what if' questions easier to tackle. The main strength of the modelling adopted derives from the inherent capability to utilise high and low level planning interchangeably in a way that human behaviour is modelled realistically whilst ensuring high computational efficiency. This is demonstrated next by presenting relevant examples.

2.3 High and Low Level Planning

High Level – Path Planning and Graph Search

With increasing complexity of the minimal geometry of the ship to thousands of doors and regions, it is very important to have an efficient path-planning process. The path-planning algorithm adopted is illustrated in Fig. 4. The shortest path from the cabin to the embarkation station is, of course, equal to the path from the embarkation station to the cabin. If all doors neighbouring the final door that leads to the embarkation station were searched and the distance information to each door were stored, doors of cabins could be straightforwardly reached within a few cases of search process. Then only the distance information from each door to the embarkation station needs to be left with the door's id. When an agent is located in a region, the distance information from each door of the region can be obtained, thus allowing the agent to simply head to the shortest path leading to the destination area. With the pre-planning of the paths to be followed completed before the actual simulation starts, re-planning during evacuation is still possible if, for example, there is a dense crowd 'blocking' the path or a blackout in the presence of fire or smoke.

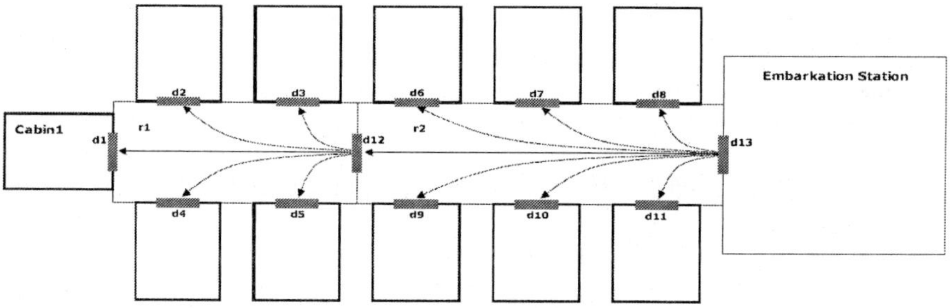

Figure 4: Simple illustration of the path-planning algorithm.

Low Level – Steering of Agents

Pursuit of a static target acts to steer the agent towards a specified position in global space. This behaviour adjusts the agent so that velocity is radially aligned towards the target. The "desired velocity" is a vector in the direction from the agent to the target. The magnitude of "desired velocity" is the agent's current speed. The steering vector is the difference between this desired velocity and the agent's current velocity, as shown in Fig. 5. Computationally, deriving the evacuation direction requires only a very simple closed form algebraic calculation. In the absence of any obstacle and other evacuees, every agent will "flow" along the evacuation direction field (passing through the gates unobstructed), hence avoidance heuristics are used to avoid collision with the neighbouring agents and obstacles present along the evacuation path.

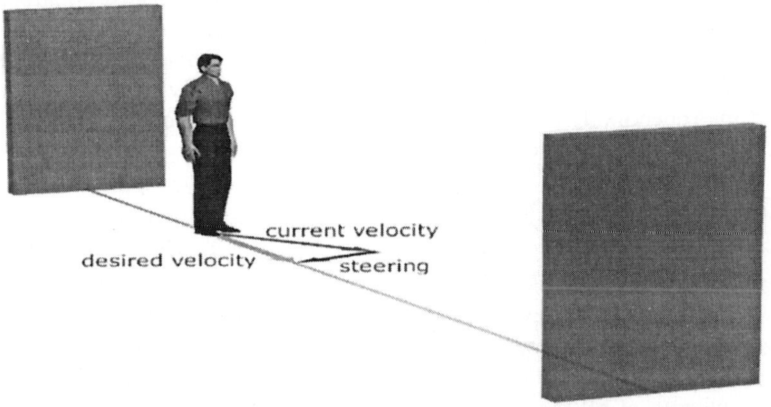

Figure 5: Pursuit of a static target.

3 Modelling Human Behaviour

3.1 Framework Adopted

As indicated in the introduction, there is a plethora of behavioural parameters that are likely to affect the evolution and the outcome of an evacuation scenario as well as vital information pertaining to the definition of the scenario itself. The latter, concerning passenger profile (number, age, gender, persons of impaired mobility and so on), passenger distribution (spatial and temporal) and crew number, distribution and functionality in any given crisis, are easier to deal with, as there is statistical information available with most reputable operators. To cater for the former, there is instead a need to adopt a framework that allows for as many behavioural parameters as deemed appropriate to be considered, at least until results deriving from sensitivity analyses of available software supported by purposely undertaken experimental data, real life data and behavioural studies indicate otherwise. This, in turn, might lead to the adoption of more simplified models and frameworks for a given purpose. For example, certification for vessel according to the IMO Interim Guidelines might require a much simpler model than for training crew in a virtual reality model. The framework adopted in the development of *Evi* treats passengers as agents moving in a "command" and "decision" structure suitable for a given emergency and in accordance with the attributes being considered, modelled as an array of "genes". These, for example, determine the behaviour of a mother searching for her child before abandoning the ship, the father taking a leadership role in a crisis, the child following parents, the members of a family forming a group and so on. "Genes" may be active or inert depending on circumstance, time and domain semantics. For example, if the current leader of a group becomes incapacitated, a new leader (someone with the right "gene") would take this role. Individual or group behaviour and requisite control is, of necessity, application specific and hence demand different levels of sophistication involving a range of control options: predefined behaviour using a script language (programmed); rule-based behaviour (autonomous), interactive control (guided) and adaptive control ("intelligent"). Emphasis in the development of *Evi* has been placed on adopting a framework model that could support all likely future developments concerning passenger evacuation in ships. Hard data has largely been obtained from open literature. An overview of the behavioural parameters currently being considered is provided in Appendix 1. Some additional relevant information on modelling human behaviour is provided next.

3.2 Speed of Advance

The speed of agents varies according to a host of parameters pertaining to ship/sea environment, the scenario in question and the passenger distribution and profile as described by an even larger number of parameters (such as passenger age, size, gender, physical location and position on vessel, signage, crew guiding and ship motion, to mention but a few). In fact, speed of advance is the compounded outcome of all that is going on onboard a ship in an emergency at sea during evacuation. As per the IMO Interim Guidelines, the speed of an agent is determined by the density of the crowd in the region. In general, crowd density is non-uniform and it may strongly depend on the size of the area considered in the density calculation. If the crowd is concentrated near a gate in a big region the remaining part of which is empty, on dividing the number of occupants by the total area of the region may give a small value of density which clearly fails to capture the situation. To overcome this drawback the concept of *perceived density* has been utilised, in which the local density in a region in front of the agent is computed and the IMO speed values assigned in accordance with this local density value. This makes the scheme conformant with IMO without sacrificing realism. The crowd density corresponding to an evacuee in an escape route describes the number of persons divided by the available escape route area pertinent to the space where the evacuee is located. As the model geometry is reproduced to exact measurements of the actual vessel, the density calculations are deemed to be very accurate. The available escape route area is determined by the actual overlapping area between the regions (e.g. corridors, stairways, etc), and a density rectangle (2.14m x 2.14m was identified as the best choice) that moves with the agent, as shown in Fig. 6.

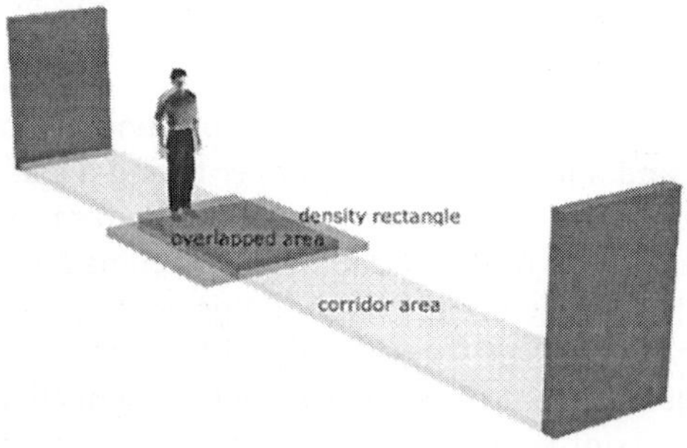

Figure 6: The concept of local density.

Additionally, when long queues are form, the effect on speed of advance is calculated rigorously on the basis of the queue length. Dependence of speed on other parameters is modelled by using multiplication factors that are reasonable functions of the parameters as explained next. These factors are adequately parameterised for calibration as experimental behavioural data becomes available.

3.3 Superposition of Behaviour

Superposition of behaviour refers to the way an action (or some parameter) is computed when some conditions hold *simultaneously*, assuming that a method is known of how to compute the same action under each of those conditions *individually*. The speed of advance of the individual agents is the most relevant example of this. As indicated above, several operating conditions, the individual's attributes and perhaps whatever happens around, affect the speed of the evacuees, and it appears that some core results in this area are due to show up in the form of data on how the different circumstantial conditions affect speed. Most of this data is in the form of speed reduction factors corresponding to different situations. The more common way of superposing quantities like reduction factors is to multiply them to obtain a resultant factor, which may be thought of as a **mobility impairment index**, ranging from zero to one and characterising the speed of every agent. Even in the case of actions that are associated with conditions probabilistically, it may be shown that the naive method of multiplying the speed reduction factors is conformant with the conditional probability based superposition, even if the speed were modelled as a normally distributed random variable.

3.4 Effect of Ship Motion

Few studies have been undertaken on the effect of inclination on the speed of people's movement but none has been particularly comprehensive and all have used young and physically fit subjects. In Evi, results from MEPdesign research [1] are used and roughly adapted to the mustering and evacuation scenario. This is likely to be modified at a later stage. A model of the effect of ship motion on speed of advance is however in place, as explained here. With time, humans can adapt to small regular motions and constant acceleration and this will tend to reduce the effect of motions and accelerations, but when the latter become unpredictable, the feedback system used by the body makes corrections too late and cannot anticipate the movements. The body concentrates instead on maintaining balance and forward motion will suffer as a result. Another way of approaching the subject is to relate the reduction in speed to the roll angle. To this end, a maximum roll angle of 20° is assumed, at which the speed reduction becomes 100%. The reduction in other angles follows the relationship described next. The effects of ship's dynamic and static heeling angles is included in the advancing speed of agents by the following analytically derived function,

$$f(\theta) = \begin{cases} \dfrac{e^{1-|\theta/\theta_{max}|} - 1}{e-1}, \forall 0 \le \theta \le \theta_{max} \\ \\ 0, \forall \theta > \theta_{max} \end{cases}$$

where θ is heel (roll) angle of the ship with θ_{max} set at 20 degrees.

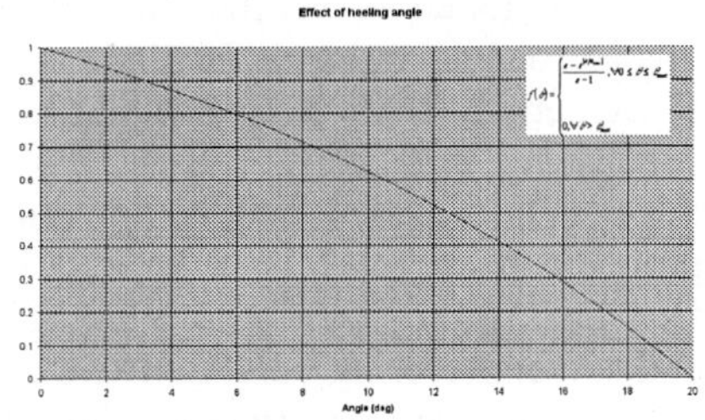

Figure 7: Heel angle factor.

The x-axis in Fig. 7 represents the heel-angle from 0 to a maximum of 20 degrees, while the y-axis represents the resulting $f(\theta)$, which is a multiplication factor, affecting the speed of advance. When modelling how speed should depend on roll motion it would be unnatural to make it dependent on the instantaneous roll angle, as the speed will return to normal even when the roll angle became zero momentarily. This indicates that the speed should in some way depend on the history of roll motion. Such dependence should account for a reasonably distant past in a way that a more recent part of the history would affect the speed more than a less recent one. Keeping this in mind, a scheme has been proposed in which every agent feels the effect of the roll angle values experienced in the immediate past (this ensures that a too distant past does not have any effect) and reduce the speed according to a weighted average of these with a decreasing function of temporal distance of the corresponding time step as weight (this ensures greater weight of more recent values).

3.5 Modelling Uncertainty

Human Behaviour Parameters

The psychological and physiological attributes of human beings are non-deterministic fuzzy quantities. Even in a contrived experiment one can hardly reproduce human actions/reactions even if all of the conditions remain the same. This inherent unpredictability of human behaviour, especially under unusual and stressful circumstances, rules out the possibility of a deterministic mathematical structure or program to model evacuation correctly. For this reason, human behaviour has to be modelled with some built in uncertainty. To this end, every parameter (continuous or discrete) is modelled as a constrained random variable with a predefined distribution. Continuous variables (attributes) are modelled as normally distributed random variables with pre-assigned mean and variance and using the Box-Müller transform a continuous distribution can be implemented in the simulation. Variables that are discrete in a deterministic context are treated as fuzzy variables and the Roulette Wheel technique is used for de-fuzzification. This is to eliminate the occurrence of unrealistic behaviour, for example, every female agent having exactly 15% lower speed of advance than a male agent or everybody of the same age reacting exactly at the same time to an alarm call.

Monte-Carlo Method

As mentioned above, the inherent uncertainty in human behaviour will give rise to a reasonable amount of variation in the result of simulation in different instances of execution. Thus the individual results obtained in each simulation will not be suitable for evaluation or sensitivity analysis and some statistical aggregate quantities evaluated over several simulation runs have to be defined that must have the property of approaching a limit as the number of ensembles grows indefinitely. Considering the above and building on the long-standing traditions of the probabilistic engineering community, the term *Evacuability(t, env, dist)*, mentioned in the introduction, is abstractly defined to be the probability of an environment being completely evacuated of human occupants no later than a time t elapsed after the alarm goes off, in a given state of the environment *env* (i.e. the layout , the conditions, and the final egress system if applicable) and a given state of initial distribution *dist* of people in the environment (including the distribution of behavioural attributes). Thus if we have data on evacuation time from several instances of simulation (given that the environment and the distribution remain the same) as a multi-set {t1, t2, t3, t4,...,tn} then by the law of large numbers *Evacuability* may be determined with an accuracy directly dependent on the number of simulation runs, as shown in Fig. 8. With this formalism a mathematically sound regulatory rule may be something like: *Evacuability (60 min., entire ship (worst anticipated conditions), worst passenger distribution) > 0.99.* This is a more scientific way of addressing regulatory issues since it acknowledges the inherent uncertainty in the system and allows for any desired degree of stringency.

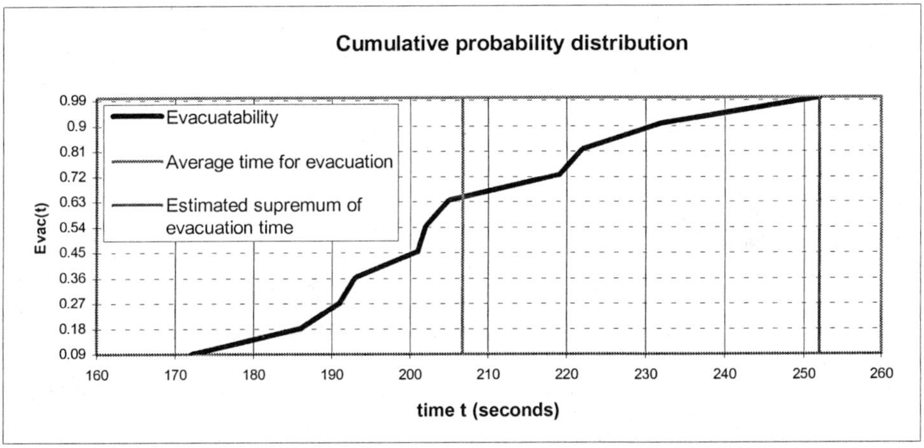

Figure 8: A typical *Evacuability* graph using the Monte Carlo method.

4 Design for Ease of Evacuation

4.1 Directions in Design

This section discusses how to use the simulation results more effectively to improve the design of the ship for ease of evacuation. One and the more apparent technique is that of model *inversion*, through sensitivity analysis with respect to different adjustable parameters. The simulation may be thought of as a function (or a forward model) that maps from the space of adjustable parameters (geometry, topology, signs, passenger distribution) to a one-dimensional real number space representing the evacuation time. The model inversion technique is aimed at arriving at some ordered multi-set of the adjustable parameters corresponding to a required value of evacuation time, respecting the constraints that exist in the adjustable parameter space whilst ensuring that other constraints imposed by considerations other than evacuation performance are also adhered to. If all the constraints were analytic and all the variables were continuous, a rigorous way of solving the problem would involve minimising a contrived grand objective function that includes all the constraints (with properly weighted penalties corresponding to their violations) and the pure objectives. But since the adjustable parameter space also contains topological parameters that are inherently discrete valued (the number of staircases between two decks, number of cross-ways connecting the corridors, number of corridors and so on) the main alternative is to undertake *sensitivity* analysis. The underlying cause of model inversion remains as important as ever, though implicitly only the goal being a bit more relaxed since the major constraint, that of evacuation time, is actually an inequality bounded at one end (the lower the better). Besides the generic method of synthesis by (sensitivity) analysis, there can be other very sound methods that deploy the prowess of microscopic simulation to a fuller extent. Microscopic simulation reveals a lot more technical information about the scenario (in addition to the information on *time* required for evacuating the entire ship or its sub-domains), which may be of great value in the design process. The first issue to address is how to extract some valuable information from the simulation. The situation is similar to that of system identification where one has to decide which outputs of a multivariate system are to be sensed ("where to put the sensor?", "what kind of sensor?" and so on). In this respect, quantities to be sensed and recorded through the course of the simulation include the following:

- the time history of queue size at the gates of interest (the ones to be designed),
- the time history of total number of crossings through each gate of interest,
- the history of total occupancy of each region of interest.

In light of the information on time histories of region occupancy, door crossings and queue lengths, the following design guidelines become obvious:

- The peak value reflected in the region occupancy history must not exceed the total capacity of the room (as determined by the area of the room multiplied by the maximum allowable crowd packing density).
- If the width values for the doors are set to large numbers, then the allowable efflux rate for the doors will be large enough to allow any number of people through and will not cause any queuing. Under this contrived situation, the maximum value of (or may be modal value of) the slope of the *gate crossings*' history curve will indicate the value of efflux rate (and thereby the width) required to prevent queuing.
- The queue length history directly reflects the bottlenecks in the simulation quantitatively without requiring someone to watch the animation to identify these and thus does not rule out the possibility of some automation in the *design for ease of evacuation.*
- It may appear that bottlenecks are always bad, but it may be argued that a narrow design of a gate may prevent an intermediate region (for which the said gate mostly plays an inlet) from being overcrowded beyond the extent

that can be that hinders guided evacuation beyond what some other constraints permit, and it may be a plausible design alternative.

4.2 Sensitivity Analyses

Do changes in input and parameter values lead to "reasonable" changes in output values, both in magnitude and direction of change? Can the main variables be examined? Does the evacuation simulation model predict average test results accurately? Sensitivity analyses constitute the normal route to finding answers to this question. To validate and verify the results of a model, use of experimental or full-scale data ought to be made. The standard deviation (root-mean-square) estimate of an uncertainty decreases with the square of the sample size and two tests under a given set of conditions are needed even to make an unsatisfying calculation. Therefore, the amount of experiments needed for validation purposes makes it very costly and in the case of an evacuation simulation this is virtually impossible to carry out. Thus, in large-scale models such as with an evacuation simulation one has inevitably to rely on professional assumptions, opinion, judgement and experience.

5 *EVI* and the Run Time Simulator

5.1 The Passenger Evacuation Simulation Model – *Evi*

Evi stands for *Evacuability* index as explained in the foregoing and is an evacuation simulation model specifically developed with ship applications in mind at the SSRC in collaboration with Deltamarin Ltd. *Evi* is available in the form of a computer program that can be coupled to any vessel environment. The vessel information required pertains to semantics, topology and geometric data, the latter varying from very simple (allowing quick calculations for high level planning) to a 3D virtual environment up to a level that replicates the actual ship with an efficiently tailored user interface and Run Time Simulator (RTS) that allow for setting up almost any evacuation scenario over a range of incidents. It is a multi-agent mesoscopic model, i.e., one that combines macroscopic and microscopic modelling to provide an intelligent multi-level planning capability. These features coupled to modelling of uncertainty in all the parameters that may affect evacuation times and the ability to play back a given scenario as video, provide a wide-ranging capability in modelling realistically the most complex of evacuation scenarios on a moving ship platform. The model has already been used to undertake systematic investigation on wide-ranging issues of passenger evacuation on the largest cruise liners and Ropax. Typically, it takes 4 weeks to complete a full investigation and to deliver a comprehensive report together with an RTS.

5.2 Run Time Simulator

The RTS allows, with the minimum of training, to set up almost any evacuation scenario whilst accounting for the dynamic behaviour of the vessel in a realistic sea environment (sea state) and human behaviour including the role of crew, which is of paramount importance in ship evacuation simulation. Fig. 9 illustrates the main features of *Evi*'s interface.

Figure 9: The Run Time Simulator.

By way of illustration, a screen shot of the main page of the *Evi*'s user interface is shown in Fig. 10 below.

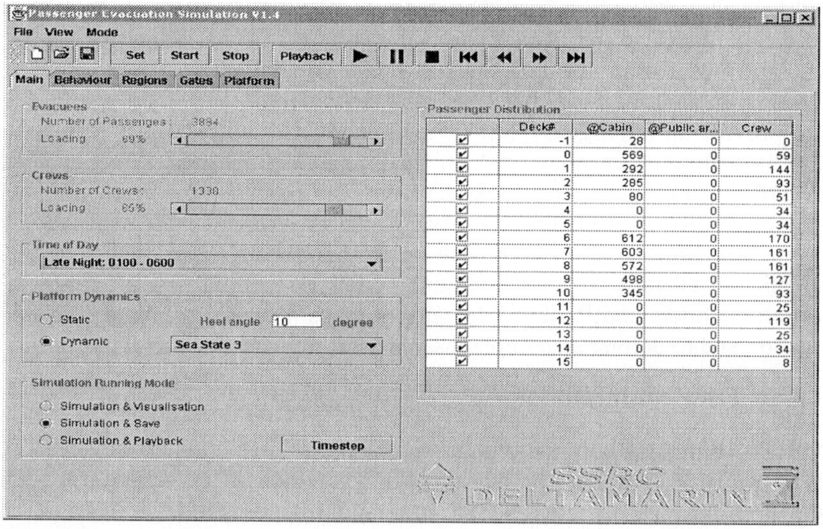

Figure 10: *Evi* – The main page.

6 Case Studies

To demonstrate the use of the simulator, three evacuation cases are considered here for a large cruise liner. Information on passenger distribution was received from the owner based on counts from 43 voyages completed by the vessel this during the year 2000 and from demographic details through CLIA (Cruise Lines International Association). All cases use 3492 passengers. The simulation runs continued until 99% of the evacuees had reached the assembly station, the obvious reason being that in some cases (like passengers getting lost), a very small number of passengers could have a very large effect on the evacuation time. The simulations were run for the full vessel - with passengers in their cabins for the first case and with the 'true' passenger distributions in the remaining two.

Case 1: This case is based on IMO Interim Guidelines, [1]. This is a standard day-case with passenger distribution according to the Escape Plans of the vessel.

Case 2: This case simulates an evacuation situation, taking place at morning time based on the cruising daily programme. The special condition during the simulation is a blocked stairway between decks 6 and 10 in Fire Zone 5. As a consequence of this, the persons onboard are forced to find an alternative route to the assembly station.

Case 3: This case imitates senior cruising at evening time. 90% of the passengers are over 60 years old and the rest are over 50. During the simulation the heeling angle increases steadily from 0 to 10 decrees over a period of ten minutes.

7 Results

Results of some real mustering exercises for large passenger ships [2] provide indications that the time it takes to assemble all passengers before embarkation varies from 7 – 28 minutes. The great variance was due to the difference in preparation of the passengers. The passengers were obviously aware of the mustering exercise and therefore some were ready with their lifejackets when the mustering alarm sounded. The average time for the reported passenger musters (discounting those that were obviously invalid) was 11.5 min. Some vessels also recorded the time taken to empty the assembly stations at the end of a passenger drill – that is the time it takes for passengers to depart the assembly stations and for the stairways to empty of other than routeing traffic. The time to empty the assembly stations consistently turned out to be 7 to 9 minutes. Members of the International Council of Cruise Liners (ICCL) have identified five incidents since 1984, in which passengers were sent to assembly stations without advance notice as a result of a collision, grounding or fire. The time for mustering all passengers and crew, including a full sweep of the ship for injured, stragglers, disabled, or uninformed, ranged from 17 to 28 minutes. In one instance where the ship had taken on a 5° list, the assembly time was increased to 25 minutes. Two of the incidents occurred in the late night/early morning timeframe when many or most passengers were in bed. Results from the case studies presently considered are shown in Figures 11-13 together with results from a "normal" case for comparison purposes. In this context, normal case is a standard case of *Evi* including the uncertainties related to age and gender effects.

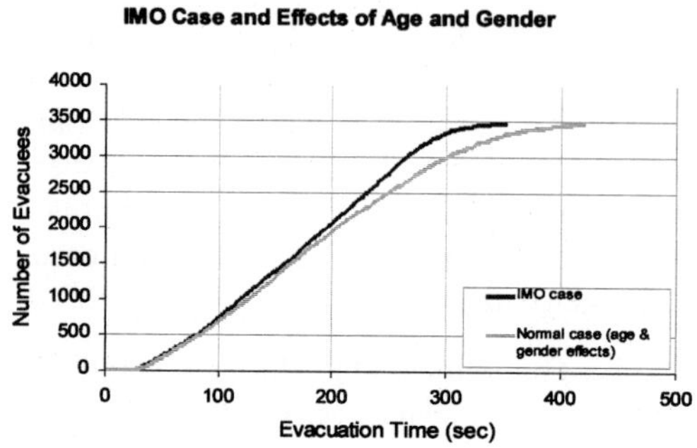

Figure 11: Case 1 – Basic IMO and normal cases.

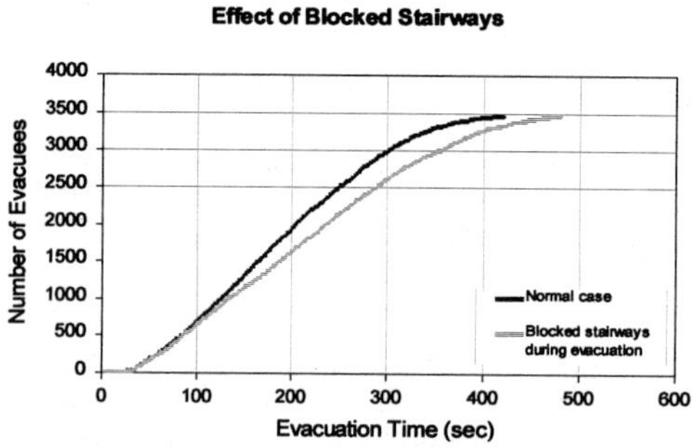

Figure 12: Case 2 – Blocked stairways in MFZ5.

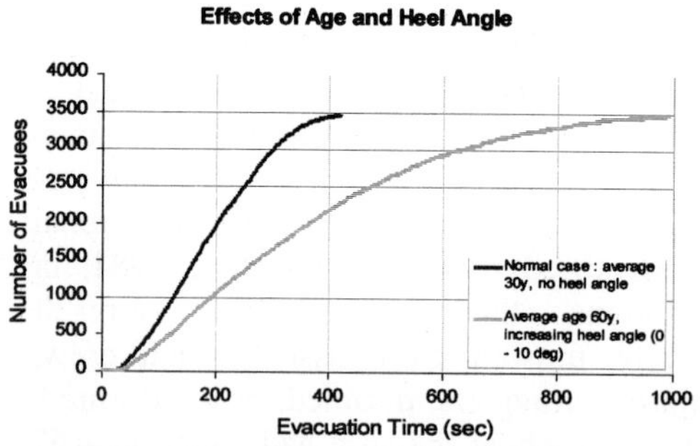

Figure 13: Case 3 – Effects of age and heeling angle.

Case 1:

Figure 11 shows a 20% increase in time due to gender and age. Both Normal and IMO cases, however, suggesting times for evacuation towards the lower end of the data provided in [2].

Case 2:

Figure 12 illustrates that the effect of blockage on the evacuation time is not as noticeable as one might expect (considering the loss of means of escape routes). The reason for this is that the distance the passengers have to travel to reach an alternative stairway (most choose staircase in Fire Zone 7) is not that much further than their original choice and does not take long to travel. In addition, when passengers arrive on the deck where their assembly station is (a short period later than a normal case) they still have to queue before arriving at their destination.

Case 3:

In this case the vessel is populated mostly by seniors (mean 60 years) who lack the agility of younger passengers. The effect of this, shown in Fig. 13, where for comparison purposes the opposite extreme is also simulated referring to a mustering scenario where all the passengers are young (thus, faster walking speeds and shorter reaction times) is to more than double the evacuation time.

8 Concluding Remarks

- A state-of-the-art passenger evacuation simulation model has been developed providing for wide-ranging capability in modelling realistically the most complex of evacuation scenarios at sea and to consider applications to the largest cruise liners and Ropax vessels to evaluate evacuation performance.

- A Run Time Simulator has also been developed based on a flexible structure that allows for realistic modelling of all the parameters likely to affect the simulation and coupled to a user interface that allows for model preparation and running in a user friendly and efficient manner.

- Following several applications, a stage of maturity with the software has now been reached where "design for ease of evacuation" could readily be undertaken and the effect of a number of behavioural parameters on evacuation time considered. The main limitation on accuracy is lack of suitable experimental and "field" data. Related to this, it has to be made clear to all concerned that the real value of this software derives from the ability to compare alternatives, i.e., relative measures of *Evacuability*, rather than absolute measures. In this respect, the unique ability of *Evi* to treat uncertainty comprehensively and to present results as statistical measures can prove immensely useful.

- Work at IMO is currently addressing the problem of evacuation of large passenger ships as a matter of priority. In this respect, it has been demonstrated that *Evi* can deal with the sheer size of the problem at hand from a computer modelling and simulation viewpoints.

- Efforts are now being directed towards specifying a number of scenarios for benchmarking purposes, following which validation and verification of these scenarios against "real" data will be sought, aggregate results concerning macroscopic modelling and controlled focused experiments to address the governing human behaviour parameters.

References

1. E. Tsychkova, *Influence of Waves and Ship Motions on Safe Evacuation of Passenger Ships*, Licentiate Thesis, (Kungliga Tekniska Högskolan (KTH), June 2000).

2. IMO Interim Guidelines for a simplified Evacuation Analysis of Ro-Ro Passenger Ships, NSC/Circ.909/04. /06.99.

3. IMO Sub-committee on fire protection: Recommendation on Evacuation Analysis for Passenger Ships and High-Speed Passenger Crafts; Paper submitted by ICCL: Passenger vessel evacuation analysis, FP 45/3/1, July (2000); Paper submitted by Italy: Report of the Correspondence Group, FP 45/3/3, October (2000); Paper submitted by Germany: Analysis of the Interim Guidelines for a simplified evacuation analysis on Ro-Ro passenger ships, FP 45/3/4, November (2000); Paper submitted by United States: Recommended guidance for computer-aided evacuation analysis tools, FP 45/3, November (2000).

APPENDIX 1: OVERVIEW OF HUMAN BEHAVIOUR PARAMETERS

Behaviour		Information Available	Implementation Details	Plans for the Future
Reaction time			Every agent is assigned a value of reaction time depending on the circumstances. At the beginning of simulation each one starts evacuating after the assigned time has elapsed.	Information on deterioration of RT of persons affected by alcohol will be taken into account later in the development. Obtain more data on RT of different types of people (e.g., gender, intoxication, etc.) from TNO and experiments with ERGO.
	Alarm	Obtained data available on people's reaction to different types of alarms.	To be implemented as multiplicative factors of the base value of the reaction time corresponding to different types of alarm.	Experiment with different types of alarms - obtain RT for all types.
	Age	Reaction Times (RT) for elderly people compared with RT of younger adults. Juniors move along with parents.	To be implemented as multiplicative factors on the base value of the reaction time corresponding to the age categories.	
	Time of day	Reaction times in accordance with IMO guidelines.	To be implemented as two different base values corresponding to day and night.	Obtain results from TNO and ERGO - find RT for different times of day.
	Location		To be implemented as a multiplicative factor.	Obtain information on RT from different locations onboard the vessel from TNO and/or ERGO. Awaiting information from Color Line.
	Role of crew	Information from human behaviour in case of fire	To be implemented as a multiplicative factor.	Specify the effect of crew guiding. Use information obtained from ship owners.
Speed of advance		Obtained data on walking speeds of humans of different age groups from a variety of studies on human mobility.	Speed variation implemented by look-up tables containing the hitherto available mobility data, as multiplicative factors that are functions of the operating parameters.	
	Number of people		Look-up table	
	Gender	Gender factor of 0.85 assigned for female and 1.0 for male	Look-up table	
	Size	Speed reduction assumptions based on mobility data of obese students in USA.	Look-up table	

(Continued next page)

(Contd.)	Age	Data from age related studies indicate muscle loss, decrease in step frequency and in step-length - speed reduction assumptions based on these factors.	Look-up table	
	Special case		Possibly look-up table	
	Ship motion	Results from MEPdesign for evacuation time in different sea states along with an exponentially fitted function depending on time history of previous roll motion	Implemented as a multiplicative factor	
	Smoke/ Black-out	Obtained results on visibility parameters and how they affect way-finding and walking speed	Implemented as a multiplicative factor	Density of smoke can affect the speed of advance in the same way density of people affects speed. Define relationships.
	Crew guiding role		Possibly look-up table	
Bonding between passengers			People in one cabin come out at the same time (i.e. sharing RT) and move at the same speed (the fast ones slow down and the slow ones speed up) and follow the same route. Similar logical groups of people are defined based on initial proximity (people in the bar) or genes (a family group).	This area is undergoing continuous development.
Route Deviation			In the initial definition of a scenario one can choose a percentage of passengers whom are likely to get lost. They will arrive at a junction and not knowing which way to go, they might go the right way or walk off in the wrong direction. They will only find their way again with the help of crew guiding or when the encounter a clear and lucid signage.	
	Signage	Way-finding errors due to signage is reported in a number of studies (e.g. MEP-Design)		
	Other passengers		Assign a 'leader' type passenger, that other passengers will tend to follow (same function as crew members).	
	Ship motion	Use data obtained for the effect on speed from ship motion.	Since ship motion has proved to reduce way-finding abilities, this factor will be linked to the results obtained for speed reduction due to ship motion.	

(Continued next page)

	Crew guiding role			
	Familiarity			
(Contd.)	Smoke/ Black-out	Visibility parameters for different types of signage obtained from various studies.		When agents encountered smoke the visibility of signs and guidance systems may become obscured. On the other hand the agent may deliberately choose to take an alternative route e.g. upon entering a corridor containing smoke, turn around and take an alternative route to the exit.
Re-routing	Number of people			
	Smoke/ Black-out			
Panicking				
Passengers returning to cabin		Require information on what type of person and how many tend to return to their cabin - returning times are obtained from the simulation itself.	Implemented by choosing a number of agents at random -instantiate equally as many path-planner objects with some random chosen cabin or region - once they have reached the cabin replace the global evacuation planner.	
Passengers carrying luggage		Agents are assigned a slower speed depending on the load they carry.	Passengers carrying luggage will keep doing so until persuaded to leave it by crew or a fellow passenger. The baggage could pose as an obstacle or be disposed of in an unobstructing manner.	Area to be further investigated and relevant data collected.

The U.S. Coast Guard Approach to Evacuation Analysis

K.C. Kiefer

U.S. Coast Guard Headquarters, Office of Design and Engineering Standards, Lifesaving and Fire Safety Standards Division (G-MSE-4), 2100 Second Street, SW, Washington, DC 20593-0001, USA

The United States Coast Guard provides maritime safety and security services, beginning with the design phase of a ship, following through to construction, and continuing with regular vessel inspections. The Coast Guard employs a wide-range of regulatory and non-regulatory measures to ensure U.S. and foreign flag vessels operating in U.S. waters are structurally sound, competently operated, and outfitted with lifesaving appliances, fire protection and other safety systems. A holistic approach to evacuation analysis is followed to determine how accidents can be avoided, and what the best evacuation procedures are when these accidents do happen. The Coast Guard is involved with a number of different national and international maritime groups to ensure the highest level of safety possible. This includes the International Maritime Organization, where the Coast Guard plays an active role in two groups currently working on passenger ship evacuation issues. Although the likelihood that a large cruise ship would have to evacuate passengers in the middle of the ocean is small, accidents show that the consequences when an event occurs can be daunting. The Coast Guard's ultimate goal is the safety of passengers, crew and the environment.

1 Role of the U.S. Coast Guard

The U.S. Coast Guard is a military, maritime service that has provided maritime safety and security services continuously for over 210 years. As lifesavers and guardians of the sea, their role is to keep passengers, crewmembers, vessels, environment, and cargoes safe. They spend over 800 million dollars annually on maritime safety alone, and their multi-faceted roles extend far beyond U.S. borders. Their goal of safety is pursued mainly through search and rescue and marine safety missions. They respond to calls for help at sea with search and rescue experts, and operate an extensive network of boat and air stations staffed with highly trained

---

Note: The opinions expressed in this paper are solely those of the author and are not to be taken as official views of the U.S. Coast Guard or the U.S. Department of Transportation.

crews. More than 50,000 calls for assistance are received annually and thousands of lives and billions of dollars in property are saved.

The primary way to ensure U.S. ports and waterways are safe is through the Coast Guard's Port State Control program. Safety and security of ships and cargoes entering U.S. ports, security of the port areas, and protection of the marine environment are the key objectives of the marine safety program. Most of the ships in U.S. ports, at any time, are foreign flag ships which are required to comply with the safety standards issued by the International Maritime Organization (IMO). Coast Guard marine inspectors examine about 25 percent of foreign ships entering U.S. ports and ensure compliance in an effort to reduce substandard shipping. The Coast Guard has the authority to board and examine ships, and, if necessary, detain them in port or order them out of port, to ensure the safe operation within our waterways.

1.1 Holistic Approach

It is well recognized that in the event of a major emergency such as a collision or fire, a ship is its own best lifeboat. Therefore, the Coast Guard's goal is to ensure the safest ship possible. The Coast Guard begins their focus on passenger vessel safety at the design phase of a ship. They then follow it through its entire construction, including approval of the construction materials and safety equipment, and continue with vessel inspections as the ship prepares to embark passengers. Coast Guard marine inspectors track most vessels from shipyard to final voyage.

1.1.1 Ship Design

Operators of U.S. and foreign flag passenger vessels (whether under construction or existing) that are going to take passengers out of a U.S. port must first submit plans to the Coast Guard for approval. This is to ensure that they comply with current safety standards. The Coast Guard reviews the ship plans for a variety of issues, including structural fire protection and means of escape. After review of the ship design, the ship may be inspected during construction. It is required to be inspected upon completion to ensure that the plans reflect the actual built ship. The primary reason for this intense focus is so this great lifeboat will adequately and safely protect its crew and passengers should an unfortunate emergency strike.

1.1.2 Regulations

The Coast Guard also creates domestic regulations and participates in the IMO so that safety on the seas will be maintained for new ships with innovative designs. The Coast Guard's foundation of the regulation process has always been — and continues to be — the intent to improve passenger and crewmember safety and environmental protection. Simply put, the Coast Guard wants fewer people to suffer injuries and less oil and hazardous materials to be spilled. In order to be proper regulators, the Coast Guard understands and works in concert with the entire maritime industry team — ship owners, operators, shore-side operators and all of the other people who are necessary to make the modern maritime industry work.

Regulatory bodies can specify minimum standards below which no one may operate, but it's a bit more challenging to point to a superior standard to which everyone should aspire. Through industry partnering and supervision, the Coast Guard supports those who operate above the standard. When all who make up the foundation act in concert, the industry is provided with a base upon which better operators can compete fairly, and marginal operators are forced to either improve or lose business opportunities.

1.1.3 Compliance

The Coast Guard enforces a wide-range of regulations to ensure U.S. and foreign flag vessels operating in U.S. waters are structurally sound, competently operated, and outfitted with lifesaving, fire protection and other safety systems. These regulations include the 1974 International Convention for Safety of Life at Sea and its Protocol of 1978, as amended (SOLAS) and the U.S. Code of Federal Regulations (CFR). Both are based on sound engineering practices and many years of practical operating experience combined with good seamanship. The Coast Guard ensures compliance with these regulations primarily through vessel plan review and a vessel inspection program that includes port state control verification examinations.

1.2 Evacuation Analysis

Even with everyone's efforts to ensure safely built ships and compliance with strict regulations, accidents still happen. The Coast Guard has prepared for the possibility of accidents with a two-fold approach.

First, should an emergency occur on the water, the Coast Guard has people dedicated 24 hours a day to search and rescue missions. No emergency is too small or large — their role as guardians and lifesavers of the sea is to protect everyone who sails on it. They will respond to every call for help.

Second, and the focus of this paper, is the Coast Guard's holistic approach to safely evacuating passengers. This method seeks to determine how accidents can be avoided, and what the best evacuation procedures are when these accidents happen, using a three-element approach. The first element addresses the events leading up to and including various accident scenarios that could require evacuation of the ship. The second element assesses the process of the actual evacuation process, including passenger decision-making and movement to assembly stations, embarkation stations and off the vessel. The third element investigates the impact of the abandon ship scenario on the passengers by including an assessment of search and rescue capabilities based on various sea and weather scenarios.

1.3 Passenger Ship Fire Investigations

The Coast Guard is the main agency responsible for conducting investigations of major marine casualties that occur within U.S. waters. During the investigation, the Coast Guard looks at many factors, including the cause of accident and emergency procedures (whether they were properly enacted and if they were adequate). Lessons learned from these investigations are then used internally to help develop new safety regulations. Even if the ship is not abandoned, the evacuation procedures (including the movement of passengers to their assembly stations) are studied in an effort to identify areas where improvements can be made. Two recent incidents are briefly highlighted below.

On July 20, 1998 the *Ecstasy* was departing from the Port of Miami, Florida with 2,557 passengers and 920 crew onboard when a fire occurred aft on the vessel. During the fire, the master ordered the general alarm sounded and the passengers and crew were directed to proceed to their assembly stations. Announcements were made that there was a fire on board, it was under control and there was no danger. Cabin stewards and lobby/corridor personnel directed passengers to their assigned assembly stations. The cabin stewards moved through each cabin, checking the closets, toilets and under the beds. Crewmembers reacted to the emergency as they were trained during drills, as this evacuation to the assembly stations turned out to be like a typical drill. However there was smoke in the aft section of the ship near an assembly station and it was determined not safe to use. The public address system was then used to change evacuation routes to an alternate area forward on the ship. Overall the evacuation went well and the passengers quickly and safely reached their assembly stations.

A similar incident occurred on September 19, 1999 when the *Tropicale*, carrying 1096 passengers and 610 crew, was transiting in international waters approximately 120 miles from Florida's west coast. An engine room fire caused the master to order passengers to their assembly stations, and announcements were made about the fire and that there was no danger. However, the passengers had to remain at assembly stations for almost twelve hours. Afterward, some passengers reported that there were communication problems and confusion while proceeding to and remaining at their assembly stations. The behavior and response of the passengers was unusual in this incident because the ship was also exposed to a hurricane. This understandably must have added to the passengers' concern of danger. Although there was some confusion and the evacuation did not go smoothly, no one was injured and an effective abandonment could have been conducted if needed.

The Coast Guard's role in these two investigations was to determine the cause of the incidents and to review the emergency and evacuation procedures. Meaningful information can be obtained from these actual emergencies. Although total abandonments are rarely required, the actual movement of passengers to their assembly stations reveals valuable insight about evacuation procedures and the behavior of passengers. Like the Coast Guard, the IMO studies actual incidents and uses them as a basis for the development of new requirements.

2 Coast Guard Evacuation Analysis Activities

Though information from actual events is helpful, unfortunately the data recovered from most casualties is too incomplete to serve as a sole source for development of improved evacuation techniques. The Coast Guard therefore relies on other sources in the development of passenger safety regulations, such as shoreside building evacuation drills and actual emergencies, building computer simulations, and shipboard evacuation drills.

2.1 Drills/Exercises

Shipboard drills and exercises provide a unique opportunity to study and practice an evacuation on a dynamic platform without the pressure of an actual emergency. The Coast Guard can take the time to ensure that all the right groups are involved, to set up specific evacuation scenarios, and to walk through the drill with a sense of calmness that all is actually safe. The greatest advantage of these drills is the training for all people involved, both on and off the ship. Bringing all the right people together in a relaxed, non-threatening event allows everyone to practice their roles as well as get a better understanding of everyone else's roles. It is a great opportunity to test evacuation procedures and recognize their strengths and weaknesses. What looks good on paper does not always work in reality, and the safe environment of a drill allows such problems to be recognized without injury to people or damage to property.

There are, however, recognizable disadvantages to drills and exercises that limit their regular performance on ships. First, drills take a lot of time to plan and enact. Ship platforms and terminals have to temporarily suspend normal operations, participants or "players" (volunteer passengers, crew, emergency response personnel, etc.) must be found and organized, and various shore-based organizations must coordinate their schedules. This planning costs time and money. Also, regardless of how many people try to make the emergency as realistic as possible, the "stand-in" passengers are aware that there is no danger. Their response understandably may not accurately reflect all the potential uncertainty that a true emergency might cause.

The Coast Guard firmly believes that the advantages of simulated evacuation exercises strongly outweigh the disadvantages, and it is for this reason that they frequently participate in them. It is only through such activity that the Coast Guard can continue to improve the safety level of passengers and crew during an actual emergency. Two recent shipboard drills, one while the vessel was underway and one dockside, are highlighted below.

2.1.1 Port Canaveral Drill

The first drill, held in Port Canaveral, Florida in the spring of 2000, involved a 65 meter (length overall), 1000 passenger offshore gaming (casino) vessel documented as a small passenger vessel under U.S. regulations. The scenario was an engine room fire that resulted in the loss of main propulsion and caused the vessel to run aground 0.9 kilometers offshore. Numerous passenger injuries and an offshore evacuation were simulated.

The purpose of this drill was to exercise the integration of numerous groups to evaluate their response to a major disaster in that area. The Coast Guard was a major participant in this drill. The following exercise objectives specifically relate to the Coast Guard's holistic evacuation analysis approach:

1. Conduct a simulated real time evacuation of 1000 passengers and crew using pre-identified Search and Rescue (SAR) platforms.

2. Identify, contact and determine availability of additional SAR platforms to assist in passenger evacuation.

3. Establish a passenger reception center to account for all passengers and crew and to accommodate passengers in need of medical attention.

4. Extinguish the fire onboard the vessel.

5. Mitigate any environmental consequences resulting from the damage to the vessel.

The drill began at 4 a.m. as the vessel, simulated to be at 100 percent capacity, returned from a gaming cruise. A fire was reported in the engine room and the crew began to isolate and fight the fire. Notifications were made and Coast Guard and Canaveral Fire Rescue resources responded to the scene and set up command posts. The evacuation of passengers to their assembly and embarkation stations and then off of the vessel was simulated since no "stand-in" passengers were actually on the vessel. The crew contained the fire; however assistance from Canaveral Fire Rescue and Coast Guard personnel was needed to extinguish it. One hundred passenger injuries were simulated by describing the injuries on index cards that were given to response personnel for action. The response personnel were evaluated on the prioritization of the injured passengers and how they were treated.

Coast Guard boats and helicopters were used to evacuate passengers from the distressed vessel. Two helicopters transported high priority victims to awaiting ambulances located at Coast Guard Station Port Canaveral. Response vessels and other vessels (Good Samaritans) that were in the area assisted with the evacuation of the vessel. A shore-side passenger reception area was established to process evacuees as they arrived on rescue vessels and helicopters. The status and location of the evacuees was documented in this area. Part of the reception area was used as a triage location to prioritize first aid for the most serious injured passengers and arrange for their transportation to hospitals. The reception area was also used to hold uninjured evacuees until bus services were arranged to transport them back to the place of embarkation.

One of the lessons learned from this exercise occurred when one of the rescue vessels docked with another vessel to transfer passengers, and found that the main decks of the two vessels did not match up. Jacob's ladders were used to solve this problem; however this slowed the evacuation process and made it more hazardous. Injured and elderly passengers could not be transferred this way so stokes litters (metal rescue baskets) were used which created another bottleneck in the already slowed evacuation process.

The drill was a great success. Coast Guard and firefighting resources were able to assist with the vessel fire, SAR resources were able to transport evacuees, and emergency medical systems (including ambulance services, hospitals, and fire departments) were able to adequately treat passengers. The vessel's fire fighting and evacuation plans were appropriate and worked well. Although the movement of passengers was simulated and therefore not recorded, this drill was successful in identifying and evaluating SAR and emergency medical resources. Overall, all participants agreed that the time and money spent on this drill was worth the results gained. It was also agreed that another drill would be held the following year to continue the cooperation and dedication to safety.

2.1.2 Port of Galveston Drill

In the fall of 2000, a dockside evacuation drill was held in Galveston, Texas aboard a 225 meter (length overall), 1800 passenger cruise ship. The Port was preparing for its first cruise ship in recent years, and the Coast Guard wanted to ensure that all related parties were adequately prepared should an emergency occur. This drill provided the opportunity for all parties to get together beforehand, to learn each other's roles and responsibilities, and to learn how this new group could work together.

Although this was not specifically an evacuation analysis drill, the drill allowed everyone to observe typical evacuation procedures. Evacuation issues being observed included movement of passengers away from the incident, subsequent movement of passengers off the vessel (as the fire got worse), and adequacy of crew response to passenger evacuation needs.

The drill scenario placed the vessel in port embarking passengers for its next voyage. Approximately 100 Coast Guard personnel volunteered to be "stand-in" passengers and were on board when the fire was detected in a passenger space. During the drill, crew response was appropriate in that the fire was isolated and passengers were escorted to their assembly stations. The drill then assumed that the fire progressed to the point where it was determined that the ship should be evacuated. The passengers and non-essential crewmembers therefore left the vessel via normal gangways and terminal routes. After the evacuation, the fire was extinguished with assistance from the local fire department.

The crewmember evacuation team was comprised of (1) cabin stewards, (2) lobby/corridor personnel and (3) assembly station personnel. The passengers easily identified them because they wore fluorescent green hats. This team was responsible for knowing the escape routes and alternate escape routes; location of adult, children and infant lifejackets; and the operation of fire screen doors and fire equipment located in their sections. Cabin stewards and lobby/corridor personnel directed passengers to their assembly stations during the beginning of the drill. Then the assembly station personnel lead the passengers to the gangways and through the terminal. One crewmember led the group and another followed the group. This controlled the flow of the passengers and ensured the evacuation was carried out in an orderly fashion. Emergency medical response teams, made up of ship medical personnel, evacuated injured passengers using stretchers.

The drill proved a great success in highlighting strengths and weaknesses of the coordinated dockside evacuation process. Overall, the ability of the different organizations to effectively communicate was determined to be the biggest challenge. Certainly in a situation such as this, complex communications would be expected to be challenging, but this was a good opportunity to experience this difficulty firsthand. It was also recognized that people who were very familiar with their own procedures did not know the procedures of other organizations with which they would have to work during an actual event. This drill provided everyone with the opportunity to learn these other procedures.

2.2 Teamwork between Coast Guard and Maritime Industry

The Coast Guard understands that safety at sea depends on everyone who uses the waterways. Everyone must continue to work together to improve safety. Therefore, the Coast Guard focuses on partnerships with various maritime industry organizations. One way to accomplish common goals is through formal partnerships that create and implement new safety measures to achieve their intended safety function without over-burdening the ship owners. Whether they are regulatory or non-regulatory, the success of these measures depends on all parties understanding them and recognizing their benefit. Two formal partnerships reflecting the Coast Guard's role in evacuation analysis are highlighted below.

2.2.1 Partnership with the International Council of Cruise Lines

The International Council of Cruise Lines (ICCL) is an organization made up of the 16 largest passenger cruise lines that call on major ports in the U.S. and abroad. Its mission is to participate in regulatory and policy development processes so that all measures adopted provide for a safe, secure and healthy cruise ship environment. ICCL's overnight cruise ship operators carry more than four million U.S. passengers on 93 ships each year.

The purpose of the Coast Guard/ICCL partnership, signed in March 1997, is to promote safety and environmental protection of large foreign flag passenger ships while strengthening communications and the working relationship between the international passenger vessel industry and the Coast Guard. This partnership is not intended to undermine the authority of the Administrations or international conventions that govern safe ship operations, nor is it intended to supplant the role of the U.S. Port State Control program. However, the partnership is intended to complement these governing bodies through an efficient means for joint Coast Guard/industry talks and action.

A prime example of this worthy partnership will be taking place this year as a mass rescue evacuation drill is planned to test the ability of the Coast Guard, industry and other organizations to adequately respond to a significant casualty on a major cruise ship with 4000 persons on board. A meeting is planned for spring 2001, where Coast Guard and ICCL will meet with other maritime industry representatives to establish a baseline of information and discuss the ability for such an exercise. The group will identify problem areas and develop strategies for follow-on table-top and field exercises. One expected final product is a major cruise ship exercise where a large number of stand-in passengers (though most likely less than 4000) will be safely evacuated.

2.2.2 Partnership with the Passenger Vessel Association

The Passenger Vessel Association (PVA) is the U.S. trade organization of the passenger vessel industry dedicated to developing a superior business environment. Founded in 1971, the PVA represents more than 500 companies that own, operate and supply U.S. Coast Guard certificated, Canadian Coast Guard and state inspected passenger vessels. PVA members carry more than 200 million passengers each year.

The partnership between the Coast Guard and PVA was signed in January 1996, and its objectives are to promote passenger, personnel, and property safety within the domestic passenger vessel industry and the protection of the environment within U.S. waters.

Although partnerships are non-regulatory in nature, they provide the Coast Guard an opportunity to work directly with a specific part of the maritime community. The partnership with PVA shows how this relationship is proving very successful with regulatory matters as well, since the Coast Guard and industry have an opportunity to work together to develop the best regulations possible. For example, the Coast Guard is currently implementing new lifesaving regulations for large domestic passenger vessels. This regulation will require these vessels to upgrade or retrofit existing lifesaving equipment. The regulation provides an alternative for some vessels to conduct a risk assessment in lieu of obtaining the required lifesaving equipment. This risk assessment must identify the risks involved with a vessel evacuation, and must show how those risks will be mitigated through a shipboard safety management and contingency plan. Evacuation plans and procedures are therefore part of the required safety assessment. The Coast Guard has hosted four public workshops around the country to discuss and further explain the expected level of detail that must be provided in the safety assessments, so that all affected members will have the most information possible.

2.3 Passenger Vessel Access Advisory Committee

The Coast Guard is also concerned with ensuring shipboard compliance with the Americans with Disabilities Act of 1990 (ADA). The ADA is a U.S. law that prohibits discrimination and ensures equal opportunity for persons with disabilities in employment, state and local government services, public accommodations, commercial facilities, and transportation. This law ensures that buildings and modes of transportation are accessible to persons with disabilities. It has been implemented successfully in buildings and in some modes of transportation. However, ADA compliance on passenger ships is not yet completed. The U.S. Architectural and Barriers Compliance Board (Access Board), in charge of developing and maintaining accessibility requirements enforced within the U.S., formed the Passenger Vessel Access Advisory Committee (PVAAC) to focus specifically on ADA compliance on passenger ships. This group, comprised of maritime industry representatives and organizations representing people with disabilities, was formed to develop guidelines to make all types of newly constructed passenger vessels accessible to persons with disabilities.

The Coast Guard joined PVAAC as a technical advisor with a multi-faceted role. First, they had to teach this group existing passenger vessel regulations so that the guidelines created would not conflict with existing ones. And second, the Coast Guard was interested in ensuring that the evacuation procedures for persons with disabilities would be practical and acceptable to passengers and ship operators.

Many issues of compliance were discussed among this group, including the great concern for safe and effective evacuation procedures for persons with disabilities. Although many cruise ships can now accommodate passengers with disabilities, specifically passengers with mobility impairments, there are few solid procedures in place for the evacuation of these passengers should an emergency occur. Passengers confined to wheelchairs must rely on crew to locate them and assist them in moving up and down stairs to the assembly stations. Crewmember assistance is also needed to help them get into lifeboats. Some vessels may not have the dedicated number of crew to provide this service. One potential solution is allowing persons with mobility impairment to use elevators during emergency evacuations. PVAAC discussed many options such as this one concerning the safe and timely evacuation of passengers with disabilities.

PVAAC submitted a report of their recommendations to the Access Board, and it is predicted that the Access Board will use these recommendations to develop regulations for passenger vessels.

2.4 International Maritime Organization

On the international front, the Coast Guard is the U.S. representative at the International Maritime Organization (IMO), a permanent international body established in 1958 to review current maritime issues and set international shipping policy. The U.S. is one of 158 Member States. IMO promotes maritime safety and pollution prevention through international treaties, agreements, and cooperative efforts. IMO also provides a mechanism for cooperation among governments concerning safe shipping regarding international trade, and encourages the highest standards of maritime safety, navigation and marine pollution control.

The Maritime Safety Committee (MSC) is the highest technical body of the IMO, and it consists of all Member States. The MSC operates with the assistance of nine Sub-Committees, including Fire Protection (FP), Radio Communications and Search and Rescue, Safety of Navigation, and Ship Design and Equipment. The Coast Guard has representatives in all nine. Currently two different groups, the FP Sub-Committee and the MSC, are working on passenger ship evacuation issues. The Coast Guard is playing an active role in both of these groups.

2.4.1 Sub-Committee on Fire Protection

Following the *Estonia* incident in September 1994 where 852 lives were lost after the vessel capsized and sank, the need to develop rapid evacuation procedures for ro-ro passenger vessels was emphasized. The IMO agreed upon a resolution to require evacuation analysis on new ro-ro passenger vessels at the design stage. This effort has led to doing the same for all new passenger ships, including high-speed passenger vessels and cruise ships through a three-year evacuation analysis project (to begin in 1999).

During the past two years, this group has focused on developing guidelines for conducting evacuation analysis on passenger vessels. One of the basic difficulties (for this group and all groups studying ship evacuations) has been the gathering of applicable information. Though there is a lot of documented material on building evacuations, it can be very difficult to apply toward ship evacuations because of the vastly different situation. For instance, abandonment in a building is the first step in the event of an emergency. It is the last resort on a ship, and because of this, emergency resources such as evacuation procedures and equipment MUST work. Otherwise, the passengers and crew have no more options.

Another fundamental difference is that buildings don't move; ships do. Ships are also subject to dynamic motion, which will definitely affect the movement of passengers and crew. And help for a ship can sometimes be hours away, as too many ship disasters (*Estonia*, and *Scandinavian Star*, to name but a few) have unfortunately proven.

The goal of conducting an evacuation analysis is to determine the amount of time needed to evacuate a vessel. Discussions on this analysis with the FP Sub-Committee have led to two distinctive analysis techniques: microscopic and macroscopic.

Because of a lack of technical information, the working group first developed a macroscopic approach. This approach is a simplified, quick, first-order approximation of the actual evacuation time. Many assumptions are made when applying this model, such as the unrealistic scenario that all passengers begin evacuating at the same time. Much of the calculation is based on known building evacuation methods with an adjustment in travel times for shipboard movement.

After the macroscopic approach was developed, the group switched to the microscopic, a more human factors approach. The working group is now monitoring the work of several members who are developing a microscopic approach that accounts for people's individual behaviors. The Australian delegation is studying the movement of people through a dynamic or moving environment. The Canadian delegation is studying the movement of people through a static listed or sloped environment. There are many additional evacuation models that account for at least some, if not many, human factors being developed, including ones by the German delegation and the United Kingdom delegation.

The Coast Guard believes that the more exact nature of the microscopic approach is the future of evacuation analysis techniques. The Coast Guard has developed a database of documents, and would like to work with other delegations to further develop a holistic approach to evacuation analysis. In choosing to support the microscopic approach, the Coast Guard recognizes that people will not all react the same. For example, parents will search the ship for their children instead of proceeding to their assembly stations (affiliative behavior). People will also sometimes return to their cabins for lifejackets and/or medications. With each of these examples, people are potentially moving against traffic and causing more congestion, as well as putting their own lives in further danger by moving away from their assembly stations.

To successfully deal with these complex evacuation problems, human factors must be acknowledged and adequately addressed. Some industry representatives are dealing with these issues effectively and have had great success. But these solutions must be shared with the entire maritime industry. The potential concerns involved with doing so reminds everyone that cruise ships are a business, and sometimes not sharing information (even if it refers to safety) can be seen as a competitive advantage.

2.4.2 Maritime Safety Committee

In 1999, IMO Secretary General William O'Neil expressed his concerns on the safety of large passenger ships. He stated in particular his concerns that the operational safety aspects in possible emergencies involving large cruise ships, which carry thousands of passengers and crewmembers, may not have kept in line with technical developments. The members of the MSC endorsed this concern and established a working group on large passenger ship safety.

The working group used a risk based, formal safety assessment approach to identify the issues that should be considered. The group had several terms of reference which are summarized as follows: (1) conduct an overview of large passenger ship current practices, existing regulations and safety philosophies; (2) identify areas of concern, using a holistic approach and taking into account the human element, relating to the ship, persons and environment; (3) examine on-going efforts at IMO, within the industry or other organizations; (4) identify potential future risks large passenger ships may face in the coming decade; and (5) prioritize the work to be undertaken and develop a preliminary work plan. Collision and grounding, evacuation and rescue, (including lifesaving systems and arrangements), fire safety, and search and rescue are some of the areas of concern identified by the group.

The group had discussions about concerns related to evacuation, abandonment, and rescue; they agreed that time was a key element in such situations. Therefore, this group decided to focus on how long a damaged passenger ship must remain afloat (given certain conditions such as sea conditions, injuries, etc) to ensure probability of an effective evacuation and rescue operation. Many points of concern were taken into account. One concern was the speed of passenger movement to assembly and embarkation stations, which depends on the design and arrangements of the corridors and stairways and the guidance provided by the crewmembers. Another point was crewmember guidance and the condition of the lifesaving equipment and arrangements, which will determine how orderly the ship is abandoned. A third concern was the rescue of the survivors, which is dependent on the ship's operational area and the environmental conditions. These factors and more will be investigated further in the holistic approach to large passenger ship safety.

The group also considered the on-going work of the Fire Protection (FP) Sub-Committee on the development of guidelines for conducting an evacuation analysis for newly constructed passenger ships. They agreed that this work had application to evacuation in their holistic approach, and that evacuation analysis on existing ships would be beneficial in identifying areas of concern to develop proactive solutions before a casualty occurs.

This group's work is ongoing through a correspondence group with a Coast Guard representative as the co-ordinator. Their efforts will be presented at the upcoming meeting of the MSC in June 2001.

3 Future Activities of the Coast Guard

What does the future hold? Hopefully, it holds safe and prosperous times for all. But valid concerns about the collective knowledge of passenger ship evacuation issues remain and need to be addressed.

3.1 Risk Management

The Coast Guard recognizes and acknowledges that the passenger vessel industry has a strong safety record. It is therefore a valid question to ask why so much time is being devoted to evacuation analysis. The answer lies in the concept of risk management. Risk is the probability of an event occurring in relation to its consequences. While the likelihood of a large passenger ship being forced to evacuate in the middle of the ocean is small, accidents such as the *Sun Vista*, *Prinsesse Ragnhild*, and *Sea Breeze I* show that the consequences when an event occur can be daunting. The ever-increasing size of passenger ships makes the evaluation of possible consequences more and more important.

Another concern is the continued increase of traffic on the waterways. Statistics show that parts of the marine transportation traffic management system are becoming antiquated. Some waterways are coming under heavier traffic use. With more than 1,000 harbor channels and 25,000 miles of inland, intracoastal, and coastal waterways in the U.S. alone, the idea of updating the entire marine transportation system is an awesome one. Also, passenger ships are continuing to gain popularity as a means of recreation. Because of this, more ships are being built to accommodate the increased flow of passengers. This influx of ships equals busier waterways. Ships are now enjoying longer voyages and traveling to new destinations. The ever increasing amount of time passengers spend on ships only adds to the likelihood for a major ship accident, and the Coast Guard is determined to be prepared should this happen.

3.2 Sub-Committee on Fire Protection

The FP correspondence group is focusing its work in four areas: (1) validate and revise, if necessary, guidelines for a simplified evacuation analysis of passenger ships (macroscopic method); (2) develop guidance on the use of microscopic models; (3) develop guidance for validation of computerized models which will be used by Administrations; and (4) identify and address evacuation problems on existing ships. This work will continue through the correspondence group and will be presented at the February 2002 meeting of the FP Sub-Committee.

3.3 Marine Safety Committee

The MSC work continues as the correspondence group develops the holistic approach, goals and objectives for dealing with safety matters relating to future large passenger ships. The group will consider how passenger ship areas of concern should be analyzed and what tools, such as formal safety assessment, human element analyzing process, cost/benefit analysis and risk assessment, should be used in the analytical process. This work will be presented at the June 2001 meeting of the MSC and will continue during that session.

4 Conclusion

The biggest roadblock to developing a comprehensive microscopic-based evacuation technique is the current lack of information specific to passenger movement on ships. One of the acknowledged difficulties in conducting ship evacuation drills to get this information is the fear from shipowners that congestion or other problems on their ships will be documented, and that they may suffer legal liability in the event of an actual accident. These fears are a major reason why there have been so few evacuation drills on large passenger ships that include the participation of the U.S. Coast Guard or other research organizations. These fears must be put to rest.

The two drills mentioned earlier (Port Canaveral and Galveston) show the great success and lessons learned that can result from a non-threatening, educational evacuation drill. The goal of everyone in the maritime industry is safety — for people, cargo, and the environment. Everyone must work together to ensure this safety, but it cannot be accomplished without a sense of teamwork and trust.

References

1. A. Grenier, LT, *Field Notes on the Ecstasy Fire*, (July 1998).

2. D. Stuart, *Escalating Risk*, Seatrade Cruise Review, (December 2000).

3. D.M. Orban, Associates, *Cruise Industry Poised for Another Record-Setting Year*, Cruise Lines International Association, (December 2000).

4. J. Reinert, LT, *After Action Report of the Evacuation Exercise involving the Port of Galveston,* Texas held on 28 September 2000, (4 December 2000).

5. K. Kiefer, LT, *Field Notes on the Tropicale Fire*, (September 1999).

6. P. Dawson, *Cruise Ships: An Evolution in Design*, Conway Maritime Press, (2000).

7. T. Attrino, *Are Big Ships Safe?*, Cruise Industry News Quarterly, (Spring 1999).

8. Economist Intelligence Unit and Business International, *Special Report on Cruise Ships in the 1990s*, (1992).

9. IMO, *Report of the Maritime Safety Committee on its Seventy-Third Session,* MSC 73/12, (12 December 2000).

10. IMO, *Report of the Working Group on Large Passenger Ship Safety,* MSC 73/WP.20, (5 December 2000).

11. IMO, *Report to the Maritime Safety Committee,* FP 45/16, (29 January 2001).

12. IMO, *1974 International Convention for Safety of Life at Sea and its Protocol of 1978,* as amended (SOLAS), (Consolidated Edition, London, 1997).

13. Joint Accident Investigation Commission of Estonia, Finland, and Sweden, *Final Report on the Capsizing on 28 September 1994 in the Baltic Sea of the Ro-Ro Passenger Vessel Estonia*, (1996).

14. U.S. Coast Guard, *Coast Guard District Seven's Year 2000 Search and Rescue Exercise*, (17 April 2000).

15. U.S. Coast Guard, *U.S. Coast Guard Marine Safety and Environmental Protection Business Plan,* FY 2001-2005, (September 2000).

16. *Surviving the Unthinkable: Can Cruise Ships Cope with Catastrophe?*, Fairplay, (29 October 1998).

17. International Council of Cruise Lines (ICCL) website, (February 2001).

18. Passenger Vessel Association (PVA) website, (December 2000).

Modelling and Simulation of Ship Motions and Vessel Traffic Services in Relation to Safe Navigation in Narrow Waterways

A.N. Ince

Istanbul Technical University Foundation, Ferit Tek Sok. 4/8 Moda, 81300 Istanbul, Turkey

This paper outlines the design of a Vessel Traffic Management and Information System (VTMIS) for the Turkish Straits and taking this as an example shows how traffic regulations may be developed for safe and efficient navigation of vessels through waterways which are narrow, winding with changing currents and are therefore difficult to navigate and prone to accidents. Ship handling and Vessel Traffic Flow simulation models are described and the simulation trials conducted under different traffic and environment condition are discussed to show the role that these navigation and simulation programmes can play in preventing marine casualities in different waterways.

1 General Aspects of Safe Navigation in Narrow Waterways

1.1 Introduction

The objective of this article is to find out, by means of modelling and simulation, the criteria to be used for the formulation of rules and procedures required for minimising the probability of accidents in waterways which are narrow, winding with changing currents and therefore are difficult to navigate and prone to accidents.

It is obvious that the risk of accident can be reduced but never altogether removed. Accidents which occur in waterways typically take the form of collision, grounding, ramming, fire on board or mechanical failures which may result in loss of life, pollution and/or irreparable damage to property and the environment.

The other papers presented in this conference treat the problems which arise during the events mentioned above or afterwards whereas this paper will discuss ship design, navigation rules and Vessel Traffic Management and Information Systems (VTMIS) which aim at preventing or reducing to a minimum accidents which occur in waterways. In fact the studies reported here have arisen in connection with the design of a VTMIS for the so-called Turkish Straits which consist of the straits of Istanbul (Bosphorus) and Çanakkale (Dardanelles) with the inland Sea of Marmara in between.

There are three main factors in the occurrence of marine accidents; ship design, environment, traffic regulations.

The important ship design parameters here relating to the vessel control are the engine and rudder performances (with and without thrusters) and the hydrodynamic and aerodynamic behaviour of the vessel.

Simulation modelling in this respect is used for studying ship manoeuvering characteristics as affected by geographic and hydrographic conditions and for the investigations of the amenability of accidents under different navigation scenarios. The master's judgment in controlling his ship is, of course, influenced by the data he has about his environment (geography, bathymetry, meteorologic conditions and locations of other vessels in the waterway). Some of these data may be obtained by means of the vessel's own sensors but the ones that cannot be obtained must be transmitted to the vessel from shore facilities such as the VTS Control Centre.

Finally, the third factor affecting accidents risks relates to traffic regulations in force in the waterway which stipulate priorities for entering the waterway and navigational advice for safe passage taking into account the size, weight and cargo (hazardous, non-hazardous) characteristics of the vessels which are to use the waterway. In this context and by means of an appropriate "Vessel Traffic Simulator" it is possible to estimate the traffic capacity of the waterway and to determine navigational rules which would assure safe and efficient passage of vessels through the waterway. Before dealing with the details of the topics above, we shall give below information about the Turkish Straits and their traffic which were used as a model for our studies.

1.2 Turkish Straits

The Istanbul Strait and the Strait of Çanakkale are two important waterways, which link the Sea of Marmara to the Aegean and Black Seas. The Strait of Istanbul has a length of about 30 km and an average width of 1.5 km narrowing to a width of 700 meters at Kandilli (Fig. 1). The strait geography is such that a sail plan made for a passage would have at least 12 important turns. The strait of Çanakkale, on the other hand, is longer (70 km) and wider, the minimum width being 1300 meters at Nara Burnu.

The annual traffic in 1999 was about 48000 vessels for the Strait of Istanbul and 41000 for the Strait of Çanakkale, 14 % of these being tankers and simular vessels carrying hazardous cargo. The present day avarage traffic for Istanbul is 150 and for Çanakkale about 100 vessels per day. A siginficant percentage of these vessels carry hazardous cargo (petroleum, chemical products etc.) and can have lengths exceeding 300 meters. Continuous tracking of these vessels is essential, particularly when their masters have little knowledge of these treacherous waters, in order to keep at a minimum the risk of accidents which result in a loss of life and damage to property and the environment. The present dense tanker traffic is expected to reach 20 tankers per day in the next five years and 30 in the following five years. The number of vessels which can be presented simultaneously in the Straits of Istanbul and Çanakkale can reach 25 and 30 vessels respectively. In addition to the transit traffic cited above for Istanbul there are some 2500 scheduled vessels criss-crossing between the two shores of the straits.

The Strait of Istanbul has, unfortunately, maintained, over many years, the character of being a high-risk waterway; about 40 accidents per 100 000 passages. The statistics of other waterways shows that the rate of accident does not change with increase of traffic unless the traffic regime is changed. The main causes of these accidents are the following: high traffic density, high strength and variability of currents, sharp bends and turns, frequent bad visibility, high density of local cross traffic, and of course faulty judgement by the masters/pilots. When we consider the expected increase of tanker traffic in the coming years the importance even absolute necessecity of taking all possible measures, including the implementation of a modern VTS which have the potential of changing the present traffic regime, becomes obvious.

1.3 The VTS System for the Turkish Straits

The information about the VTS system being implemented by Turkey is given below for use in our later discussions. Three VTS (Vessel Traffic Services) systems are presently being implemented for the three VTS areas of the Istanbul and Çanakkale Straits and the Sea of Marmara. These systems which use the most up-to-date technology are designed by us to provide maximum safety and efficency of navigation and protection of life and environment.

For the provision of an effective Vessel Traffic Service it is necessary to collect data continuously about vessels, sea state, meteorological and other environmental conditions. For these data to be collected, evaluated and distributed to all those concerned it is necessary to have radars, hydrologic and meteorologic sensors, imaging day/night cameras, dGPS stations and transponders, VHF and MF/HF communications, VHF direction finders (DF), RACON and other navigational aids.

Radars are the most essential sensors for the tracking of non-cooperative targets which are least affected by environmental conditions. The international Association of Lighthouse Authorities (IALA) stipulates that microwave radars used for surveillance of waterways should have range and cross-range accuracies as well as range and cross-range resulations of about 20 meters within a range of up to 3 nm. For this reason the radars are to be deployed at appropriate sites on both shores of the Straits so as to provide continuous coverage of all areas with one or two radars, depending on the density of traffic and accidents. It has been ascertained that the deployment of dual radars on 8 sites in Istanbul and 5 in the Çanakkale strait would provide the coverage needed with 99.9 % availability (see Fig. 1). There will also be two GPS reference stations at the Black Sea and Aegean Sea entrances together with two dGPS broadcast stations to enable vessels to determine their own positions (with an accurary of about 5 meters) and using on-board transponders to rebroadcast this together with other data such as identity to the VTS Centre on shore. If a vessel were to be equiped with two transponders, one at each end of the vessel, it would be possible for it to determine its own rate of turn. The Turkish VTMIS will also have a pilot carry-aboard laptop package (computer and dGPS transponder) which combines GPS satellite data with dGPS corrections and VTMIS traffic and weather data including automatic vessel identifications using AIS technology.

In addition to the elements mentioned above, the VTMIS will have an "expert system" and a "hydrodynamic predictive model" which will be used to provide various vessel traffic services such as vessel passage schedules and safe vessel speeds. The expert system is a decision aid for preventing critical navigational situations arising in the straits and will advise the traffic controller as to which vessels wishing to transit will have to wait, their route and speed and what would happen if the advice given were not carried out by the ship master concerned.

Surface sea currents affect navigation and therefore safety and their presence and future intensities/directions should therefore be measured and predicted for safe and efficient navigation as well as for pollution control. The hydrodynamic Predictive Model will be used to predict currents up to four hours in advance anywhere in the waterway for the preparation of sail plans for vessels thus minimizing the risk of accidents which are mainly due to currents, the main cause of accidents. An adequate number and type of sensors (current, water level, atmospheric pressure, wind, water salinity and temperature) will be deployed in the straits, as show in Fig. 2, in order to provide input data for the model. Configuration of the VTMIS is shown in Fig. 3.

In addition to the above components, the Turkish VTMIS will have emergency services utilizing appropriately equipped vessels and helicopters for fire fighting, pollution control and search and rescue operations which will go into action when accidents occur.

1.4 Issues to be Resolved for Safe and Efficient Navigation with VTS

Despite all efforts and studies by several national and international organisations such as IMO, IHO and IALA there are still several subjects which need to be investigated further by the international maritime community. Recommendations and/or guidelines for use by VTS authorities and all mariners are to be developed in order to achieve safer and more efficient navigation through difficult waterways characterised by limited and varying width, bends, strong and unpredictable currents and cross currents. Some of these issues are listed below. These have arisen out of studies by the author in connection with the design of the Turkish VTMIS. The issues in question have been submitted to and accepted for study by IALA [1].

The topics are given below:

1. Criteria for setting limits for: maximum sailing speed, minimum sailing speed

Considering safety, steerageway, overtaking, currents, delays caused to traffic, disturbance to small vessels and shore assets.

2. Responsibility/liability/ethics about delays caused to traffic by slow moving, large and/or old/old fashioned vessels.

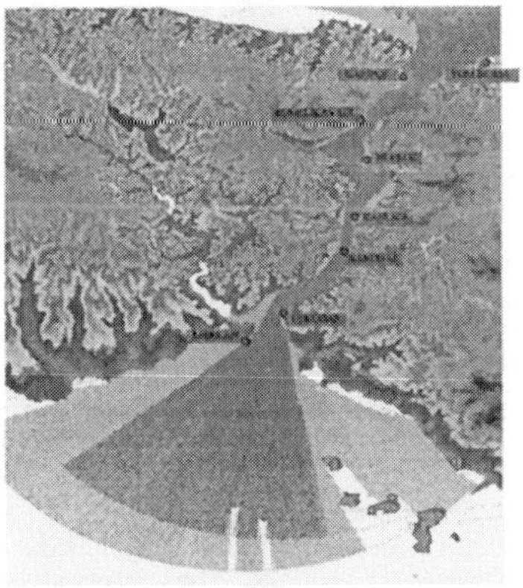

Figure 1: Radar coverage for the Strait of Istanbul.

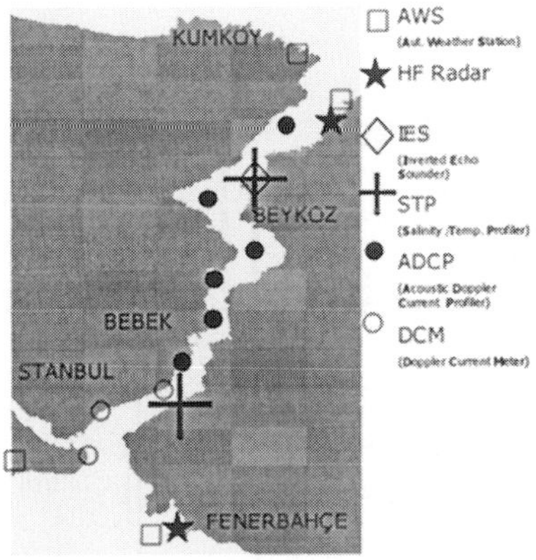

Figure 2: Hydrographic sensor types and locations.

3. Rules for overtaking.
4. Rules of priority for entering a waterway (e.g., FIFO, LIFO, fast before slow, small before large etc.).
5. Methods of measurement of current and hydrodynamical modelling for prediction/forcasting of currents driven by tides, winds and gravity for safe and efficient navigation, SAR and pollution control.
6. How to specify “safety” for use in the design of VTS or how one can or should handle the safety factor (safety criteria) in the design of navigational systems.
7. Quantitative rules and criteria to be used in defining “Traffic Separation Schemes” and “Precautionary Areas” considering the geography, currents and bathymetry of the waterway.
8. The use of factors such as “Ship Domain Theory”, turning radius, distance between ships, speed, vessel handling capabilities and environmental conditions in the determination of traffic regulations for navigation in waterways to be defined.
9. The role of voyage planning in reducing the risk of casualties.
10. Development of new guidelines and compensation for the inadequaties of the present IMO Colreg 72 rules [2] for safe navigation in narrow (to be defined) waterways with turns and unpredictable currents/cross-currents.
11. Study of ship handling capabilities with reference to risk of accidents.
12. As world trade continues to increase particularly in the energy sector (oil and gas) the question arises, as it does for the Turkish Straits, as to how much oil/gas can be transported through straits etc. considering that all channels

(telephone, data, road or waterway) have limited traffic carrying capacities. A reliable method of calculating this capability is required taking into account all relevant factors related to vessels, cargoes and the environment.

Studies carried out by the author so far for IALA have produced some preliminary results which shed some light on the topics above and are presented in Ref [3].

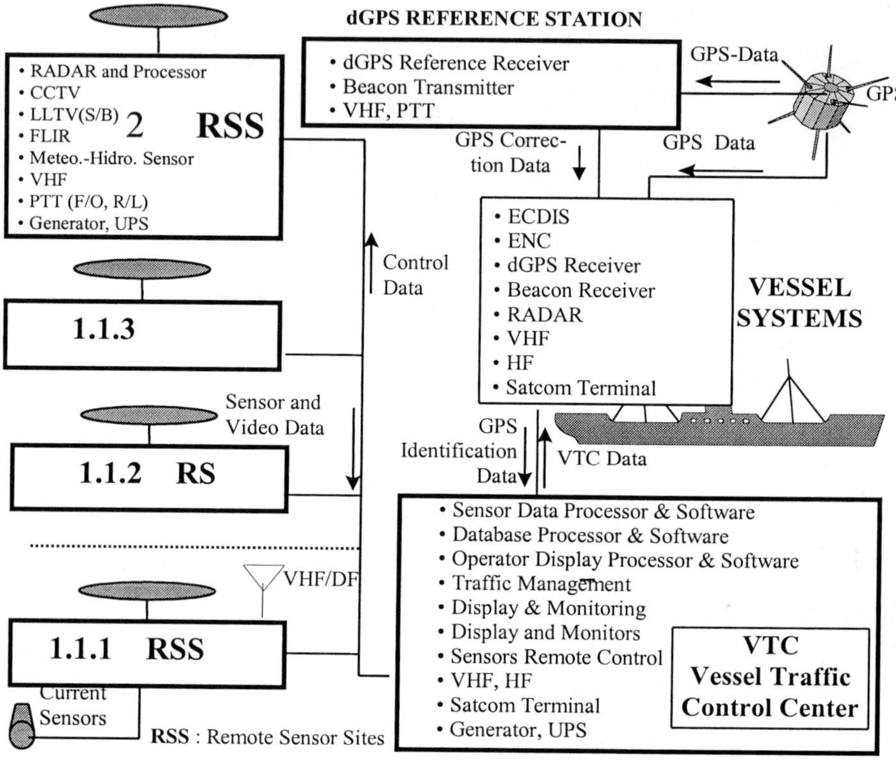

Figure 3: VTMIS Configuration for the turkish straits.

2 Modelling and Simulation of Ship Motions

2.1 General

Marine Navigational Simulators are found to be useful, even essential tools for the general training of seamen in order to prepare them to cope with unusual situations during sailing and more generally with maritime activities of all kinds. Our investigations show that most of the world's maritime organizations use simulation programmes of various kinds and that in recent times when new ships are delivered to their owners these are often accompanied by their specific navigation simulation programmes the accuracy of which can be as high as 96 %.

As in all advanced simulation applications, marine navigation simulation also relies on computer support. In navigation simulations the computer, apart from specifying the variations of the physical parameters of the object or event to be simulated, also provides graphics and image display support. A simulation programme prepared in this fashion, works out all the calculations prescribed in the mathematical model. The mathematical models used in simulation programmes take into account the characteristics of the vessel and the environment and calculates the motion of the vessel under all conditions and transfers the calculated numerical values and the video variations into the display system used.

A simple model of ship motions would ignore the environmental effects such as those due to wind, waves, currents and bathymetric variations. A similar simplified approach may be taken also as far as the vessel itself is concerned and where only its hydrodynamic characteristics are considered.

The hydrodynamic characteristics of a vessel can be specified using analytic and empirical methods and are tabulated as parameters of a system of equations. A table prepared in this way contains all the basic information needed for the determination of the accelerated translational and the rotational movements of the vessel. The second aspect to be modeled would be the propeller or propellers. There are tables available which give the performance characteristics of the various types of propellers used (with 3, 4 or more blades and fixed or variable pitch). When selecting the rotational speed and power characteristics of the propeller it is necessary to know the characteristics of the engine used in the vessel or alternatively to model the engine separately or together with the propeller. The third important aspect to be modelled is the rudder. Modelling of the rudder is accomplished by determining the hydrodynamic forces as defined by the rudder hydrodynamical parameters as a function of the rudder profile characteristics and rudder angle.

While the modelling of ship motions requires as a minimum the modelling of the ship's hull, engine, propeller and rudder as explained above, professional simulation programmes do, infact, incorporate several other modules as shown in Table 1.

2.2 Operational Properties of Simulation Programmes

In any simulation programme, there exist ship control, external effects and navigational aids for the user. The elements for ship control are: three dimensional view of the environment (in 60° steps), options for binoculars, options for the engine selection, change of rudder angle and selection of thrusters (if exist). The controllable external effects are: direction and speed of wind and current. Radar and video plotters are used as the navigational aids. Viewing facilities, provided by the programme are shown in Fig. 4.

2.3 The Role of Simulation Programmes in Preventing Casualties

The prediction of ship behaviour in sea regions with critical navigation conditions must take into account numerous and instantly changing external effects, which influence the ship motion. The simulation programmes developed in fact do this and allow the control of both ship and envirnmental effects, thus enabling the conduct of various trials, almost impossible to do in real life, on the table.

MAIN MODEL			
Control Program, Hull, Rudder(s) and Rotatable Nozzle(s), Propeller(s), Main Engine, Thruster(s), Anchors, Mooring Lines and Fenders, Current, Wind and Waves, Shallow Water, Banks, Channels, Ship to Ship Interaction Effects			
ADDITIONAL MODULES			
BRIDGE	**VTS**	**TRAINING**	**DATA BASE**
ARPA Radar Module, ECDIS (Electronic Chart Display and Information System) Mod., Navigational AIDS Module, Fishing Simulator Module, GMDSS (Global Maritime Distress & Safety System) Module, Real Communication Equipment, Visualisation System, Sound System	VTS Operator Module, GMDSS Module, Visualisation System	ASD (Active Situation Display) Module, Selective radar Module, GMDSS Instructor Module, VTMS Instructor Module, Visualisation Slave Monitors, Debriefing System	Common Database System, Exercise Editor Module, 3D Scene Editor Module, Radar Scene Editor Module, Chart Editor Module, Ship Modelling Data Generator

Table 1: Mathematical models used in navigation simulation.

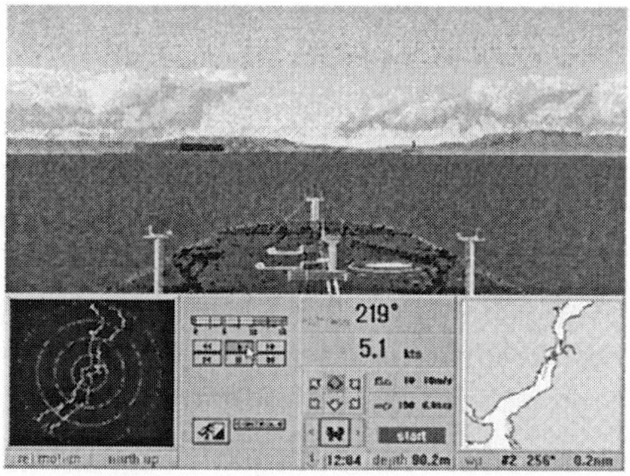

Figure 4: Viewing facilities provided by the navigational simulation programme.

As a result, it is possible to determine the measures to be taken by, or to be brought to the attention of the ship's master sailing in critical navigational conditions, by the simulation trials for the designated ships and environmental parameters, taking into account the casualties occurred in the past and possible ones in the future.

In addition to the critical navigational conditions, engine and rudder malfunctions, fire etc. may be the cause of casualties. All of these may be conceived as "Ship Safety State Model". In this model, the state entities (risk entities) which can be initiator or contributer to accidents would be as follows:

- Bridge operations (commands, process, staff),
- Route (channel width, other ships, depth, heading, docks),
- Ship engine,
- Ship control (response to rudder commands),
- Environment (wind, current, visibility, temperature),
- Ship cargo (fire potential, loading),
- Ship structure (corrosion, fatigue, design, equipment and maintenance).

The state entities which may change the above states by interaction may be :

- Ship data (radar, charts),
- Condition of ship (corrosion, design, equipment, maintenance),
- Coast Guard (data, inspection, navigation aids, radio, radar, SAR),
- Ship ownership (management, access to capital, experience),
- Logistics (food, fuel) and personnel (number, training, experience).

The factors listed in the first group above are all represented in our simulation programmes and trials.

2.4 Simulation Trials

For the simulation trials, Cape Yeniköy, one of the most critical regions for the navigation in the Strait of Istanbul, is chosen, taking into account its geography and the casualties occurred in the past. Various trials have been conducted, by changing ship parameters and environmental conditions in each trial.

Cape Yeniköy is one of the sharpest turning points in the Strait of Istanbul (Fig. 5). There is a shallow area (less than 20 m deep) just near the cape. Moreover, the width of the designated Traffic Separation Lane for this region seems quite “narrow”. The effects of the above negative factors are made worse when weather and current conditions deteriorate.

Investigation of past casualties shows that, in addition to the navigational parameters which are common for most casualties, the ship parameters are also important. There are therefore numerous parameters which affect ship maneouvering. The most important ones among these are: ship’s hydrodynamic performance, length, structure, engine power, propeller and rudder characteristics, thruster etc. In this study, the simulation trials have been conducted for three different types of ship with comparable characteristics. The general characteristics of the ships are given in Table 2.

Another important factor in examining the critical navigation conditions relates to wind and current which, under the conditions necessitating sailing with low speed, may seriously hamper the maintenance of ship's designated route thus becoming the main reason for ship accidents. Wind and current conditions are taken at three different levels in the simulation trials. In the first group of trials "no wind and current" condition is assumed, afterwards, 3 knots current value is inserted and finally a 6- knot current with 30 – knot windspeed value is applied. The current patterns are set parallel to the waterway and the wind direction is defined as NorthWest.

	Ships		
Characteristic	Container	Oil Tanker	
Length (m)	204	243	261
Displacement (t)	32025	7710 0	159584
Ships Speed (kt)	19.5	15	15
Turning Radius/L Deep-shallow	3-6.1	6.5-10	4.5-8
Thruster	Yes	No	No

Table 2: Vessel characteristics used in navigational simulation.

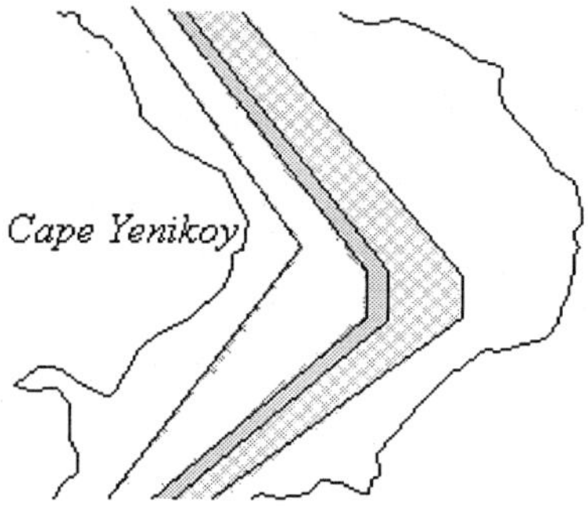

Figure 5: Geography and Traffic Separation Scheme (TSS) at Yenikoy.

2.5 Conclusions of the Trials

In the trials conducted with different types of vessels, even though the rudder was commanded hard-starboard always at the same location when approaching Cape Yenikoy (Fig. 5), the tracks obtained with different vessels were different as shown in Tables 3, 4 and 5 and Fig.s 6, 7 and 8. The information thus provided by the simulation trials as above, for the designated ships and environmental conditions, allows foresighted planning and navigation which would prevent dangerous situations from arising in the future.

Ship Speed (kt)	Environmental Conditions
	Current 6 kt + Wind 30 kt
	Fully loaded
4	*Successful*
5	*Successful*
10	*Successful*

Table 3: Trial results with a 204 m container ship.

Ship Speed (kt)	Environmental Conditions	
	Current 6 kt	Current + Wind 6 kt 30 kt
4	*Crossing over to the other lane*	
5	*Successful*	*Crossing over to the other lane*
10	*Successful*	*Successful*

Table 4: Trial results with a 243 m tanker.

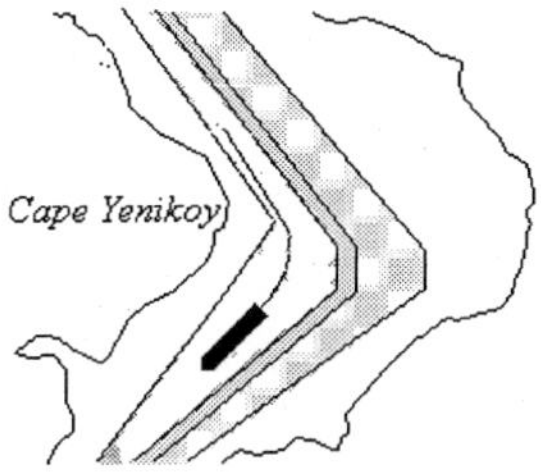

Figure 6: A 204 m container ship turning Cape Yenikoy.

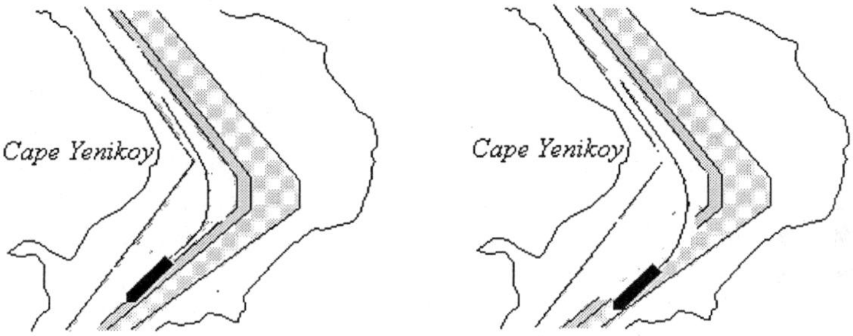

(a) (b)

Figure 7: A 243 m tanker, a) having difficulty in turning, b) crossing over to the other lane.

Ship Speed (kt)	Environmental Conditions			
	No currents No wind	Current 3 kt	Current 6 kt	Current + Wind 6 kt 30 kt
4	*Successful*	*Crossing over to the other lane*	*Ramming to the shore opposite*	*Ramming to the shore opposite*
5	*Successful*	*Crossing over to the other lane*	*Crossing over to the other lane*	*Ramming to the shore opposite*
10	*Successful*	*Successful*	*Crossing over to the other lane*	*Crossing over to the other lane*

Table 5: Trial Results with a 261 m Tanker.

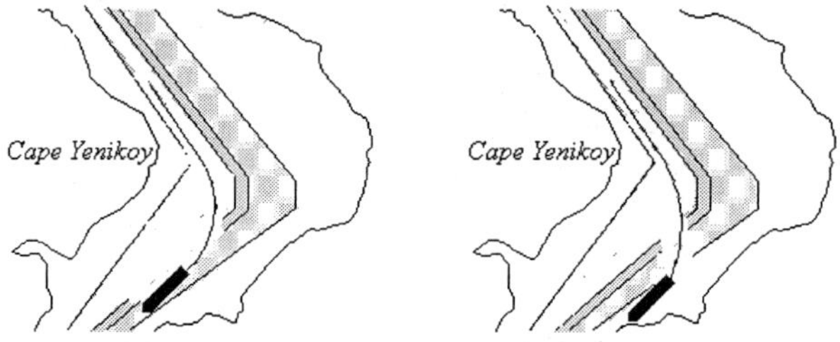

(a) (b)

Figure 8: A 261 m tanker, a) crossing over to the other lane, b) Ramming to the shore opposite.

3 Modelling and Simulation of Vessel Traffic Flow

Objective of the traffic flow simulation is to study the traffic system in a narrow and congested waterway, as exemplified by the Turkish straits, formed by the transiting vessels interacting with each other, with traffic regulations and with the environmental factors (geographic, meteorologic, hydrologic).Traffic Flow simulation helps to conduct experiments on the narrow passages and allow the following questions to be answered:

- What is the quality of service provided?
  - How many vessels can transit in a limited time period?
  - How many vessels have to wait in the queue and for how many hours?
  - What is the accident risk associated with the traffic regulations in force?
- What if traffic service demand increases in future and what kind of possible system weaknesses can be result?
  - What happens if the number of vessel increases?
  - What are the effects of the increase of the number of specific vessel types?The simulator also provides a test bed environment which allows the development of optimum navigational rules before they are formalized for the waterway, and the estimation of additional system improvements which may be brought about developments.

A simulation tool is often required for the design of new VTS systems and/or improving the safety and efficiency of navigation in existing VTS areas including ports, harbor approaches, straits and channels.

The following factors bear on these problems:

- geographical restrictions imposed by the waterways: width, turns and bathymetry.
- environmental factors: winds and currents, visibility, precipitation, and waves.
- vessel traffic characteristics: density and variance, composition, arrival statistics.
- vessel characteristics: length, weight, speed, manoeuvring capabilities, cargo and associated risks.
- "service regime": set of national and international rules for regulating traffic.
- priority rules: used to determine the order in which the vessels waiting in the queue enter the waterway; FIFO, smaller before large, hazardous before non-hazardous, etc.

The simulator that is needed here must be designed in such a way to enable the maritime authorities to address the following specific issues:

i. Determine the inter-relationships between the above given factors in a quantitative manner.
ii. Determine the optimum "service regime", a balance between safety and efficiency.
iii. Generate simulated statistical data on traffic flow to be used for planning of future upgrades of the service regime, system software/hardware features and for risk assessment.

3.1 Model Outline

"GETSIM" (Vessel Traffic Simulator) which has been developed and used for the Turkish Straits project has the logical flow diagram depicted in Fig. 9. The functions performed by different blocks shown in the figure are described below:

Traffic Statistics

The flow of maritime traffic, as the flow of any traffic, is determined primarily by the arrival statistics and the characteristics of the waterway. Normally, some statistical data on certain aspects of the maritime traffic in the area of interest would be available. We shall assume that the following are known:

- Statistical data on ship arrival times,
- Statistical data on ship parameters, including:
  - dimensions (length, width, height),
  - speed (maximum, cruising),
  - manoeuvring capabilities (trusters, one or two propellers),
  - cargo (type, hazard class).

In general, some processing of the available statistical data and in some cases additional observations would be needed in order to determine and to validate the traffic statistics to be used in the model.

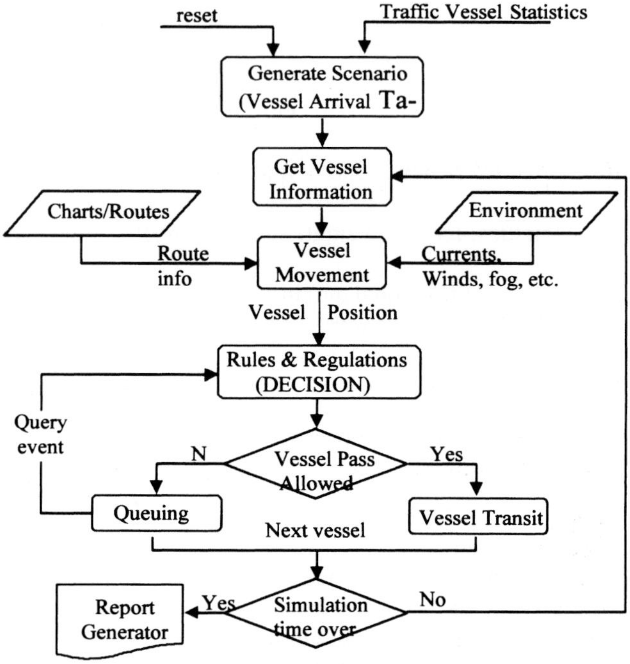

Figure 9: The logical flow diagram of GETSIM.

The following conditions are assumed to apply for the specific application of GETSIM discussed in this section.

- Arrivals per unit time of the transiting vessels are closely represented by Poisson distribution with given average arrival rates in each direction (see Fig. 10).
- Dimensions and speeds of ships can be assumed to obey a truncated Gaussian distribution with specified average values and standard deviations.
- The probabilities related to the composition of the traffic in terms of ship types is available. This information can be used to distinguish between ships carrying hazardous cargo, and those that do not.
- Information is available on the flags, and sometimes also on ship id.s, which can be used to assess the manoeuvring capabilities of the various ship classes in a statistical manner.

Environmental Factors

The model uses the geography of the area, and the statistics on the temporal and spatial variation of the currents, winds, and visibility conditions in the area.

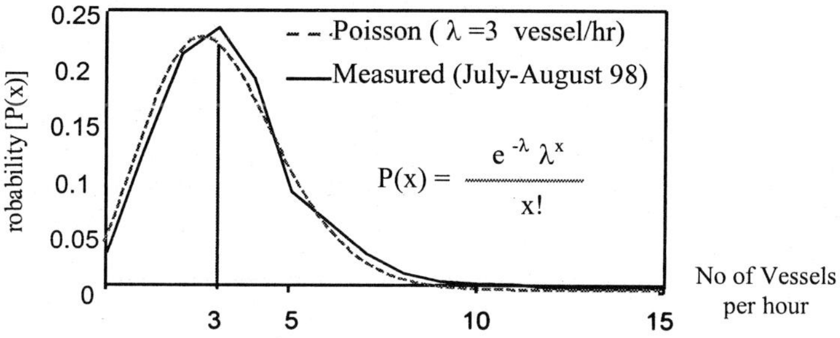

Figure 10: Validated vessel arrival statistics for the Strait of Istanbul.

Currents and winds are used to modify ship speeds on the various legs along its planned course and are assumed to be constant during a simulation run. Seasonal visibility statistics are used to generate random time intervals within which different rules may apply to transiting vessels.

Geographical data relate to the map data. The model is designed to operate on ENC (Electronic Navigational Chart) data but can also accept raster map data. Facilities are provided to define water column depths, man-made and natural obstructions and hence the limits of navigable waters for different ship types. Traffic planning and monitoring aids can also be obtained including Traffic Separation Schemes (TSS) and high risk zones, i.e. areas such as sharp bends, narrow stretches and areas with high currents where overtaking and encounters may be very hazardous.

Rules and Regulations

The model accepts various rules of navigation and enforces them to all vessels, thereby simulating a traffic situation wherein all vessels comply with all the rules applicable to them. These rules and regulations which constitute the "service regime" may include:

- Service discipline. This is related to the priorities given to arriving vessels when ordering them in the queue of vessels waiting their turn to enter the waterways and may include such rules as FIFO (first in first out, or first come first served), LIFO (last in first out), large before small, etc.
- Distance between ships sailing in the same direction.
- Maximum sailing speeds in various parts of the area, and corresponding minimum sailing speeds.
- Rules for overtaking.

- Rules for ship encounters, as a function of ship length, cargo (hazardous, non-hazardous), geographical area of the encounter, and the prevailing environmental conditions in the area, etc.
- TSS limits.
- Types of limitations on the flow of traffic and the conditions under which they apply. These limitations may include:
  - Closing one or both directions (total closure) to traffic.
  - Modifying the rules which apply under normal conditions, such as setting different criteria for distances between ships, ship speeds, and for overtaking and encounters, depending on the environmental conditions (visibility, fog, strong unpredictable currents) and on the characteristics of the vessels (large vessels carrying dangerous cargo, etc.).

The model performs all the necessary logical controls in order to assure that the motions of each vessel in the traffic comply with the appropriate subset of these rules.

Input Queue

The construction of the input queue is described below assuming that the arrival statistics conform to a Poisson distribution and the distribution of the vessels among various ship types as well as their lengths and speeds approximates to a truncated Gaussian distribution as noted above. The model takes the simulation time (T) and the number of ships (N) to be generated in a specific simulation run as inputs, together with the probabilities of ship types and the averages, standard deviations, and the truncation criteria of ship speeds and ship lengths, which are all derived from existing statistical data and performs the following operations:

- A ship is scheduled to arrive at simulation time zero. Subsequently the inter-arrival time differences between the arrival times of, say n-th and (n+1) th vessels is obtained from an exponential distribution

$$\Delta t_n = \frac{-1}{\lambda}\log_e U \qquad ; \qquad n=1,.2,....N$$

  where U stands for an uniformly distributed random number in the interval 0,1, and λ represents the average arrival rate of the Poisson distribution i.e., λ=N/T. These inter-arrival times are then normalized so that their sum complies with the total simulation time T. The time of arrival of the n-th vessel, (t_n), can then be expressed as,

$$t_n = T\sum_{1}^{n}\Delta t_n / \sum_{1}^{N}\Delta t_n$$

- For all vessel arrival times t_n, as determined above, the model assigns a vessel with an average ship length obtained by utilizing the frequencies of occurrence of that average length as determined from existing statistical data, and randomizes this length using a truncated normal distribution. The following algorithm is used for this purpose:

Generate two random numbers R_1, R_2, from a uniform distribution in (-1,1), calculate :

$$\Delta L = \sigma \; R_1 \sqrt{\frac{-2\log_e(s)}{s}} \quad ; \quad s = R_1^2 + R_2^2$$

where σ is the standard deviation of the distribution. If ΔL is not less than a user specified truncation limit (default value used is 2 σ) repeat the above steps as necessary until this condition is met. Calculate the randomized value of ship length as, $L = \overline{L} + \Delta L$. The model then assigns a type (tanker, container ship, cruiser, etc.) for each ship, using the frequencies provided for these ship types as derived from the existing statistical data. The arrival times, lengths, and types of vessels are written in a file for further processing.

Ship Routes

Ships are assumed to move along prescribed routes, defined as a sequence of waypoints. The model provides facilities to define the waypoints and assign different random components to its average speeds at different waypoints which are determined as a function of the prevailing local current and wind conditions. These random components are updated at user defined intervals.

Service Regime

Together with the ship arrival statistics, the service regime, which is the totality of rules and regulations imposed selectively on all transiting vessels is the main factor in determining the traffic delays and the throughput of the waterways. Arriving vessels, in both directions of the waterways either pass freely through the waterways, or they queue up (on both sides) and wait for their turn to enter, as determined by the service regime imposed. The waiting times of vessels in the queue will be referred to as "delay", since they add up to the normal transiting time of each vessel. Clearly, delay is a random variable which is a function of the vessel's arrival time, type, speed, and of the service regime. On the other hand, the aim in introducing a service regime is to reduce the risk of accidents, while at the same time increase the efficiency of traffic flow. The efficiency of traffic flow can best be described by the "throughput" which is the average of the vessel-transiting rate. Since there is limit to the traffic handling capacity of all waterways the throughput has an upper bound which is a function of the service regime as well as the permissible average delay. It is well known from queuing theory [4] that throughput of a channel can, in general, be increased by increasing the average delay which in turn will result in longer queues. The determination of a balance between delay and optimal throughput is one of the main problems of traffic engineering and is one of the main issue which GETSIM is designed to address.

In the simulation, the service regime is modelled through a sequence of logical operations, which perform various controls, in order to ensure that the specified rules and regulations are properly exercised. The logical operations performed by the model can be outlined as follows:

At the start of the simulation, a sail plan for the first vessel is generated and its transit scheduled, and then its position on the assigned ship route is updated until it exits from the area of interest. When this ship leaves the area of interest the ship statistics (type, length, time of exit, time spend in the queue, i.e., delay) are recorded for the purposes of generating reports and the ship is dropped from the simulation.

At subsequent simulation times, when the simulation time reaches the arrival time of the next vessel, and the input queue is empty, i.e., no ships are waiting for service, all applicable rules and regulations (service regime) are imposed and considering all other vessels presently in transit, a decision is made as to whether or not the arriving vessel can be allowed to enter the waterways. If yes, the vessel proceeds as explained above, else the ship is put into the input queue.

If the input queue is not empty at the time of arrival of the ship, then using the specified "service priorities" the entrance order of the waiting ships is updated.

When the service regime allows the entrance of a vessel to the waterways from either direction, then the pending ship from the respective queue is scheduled for entrance and the order of the waiting ships in the queue is updated accordingly.

3.2 Sample Outputs

Some examples of the outputs that can be generated using GETSIM assuming the specific geography, traffic parameters and service regime which are descriptive of the Straight of Istanbul. The environmental parameters will, however, be grossly simplified for the sake of brevity. The input data used in these simulation exercises can then be listed as follows:

i. Play area: The Strait of Istanbul

   The geography of the Strait is illustrated in Fig. 11 together with the two subareas referred to as Critical Area 1 (CA1), and Critical Area 2 (CA2), since different rules may apply to ships when sailing in these areas.

ii. Traffic:

   Average transit traffic density : 4200 vessels/month (both ways)

   Traffic composition: The relative normalized frequencies (probabilities) of the various ship types/lengths, for which different rules may apply, together with the abbreviations to be used to represent these different cases in describing the service regime are as shown in Table 6.

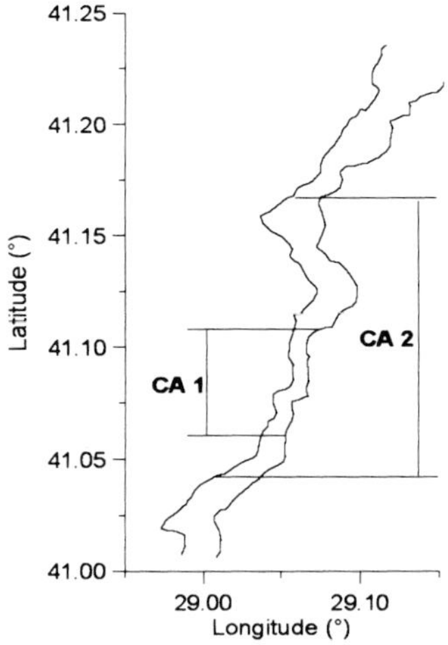

Figure 11: The Strait of Istanbul and critical areas.

SHIP TYPE	SHIP LENGTH (m)	PROBABILITY	ABBREVIATION
Any	$L \geq 300$	0,05	LS
Tanker	$300 \geq L > 250$	0,55	LT
Tanker	$250 \geq L > 150$	3,60	MT
Tanker	$150 \geq L$	5,80	ST
Non-Tanker	$300 \geq L > 150$	25,8	MS
Non-Tanker	$150 \geq L$	64,2	SS

Table 6: Probabilities of ship types/lengths.

iii. Ship routes – Transiting Speeds: Ship movements will be accounted for in a simplified manner in these simulation exercises, namely the ships will be assumed to move at constant speed along predetermined routes at constant (maximum 10 knots) speed with respect to ground.

iv. Environmental Effects: As noted above, for brevity the environmental factors, especially those related to the representation of the very complicated nature of the currents in the strait are introduced in a greatly simplified manner. The average effects of the prevailing currents are accounted for, in a grossly approximated manner, by making allowance for different ship speeds when sailing in opposite directions, i.e., north-to-south ($N \rightarrow S$) and south-to-north ($S \rightarrow N$), in order to represent an average effect. The probability for occurrence and duration of the three visibility conditions for which different transiting rules may apply are listed in Table 7. Durations are assumed to obey Gaussian distribution with the indicated standard deviation σ, truncated at the $\pm 2\sigma$ limits.

Visibility [nm]	Probability [%]	Duration [hr]	σ [hr]
< 1,0	0,066	6	2
< 0,5	0,045	6	2

Table 7: Probabilities and Durations of Various Visibility Conditions

v. Service Regime: The service regime described here derives from the actual service regime in force in the Istanbul strait.However, certain omissions (e.g., rules depending on current strength), modifications and simplifications are introduced for convenience. In particular, only the following six rules of navigation are outlined below which are a small subset of the whole of the applicable rules:

Rule 1: Large ships " LS" (Ship length > 300 m) No other ships can use the strait before the LS completes its transit, i.e. strait is closed to all other traffic in both directions.LS can transit day and night. If, however, the LS carries dangerous cargo (a tanker) then it can transit only in daylight.

Rule 2: Large Tankers "LT" (300 m ≥ vessel-length ≥ 250 m) Vessel completes its transit in day-time, and the strait is closed in the opposite direction.

Rule 3: Medium - large Tanker "MT" (250 m ≥ vessel-length ≥ 150 m) The vessel is not allowed to encounter any vessel in both critical areas (CA1 and CA2 , in the Fig. 11), additionally, tankers longer than 150 m are not allowed to enter to the strait from either direction until MT completes its transit.

Rule 4: Medium - large Ships "MS" (Ships carrying non-hazardous cargo, with length greater than 150 m)Encounters are not permitted within CA1 with similar ships (and with ships specified in Rules 1 through 3) .

Rule 5: Limited visibility (visibility < 1.0 nm) If visibility drops below 1 nm, the strait will be open for vessel traffic only in one direction, until visibility conditions improve.

Rule 6: Bad visibility(visibility < 0,5 nm) If visibility drops below 0.5 nm, the Strait will be closed for all vessel traffic until visibility conditions improve. Some examples of the outputs that can be generated by the report generator are given in Fig. 12 to 15 and Table 8.

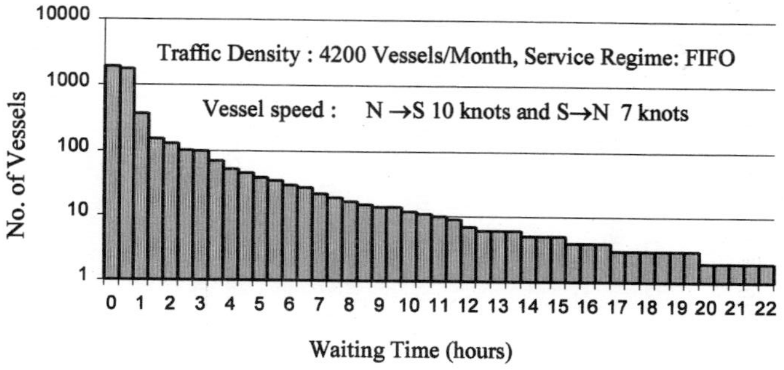

Figure 12: Monthly vessel delay distribution.

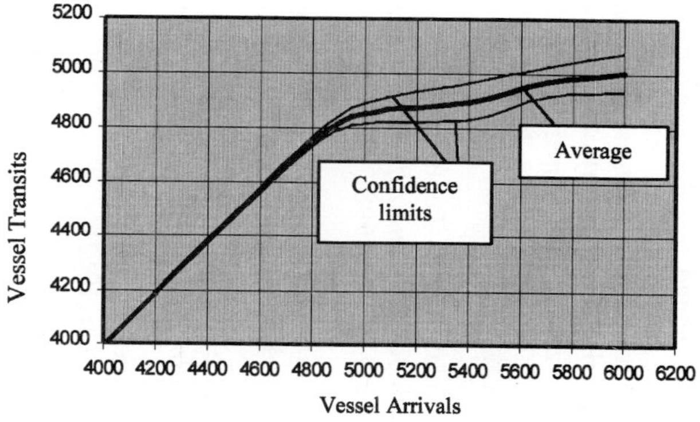

Figure 13: Monthly average throughput and 95% confidence limits.

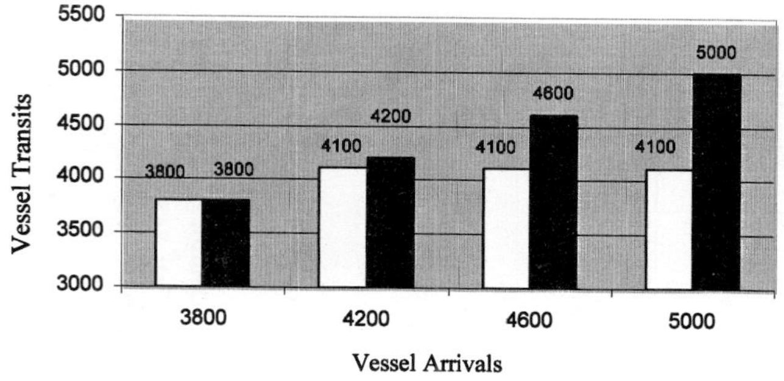

Figure 14: Monthly vessel transit statistics for two sets of vessel speeds.

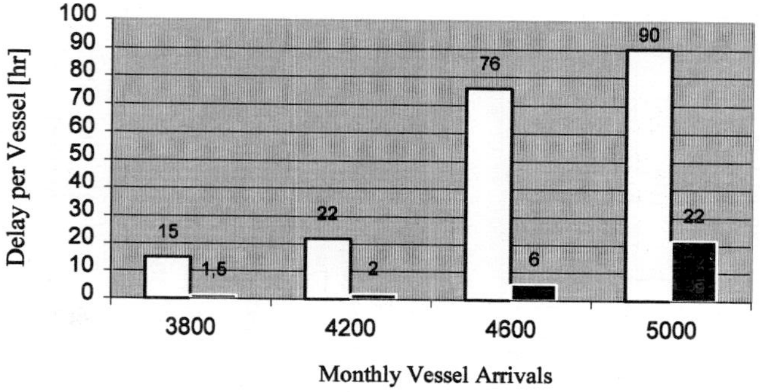

Figure 15: Variation of average delay as a function of vessel arrival rate, and vessel speeds.

Tanker Ratio (%)	No of Ves Passed (%)	Average waiting (hr)	Closure (hr)		Encounter prevented [%]
			One Dir	Two Dir	
10 (today)	100	2	12	83	48
15	85	60	15	132	48
20	70	146	22	158	48
25	54	283	31	205	48
30	47	404	45	240	48

Table 8: Increase of waiting and strait closure times with increasing number of hazardous cargo carrying vessels.

3.3 Conclusions

The following important conclusions may be drawn from the above simulation studies for the Strait of Istanbul.

- As for all channels also the capacity of the Strait (channel) of Istanbul is limited and there is a trade off between the number of vessels which can transit and the queueing delay. Regardless of how many vessels arrive for passage through the Strait the number that can actually pass without incurring unacceptable delays does not exceed 5000 vessels per month.
- The casualties in Istanbul are expected to be reduced to levels below one fifth of its present level. This is made possible by the traffic regulations which are made enforceable by a modern VTMIS which is at present being implemented.
- The traffic regulations are expected to reduce the probability of encounters (i.e. risk of collision) of vessels carrying hazardous cargo in critical areas from zero to 50 % depending on the extent of these areas.

References

1. A.N. Ince and E. Topuz, *Issues to be Resolved by International Maritime Community for Safe and Efficent Navigation with VTS*, Private Communication to AILA, (Paris, March 1998).

2. *Convention on the International Regulation for Preventing Collisions at Sea, 1972*, (IMO, London, 1990).

3. A.N. Ince, E. Topuz, E. Panayirci, and C. Isik,, *Principles of Integrated Maritime Surveillance Systems*, (Kluwer Academic Publishers, 2000).

4. E.V. Benes, *General Stochastis Processes in the Theory of Queues*, (Addison-Wesley Pub. Co.,1962).

Optimisation of the Evacuation Plan of an Offshore Vessel

M. Flier

Global Maritime BV, Rotterdam, The Netherlands

1 Background

Vessel evacuation simulation programs have been developed as decision support programs for simulations of RO-RO passenger vessels, High Speed Craft and Cruise liners driven by the SOLAS guidelines for emergency management. Furthermore these programs might be helpful in assessing the evacuation aspects of new ship designs and midlife conversions. Such an evacuation program as developed by the Mercator University of Duisburg was used by Global Maritime for an evacuation analysis trial on a large offshore vessel. The objective was to determine the usability of such programs in their present state to assess and optimise the existing safety plan of this vessel. Furthermore the longterm usability, as a bridge management tool, was investigated.

An offshore platform was chosen as the crew of offshore installations may sometimes reach up to around 700 people and these vessels are often converted (significantly affecting the layout of the vessel) for specific projects.

2 Objectives

The focus of the study was laid upon the modeling of the actual evacuation plan, assuming that the crew and personnel of the subcontractors on the vessel actually follow this plan during a required disembarkation of the vessel. The objective of this study was therefore to determine if any layout bottlenecks prevent an optimal evacuation time. In other words the project aims at identifying bottlenecks in the vessel's layout and verifies if any changes in the layout of the vessel could decrease the evacuation time.

An added objective was to determine whether or not safety-conscious operators would investigate the software and if so whether they would be interested in applying simulation programs.

3 DCV Thialf (Heerema Marine Contractors)

The vessel that was chosen to perform this exercise on is the semi-submersible crane vessel Thialf, one of the largest offshore platforms in service today, with a maximum crew of 736 people. Its dual cranes provide for depth reach lowering capability as well as heavy lift capacity up to 14,200 tonnes and has an overall length of 201.6 metres (661 ft) and a width of 88.4 metres (290 ft). The vessel can be equipped with either the existing 200 mT J-lay system or with the new 525 mT J-lay system. This multi-functional dynamic positioned Deepwater Construction Vessel is tailored for the installation of foundations, moorings, SPAR's, TLP's and integrated topsides, as well as pipelines and flowlines.

Figure 1: Shown on the picture: DCV Thialf during platform installation

4 Scenarios

The scenarios that were developed for the project are divided into two main operational categories: normal operations (without obstructions in evacuation routes) and emergency situations (with obstructions in evacuation routes). For the normal operations, scenarios a normal operational day with 224 crewmembers on the platform was studied and the 'worst case' situations with regards to concentrations of people were identified. One of the scenarios, number 4, was simulated with the maximum crew of 736 on board. One emergency situation was simulated: a helicopter crash on the portside accommodations of the vessel blocking evacuation routes in this area. Any bottlenecks identified were studied further and where necessary solutions to the problems were simulated again to test their effectiveness. This approach is significantly different from the current requirements for RO-RO ferries and High Speed Craft, however these regulations are regarded as insufficient. It should be noted that there are two shifts on this platform, both working 12 hours each day. This resulted in the following scenarios:

No	Scenario description	No. crew	Time	Distribution over vessel
1	shift handover	224	11.50	85% workplace 15% messroom 1% cabins
2	messroom	224	12.20	50% workplace 39% messroom 6% cabins 6% recreation
3	after dinner	224	21.00	36% workplace 33% cabins 33% recreation
4	cabins	224	02.00	37% workplace 47% cabins 16% recreation
5	project start-up	224	21.00	36% workplace 64% auditorium
6	maximum capacity	736	02.00	37% workplace 47% cabins 17% recreation
7	helicopter crash	736	02.00	37% workplace 47% cabins 17% recreation

Table 1: Scenarios.

5 Simulation

The tool used to perform the simulation was 'Flightsim', developed by the Gerhard-Mercator-University of Duisburg in Germany. The benefits and limitations of the tool were verified and it was decided that it was possible to model the evacuation plan of the vessel in the program and complete a realistic simulation. Furthermore the difference between simulating passenger vessels and offshore vessels was identified leading to two main differences that have an effect on the evacuation:

- The level of emergency preparedness training.
- The familiarity with the vessel.

These differences were incorporated through the parameter setting options in the program.

The program has two default parameter settings defining ideal or realistic behaviour. Simulations in the realistic mode allow adjustment of the following parameters, that are subsequently randomly distributed over the persons in the simulation: maximum speed, time out duration, see forward range, start-up delay, dawdling, and turning probability.

6 Analysis of Results

The results of the simulations were visualised in animations and statistical data. The graphs and data show, for each deck separately, the amount of people standing still, moving with restricted speed or moving at full speed. In this manner the bottlenecks were identified and studied in more detail. The following data is derived from the simulation:

- Total evacuation time per simulation.
- Mean evacuation time.

Furthermore more detailed results from the simulation with the evacuation time closest to the mean evacuation time of all simulations are presented:

- Evacuation time for each individual.
- Walking speed distribution for each individual.
- Starting point and musterpoint for each individual.
- Parameters assigned to each individual.
- Total distance covered for each individual.

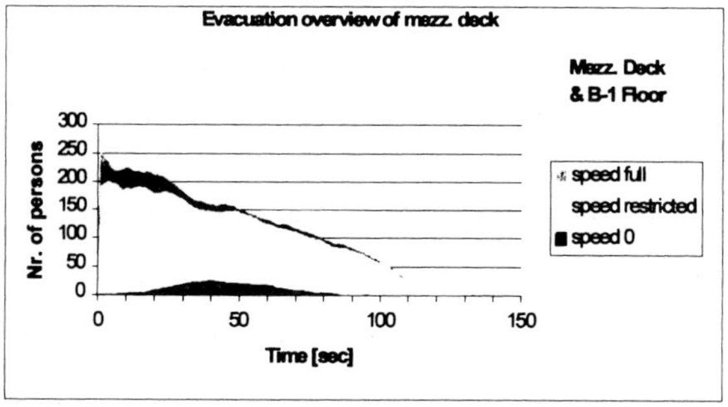

Figure 2: One of the results from the simulation: The graph shows the number of persons on a certain deck versus the time. Furthermore it shows the number of people walking at full or restricted speed and the number of people standing still.

7 Conclusions

The evacuation simulation analysis led to the following conclusions:

- No serious evacuation problems were found on this vessel, which was not to be expected because of the size of the vessel and the size of the crew. However the exercise proved successful in identifying the critical routes on the vessel and provided useful information to optimise the evacuation plan.
- The crewmembers working in the cranes and the aft engine rooms arrive last at the musterstations in all scenarios. Any bottlenecks that are removed in the

accommodations will not affect the total evacuation time of all crewmembers. It should be noted here that the columns and floaters were not modeled, else any crewmembers in these areas would be the last to arrive at the mustersta- tion.

- Of the two exits in the messroom, one is clearly the fastest exit leading to the muster stations. A more detailed study was completed to determine the fastest set-up under which the messroom can be evacuated. This optimisation study resulted in an ideal (and practical) exit configuration and desired routes leading to a 40% decrease of the average evacuation time for the messroom.
- The open deck spaces on port- and starboardside of the accommodation are critical in an evacuation and are to be kept free at all times.
- The four exits of the auditorium are very effective, if the evacuation is completed according to the safety plan.
- From the comparison between a full capacity against normal operational crew (increase in persons 228%) it was determined that the total evacuation time increases with 24%. On average there is a negligible increase in the time spent standing still per deck. It may be concluded that the evacuation routes are sufficient to accommodate 736 persons on board the vessel, however there was no comparison material available to check against other simulation exercises.

8 Limitations of Evacuation Analysis Software

As the limitations were stated before, identified at the start of the exercise enabling the project team to complete a successful evacuation simulation. The following limitations were identified:

- It is not possible to simulate real incidents and people's reactions to those incidents. For the optimisation of the safety plan as done in this exercise, this is however not essential.
- The grid of 40x40cm is a limitation on the possibility to input the vessel geometrical layout. A study into the effect on the accuracy of the results is required.
- The input of the tool does not allow to read of AutoCAD (dxf) files. This would enhance the speed of the input of vessel layout, as well as creating greater flexibility to implement design modifications.
- The statistical and visual output of the program is still limited.

Some of the remaining limitations that were identified have by now been corrected in the latest version of the 'Flightsim' tool.

9 Utilisation of Evacuation Analysis Software

The exercise proved that it is possible to design and optimise the safety/evacuation plan of a vessel or offshore installation. This exercise was completed on an existing vessel, however the simulation should really be utilised during the design stage of vessel or fixed offshore installation newbuild projects or conversions as this leaves room to implement changes in the geometrical layout of the vessel.

The second use for the simulation software is a more longterm utilisation. The evacuation simulation tool could be combined with fire control and monitoring systems and leak stability software to create an expert system used as a management decision support tool on the bridge of ferries and cruise vessels. An expert system like this should compare the time needed to evacuate the vessel with the time left to evacuate the vessel, the remaining 'floating time'. In other words the fire control and monitoring system and the leak stability system calculate the progression of an incident and assist the master in his difficult decision *when* to give the sign to disembark the vessel. This utilisation requires an intensive further development of evacuation simulation tools.

Influence of Ship Listing and Ship Motion on Walking Speed

W. Bles, S. Nooy, and L.C. Boer

TNO Human Factors, Soesterberg, The Netherlands

The walking speed of passengers is important in estimating the time required to evacuate a ship. Therefore we investigated the walking speed on stairs and in corridors during ship listing and dynamic ship motion conditions, using the TNO Ship Motion Simulator. Data were obtained from subjects between 18 and 83 years of age. Ship listing reduced walking speed up to about 35%, dependent on the direction of tilt. Seniors were about 15% slower than other age categories. Ascending stairs slowed down walking speed up to 40% for all age categories, descending up to 30%. Another finding was that people needed the handrail during ship listing. This imposes a considerable reduction on the specific flow through the escape ways. Dynamic ship movements slowed down walking speed up to 15 % with increasing frequency and with increasing amplitude. At large eccentric positions, parasitic linear accelerations will be the main factor reducing specific flow: passengers need handrails if these accelerations exceed 0.5 m/s^2. For the purpose of predicting the time required for evacuations, the reduction in passenger walking speed and the reduced capacity of escape ways should be taken into account in the IMO guidelines.

1 Introduction and History

The capsizing of the HERALD OF FREE ENTERPRISE in 1987 (death toll 193) and the sinking of the ESTONIA in 1994 (death toll 852) has shown that safety was terribly insufficient. That is why, since these disasters, actions to improve safety on board ship have gathered momentum. This also explains much of the thinking behind the BriteEuram project "Mustering and Evacuation of Passengers: Scientific Basis for Design" or MEPdesign for short. Within MEPdesign different aspects of the mustering and evacuation process were examined, and our task was to collect data on how passenger (walking) speed is influenced by the unstable ship environment. We obtained data on walking speed in corridors, in open spaces, and on stairs (1). The present paper describes results from an experiment which was subsequently carried out to provide data which were still missing from the original experiments in MEPdesign. Now data is available for static ship listing and dynamic ship motion conditions for some 150 subjects of all age categories concerning walking speed in corridors, while ascending or descending stairs, and concerning additional delays caused by opening doors. Together, the data cover the whole path of passengers moving to their muster station, and enable reliable simulation of propagation speed during mustering.

Figure 1: The TNO Ship Motion Simulator.

2 Methods and Experimental Procedures

The TNO Ship Motion Simulator (SMS) is a cabin of 4 m*2.4 m*2.3 m on top of a hydraulic system allowing for heave (peak-peak 1m), pitch (p-p 40°) and roll (p-p 30°) motion (see Fig. 1). The walls of the cabin were covered with soft padded material to prevent injuries if people would lose their balance. For added safety, subjects were always accompanied by an assistant in the SMS. The control room monitored the SMS cabin with video-cameras and also had audio-communication.

Placement of a wall segment in the middle of the SMS cabin created a walking concourse for the experiments carried out in MEPdesign. By keeping the width of the corridors at 1.1 m, straight corridors were created with a length of 1.6 m. In the present experiments only one leg was used to obtain data on walking in corridors. Because of the motion characteristics of the SMS, trim angles up to 20° could be generated for walking uphill and downhill, and list angles up to 15° for walking in a laterally tilted corridor. At the corners of the corridors, invisible sensors (LEDs) were placed at a height of about 1.5 m, allowing for automatic time registration. Walking speed was computed as the distance between two sensors divided by the time required to cover that distance. In the MEPdesign experiments it was shown that walking speed as measured in the small SMS cabin is not different from walking in larger mock-ups, but one should realise that at passing the first sensor the steady state walking phase was not reached, just as at passing the last sensor the deceleration had already started (1).

To study passengers on stairs, one leg of the concourse in the SMS was closed, and a 6-step staircase was build in it. The ceiling of the SMS was raised to allow upright standing throughout the whole stair. The rise angle was 40°, the steps had a depth of 23 cm. Handrails were present on both sides, contrary to the MEPdesign experiment, where only one handrail was present. The stair ended on a platform of 1.1 x 1.1 m. For recording purposes, all five middle steps had LED sensors at a height of 0.05 m above the step. Subjects were told to use all steps, allowing us to compute the average climbing speed (speed projected on the deck).

Sixty subjects participated, divided over three age categories: young (18-40), middle-aged (41-60) and senior (61-83). All were in good health, and able to walk unsupported. In all experiments subjects were encouraged to use the handrail, especially when climbing the stair. Subjects were allowed to rest between conditions to assure that fatigue was not introduced. All experiments were approved by the Subject Ethical Committee of TNO Human Factors. They signed an informed consent.

3 Experimental Data

3.1 Ship Listing and Walking Speed

In Fig. 2 the data of all the experiments in the SMS are summarised for walking uphill and walking downhill, in Fig. 3 for walking in a laterally tilted corridor. Analysis of variance revealed that tilt angle had a significant ($p<0.01$) effect on the walking speed, as well as age ($p<0.01$). On average the senior subjects were 15% slower than the other age groups. Apart from the fact that some of the seniors could not walk that fast, no one of that category took any risk and behaved consistently on the safe side. The effect of list angle is largest when walking uphill (about 35% reduction), but walking downhill slows down the walking speed as well with increasing tilt angles, probably because people are careful not to slip or to start running (about 10-15% reduction). These data are consistent with the data from MEPdesign. In Fig. 4 the different conditions involving the stair have been summarised together with the terminology used in the text: Conditions A, B, C and D concern ascending stairs, conditions E, F, G and H descending stairs.

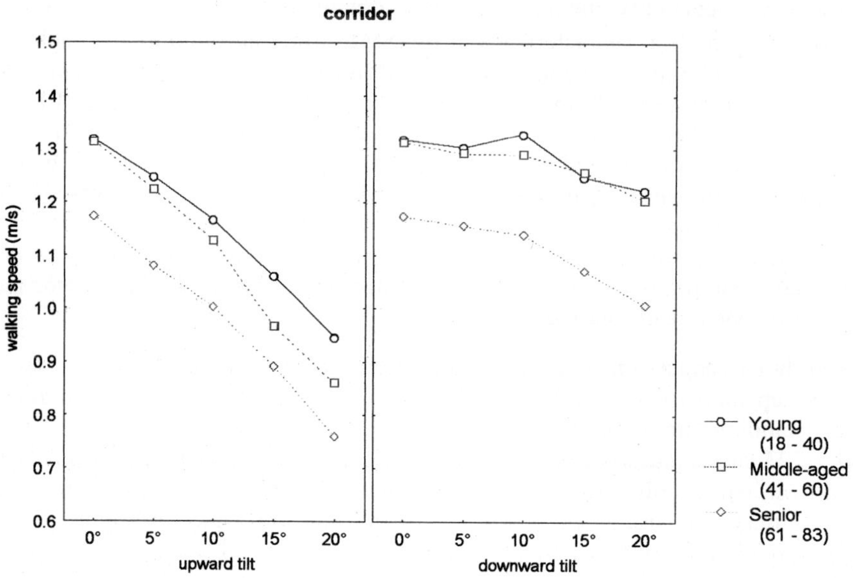

Figure 2: Mean walking speed as a function of up- and downward tilt of the corridor.

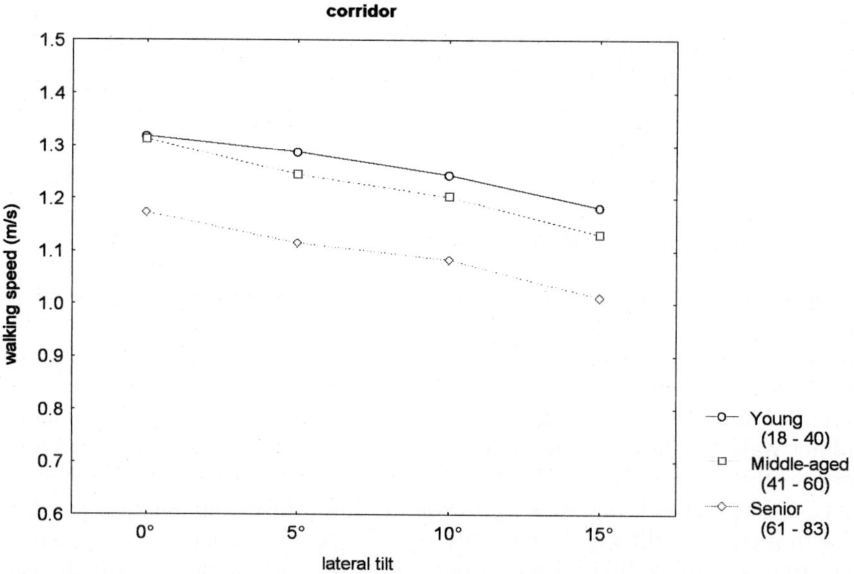

Figure 3: Mean walking speed as a function of lateral tilt of the corridor.

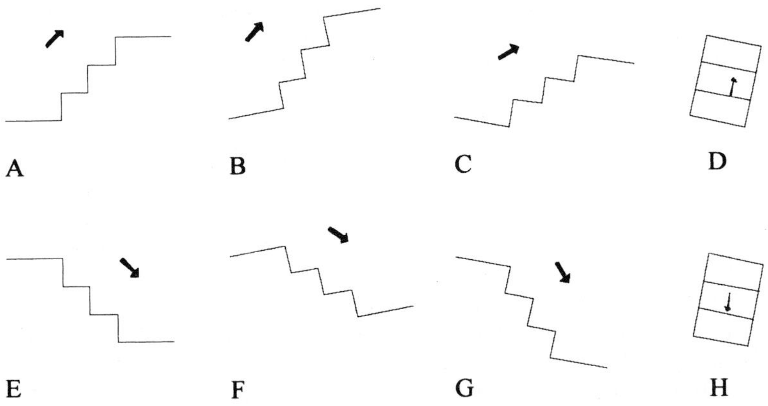

Figure 4: A: ascending a 'level' stair. B: ascending a stair tilted upward (the angle will be indicated with a '+' sign). C: ascending a stair tilted downward (angle indicated with a '-' sign). D: ascending a stair tilted laterally. E: descending a 'level' stair. F: descending a stair tilted downward. G: descending a stair tilted upward. H: descending a stair tilted laterally.

In the MEPdesign experiment (1) we found that climbing stairs was much more difficult than expected, especially ascending stairs tilted upward and descending stairs tilted downward. In the latter condition tall subjects had to be careful not to hit the ceiling overhead, because they walked more perpendicular to the stair, which shortens the actual distance between stair and ceiling. Subjects often missed the first step or put only the toe of the shoe on the first step, leading to some unstable behaviour when ascending an upward tilted stair. The handrail was an absolute necessity in these conditions. With angles of more than +10° the majority of the subjects pulled themselves up with two hands on the handrail. Most subjects were surprised by the difficulty of the task, and it was agreed that an important factor provoking these problems was that the whole surround was listed together with the stairs. Because of those problems senior subjects were not examined in the MEP-design experiment.

Since we wanted to include seniors as well in our present experiment and also include dynamic motion patterns, we adjusted a second handrail, allowing for manual support on both sides. Although the same problems remained as mentioned for the MEPdesign experiment, the subjects were much more stable on the stair, no twisting of the body was observed anymore (clinging to the handrail on one side).

In Fig. 5 the average climbing speed (projected onto the floor of the corridor) is given for ascending and descending upward and downward tilted stairs. It shows that the average walking speed for ascending stairs which are tilted downward even increases compared to ascending a level stair. People reported that it was much easier than ascending a level stair. With upward tilted stairs walking velocity decreases with about 35% at a tilt angle of the stair of 20°. Subjects found this more difficult, but the decrease here is less than the 45% as found in the MEPdesign study, where we had only one handrail. Senior subjects behaved as the young and middle-aged people, except that they were systematically about 10% slower ($p<0.01$).

Descending a downward tilted stair leads to a reduction of the walking speed with about 20%, which is considerably less than found in the MEPdesign study (45%), most probably due to the second handrail again. The reduction in descending speed is even more for the seniors (about 30%), and their overall velocity is systematically ($p<0.01$) lower than the velocity of the other age groups (about 20%). Just as in the MEPdesign study, part of the slowing down is explained by the fact that people had to be careful not to hit the ceiling with their head, because their position was more perpendicular to the stair, decreasing the distance between the stair and the ceiling over the stair. As shown in Fig. 5 descending an upward tilted stair apparently goes easier with small inclinations, but with larger list angles people are more careful and slow down. Again, senior people were slower on the average than subjects of the other age categories ($p<0.01$).

In Fig. 6 the average walking speed is presented for ascending and for descending on a laterally tilted stair. The reduction in walking speed is now only 10% at the most. Subjects found these conditions much easier than the conditions with walking on an upward or downward tilted stair.

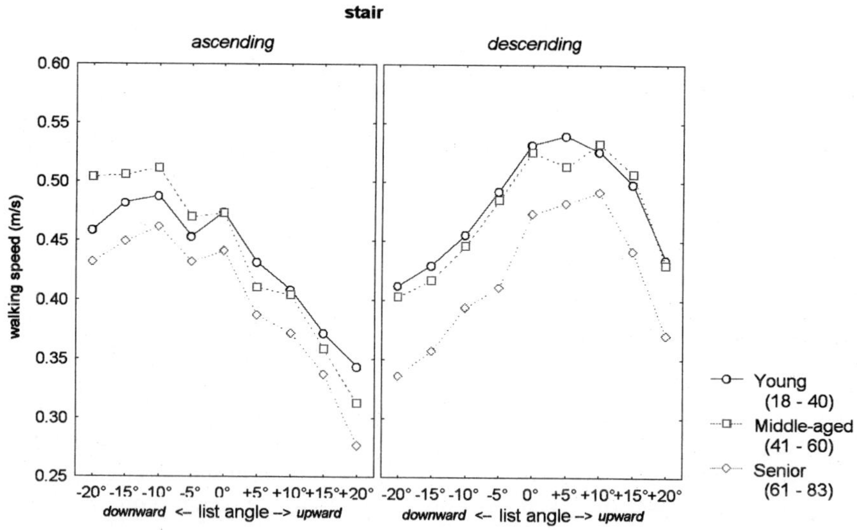

Figure 5: Average climbing speed for ascending and descending stair as a function of the angle of an upward (+) or downward (-) tilted stair. Cf. Fig. 4.

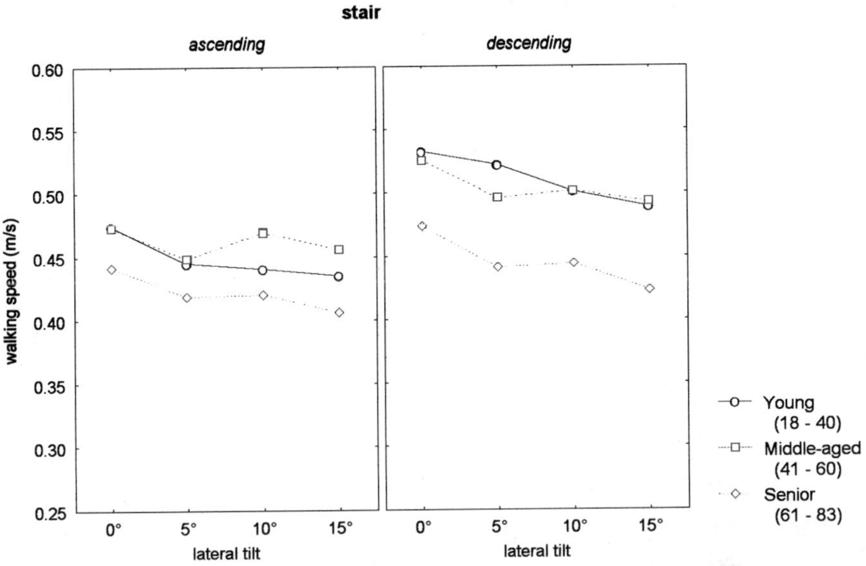

Figure 6: Average climbing speed for ascending and descending a laterally tilted stair as a function of tilt angle of the stair. Data are averaged for tilt to the left and to the right.

3.2 Ship Motion and Walking Speed

In the MEPdesign study we studied the effects of ship motion on walking speed for sinusoidal pitch or roll motion patterns, while the subjects were walking around in the SMS along the concourse as described in section 2. The reason for choosing artificial motion patterns in stead of real ship motion patterns was due to the fact that walking had to be minimised as much as possible, because of the motion sickness provocative character of the walking around and on the other hand the need for as many motion conditions as possible.

For the present experiment we diminished the number of motion conditions: we only examined the most extreme conditions. This was necessary because now we had the senior subjects participating and we had not only walking as in the MEP-design study, but we had also ascending and descending stairs, and opening doors. So we chose five motion profiles: A (frequency 0.89 rad/s, ampl. 2.5°), B (0.89 rad/s, 5°), C (0.89 rad/s, 10°), D (0,45 rad/s, 15°) and E (0.63 rad/s, 15°). Since the stair prohibited walking the concourse and since we wanted to prohibit that subjects had to turn around too many times in order to prevent motion sickness, subjects walked the corridor, or climbed the stair only once per condition. By dividing the start moments equally over the sinusoid, and measuring the exact relationship between the walking period in relation to the phase of the stimulus, we could show that the mean walking speeds are representative for the mean walking speed over a full stimulus cycle (2).

One member of the experimenter team was always present in the cabin together with the subject, but no major effort was requested during the measurements to prevent subjects from falling. Each condition started with to and fro walking in the corridor. Subsequently they ascended the stair and descended the stair, followed by walking the corridor again, but now with the closed (inward swinging) heavy door, and finally back with the again closed (outward swinging) heavy door. Each of these six legs was started on command by a sound signal. Each subject was walking with a sound signal with a fixed phase lag with respect to the SMS stimulus sinusoid. Each age population was equally distributed over the different phase lags.

Just as in the MEPdesign study we depict the walking speeds as a function of frequency and amplitude of the sinusoidal stimulus pattern.

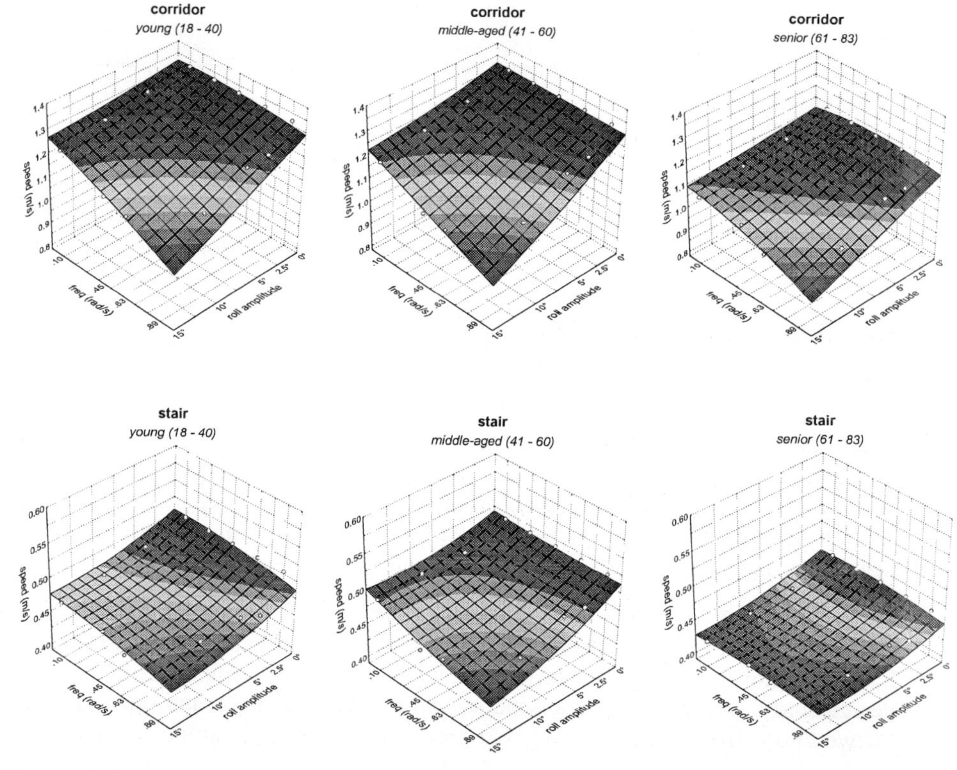

Figure 7: Mean walking speed in corridors (upper row) and on stairs (lower row) for all age categories as a function of dynamic roll motion. See text for further explanation.

Up till a certain frequency no differences in walking speed are expected, because at low frequencies dynamic motion can be assumed to be a sequence of static tilt walking conditions. Therefore we have added the (theoretical) walking speeds in the concourse at the very low frequency of 0.1 rad/s. In this approach we have assigned the static amplitude of 0° to each value of the sinusoidal stimulus between 2.5 and –2.5°, the static tilt amplitude of 5° to all values between 2.5 and 7.5°, etc. Knowing the time spent at a certain tilt angle, the speed could be computed for the total stimulus, taking into account the differences in walking speed for each tilt angle for the different age categories.

In Fig. 7 we have also added the value of walking speed for the different frequencies at the amplitude of 0°, because those values should all be the same: the average corridor walking speed in a stationary level surrounding. This data is also available from the experiment described in section 3.1.

The results on roll motion in the upper row of Fig. 7 show for all age categories a decrease in walking speed in corridors with increasing frequencies and with increasing amplitudes. In general, seniors are 10-15% slower than the other subjects.

The results on pitch motion are not shown: They show similar effects as for roll motion, be it to a lesser extent (2). This is understandable, since the roll motion is always perpendicular to the walking direction: An increase of either frequency or amplitude results in an increase of the lateral forces exerted on the body during walking, which makes propagation increasingly more difficult. This is not the case with pitch motion: In one part of the sinus propagation is hampered (walking upward), in the other part propagation is facilitated (walking downward).

All together the data show that the decrease in walking speed for the dynamic motion conditions is less than for static list angles. Again, this is understandable because in static list conditions, the angle is always the same, whereas in the dynamic motion conditions the corridor is almost level during a substantial part of the sinus.

The 3D-plots just serve to visualise the main effects, the interpretation is pretty complicated by the fact that the corridor remains a small corridor, the stair only a six step stair, etc. Moreover, the distance to the rotation axis is the main determinant of the resulting walking speed: One should note that this distance varies for each step of the stair, and virtually also for each place in the corridor, invoking linear accelerations of different magnitudes for each location during the sinusoidal motion. These complications will be discussed in the discussion section 5.

In the lower row of Fig. 7 the effects of roll motion on the climbing speed on stairs are shown for the three different age categories. For the young and middle-aged subjects we see similar behaviour compared to walking in corridors, i.e. a slowing down with increasing frequency and with increasing amplitude with maximally about 20 %. For the seniors these effects are negligible: they are already slow. The effects for ascending and descending were not too different, reason to show in Figs. 7 the mean climbing velocity of both ascending and descending.

For pitch motion (not shown here) there was hardly a difference between the semi-static condition and the experimental conditions. This was true for all age categories.

4 Discussion and Conclusions

4.1 Comparison with MEPdesign Data

The present experiments and the MEPdesign experiments showed the same decline of the walking speed with increasing tilt amplitude for walking in an upward or in a downward tilted corridor. Also for lateral tilt the effects were the same. Both studies showed the same decrease of walking speed for seniors compared to the other age categories: some 15%. However, an interesting finding was that in all corridor conditions, walking speed was systematically about 10% *faster* in the present experiment compared to the MEPdesign study. This could be due to the sound signal that commanded the subjects in the present experiment to start walking, as opposed to a self-initiated start in MEPdesign. Alternatively, this could be due to the current subjects walking the corridor only once, whereas the subjects in the MEPdesign study walked the corridors five times (three for the seniors). The first walk was usually fastest and the last one slowest, with about equal speeds for the walks in between. In the MEPdesign study the data from the middle walks were taken as the mean.

Analysis of the walking speed for those subjects who participated in the SMS and the mock-ups, showed that the walking speed was the same for the young and middle-aged groups (5.7 km/h on the average, i.e. 1.58 m/s), and 12,5 % slower for the seniors (5.0 km/h on the average, i.e. 1.4 m/s) (see (2), Fig. 21). The concourse walking used in the MEPdesign experiment shows smaller walking speeds because of the presence of corners in the concourse, which ask for deceleration and acceleration.

The stair of the current study had handrails at both sides whereas the stairs of the MEPdesign experiment had only one handrail. This probably explains the decrease of ascending an upward tilted stair of 15% / 10° in the current study, as opposed to 20% / 10° in the MEPdesign study. One or two handrails was not a factor ascending a downward tilted stair. For descending stairs both studies found a similar decrease in walking speed of 15% / 10°, irrespective of upward or downward tilt of the stair.

For ascending and descending laterally tilted stairs both studies found a decrease of 5% / 10°.

In the dynamic conditions the subjects' climbing speed is less influenced than for static conditions, but handrails are a prerequisite. Inspection of the video tapes showed that subjects were more stable on the stair if they used both handrails. Because climbing tilted stairs is virtually impossible without handrails, evacuation capacity depends on the number of handrails present on the stairs.

4.2 Miscellaneous

Age effects. In general, seniors were slower than people in their young and middle-aged years. Moreover, the reduction in propagation speed is underestimated because of the selection process implied in the recruiting procedure. We forewarned all subjects about the conditions of the experiment, and allowed no people in ill health. In short, were our seniors were in good health and able to walk unsupported. Moreover we allowed our subjects to quit the experiment if they felt like it. Some seniors did so, similar to what happened in the MEPdesign experiments. Reasons included vertigo and motion sickness. This process of self-selection wed out the relatively unfit. In consequence, the current data are representative for senior passengers on ferries. For senior passengers of a cruise liner, larger effects of old age are expected.

Motion sickness. It was unfortunate during the experiments of MEPdesign that some subjects (about 5% of the total group of subjects) got sick, even in the static list conditions. It was obvious that the most provocative effect was not the walking itself, but the turning around at the end of the corridor. This was dependent on the speed of turning around, the faster the more provocative. This is understandable in terms of the subjective vertical mismatch motion sickness model (3). In the present experiment the incidence of motion sickness was less, probably because each condition consisted of walking one leg only, no need to turn around fast in order to proceed walking the same leg again and again. Motion sickness influenced the experiment only for losing data of one senior subject, who stopped the experiment halfway.

Doors. It is obvious that doors in a listed environment may hamper propagation. For this reason we studied also the delay times caused by opening doors as a function of tilt angle. These data are described in detail elsewhere (2), but the main findings will be shown here. In the corridor of the SMS we had also a door (weight about 30 kg), with the hinges connected to the middle wall segment. In Figs 8 and 9 we show the time delays during propagation caused by opening the door as a function of tilt angle of the corridor. The delay time is defined as the extra time relative to unobstructed walking (door already open) as discussed in section 3. Effects depend on direction: Doors witch hinge only to one side, have different delaying effects on subjects that want to pass from one or the other side. One expects that the delay should be most prominent when one has to pull the door towards oneself to open it (denoted as 'inward' swinging), whereas pushing should hardly hamper the propagation (denoted as 'outward' swinging). This was found indeed: Comparison of left and right frames in Figs. 8 and 9 shows that an inward swinging door causes a delay of 1.5 s on the average, an outward swinging door only 0.2 s. It is also obvious that, depending on the side of the hinge and the direction of the list angle, the weight of the door helps the subject to open the door ('door going down'), or just increases the task ('door going up'). However, pulling the heavy door up against gravity as shown in Fig. 9 left frame for a list angle of – 20 ° poses serious problems towards the seniors; their time delay doubles in comparison to what happens in case of a light door. Given the fact that fire doors on board of ships may weigh as much as 75 kg, one may expect that this may prevent some seniors from passing these doors under these conditions, when there is no one from the crew available. This is clear from comparison of the delays at the + and – angles in Figs 8 and 9. Senior people needed in all conditions with inward swinging doors systematically more time to pass the door.

In many cases there was also contact between the shoes of the subjects and the door on opening, especially when the door swung inward when going downhill. With an inward swinging door going down in a laterally tilted corridor similar observations were made. This happened because opening the heavy door in these situations caused an unexpected swing of the door which was difficult to control by untrained (bantam weight) subjects.

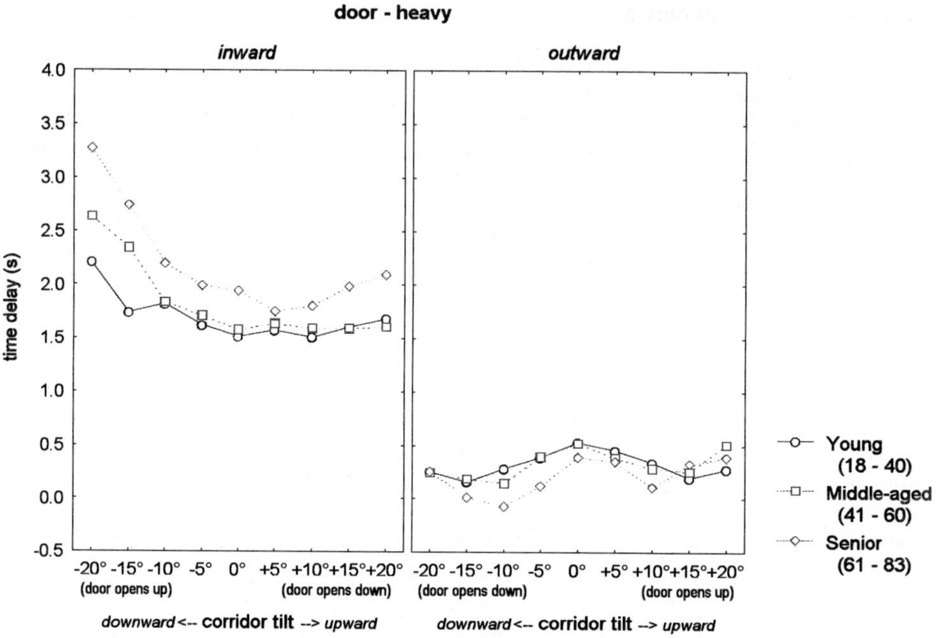

Figure 8: Time delays for opening an inward or outward swinging heavy door as a function of upward or downward tilt of the corridor, i.e. of the whole SMS.

The experiments concerning opening doors during dynamic ship motion provided data which showed that the time delays are of the same order of magnitude compared to the time delays found for static ship list.

Eccentric roll and pitch motion. Ship motion has been addressed in this study with sinusoidal stimuli, varying the frequency and the amplitude of the sinusoid. However, it is obvious that one could walk unsupported in a laterally tilting corridor if the rotation axis is at floor level. However, if the rotation axis is five floors down, walking may not be easy anymore, because now parasitic linear and vertical accelerations are involved. If these accelerations exceed 0.5 m/s² (4), people will have problems keeping balance. Considering that people at the upper decks are around 10 meters away from the rotation axis, a roll amplitude of a few degrees bring the passengers to the handrails already to prevent falling or bumping into each other.

Experimental listing conditions. Some thoughts can be offered on what can be expected for steeper listing angles than the ones used here (20° in the simulator, 15° in the mock-ups). In the MEPdesign experiments two "young" subjects on flat leather soles already started slipping at these angles. We suppose that at 30° list normal walking without support of a handrail is hardly possible. It was interesting that so many subjects were attracted by the visual vertical when walking unsupported in an environment which is completely tilted with respect to the true vertical. The result is that they are "driven" to the lower side. This means that the use of open spaces is limited to the lower side, minimising the capacity of the evacuation route during listing. This also shows that researchers using a listed floor in otherwise upright surroundings, need caution when interpreting their results.

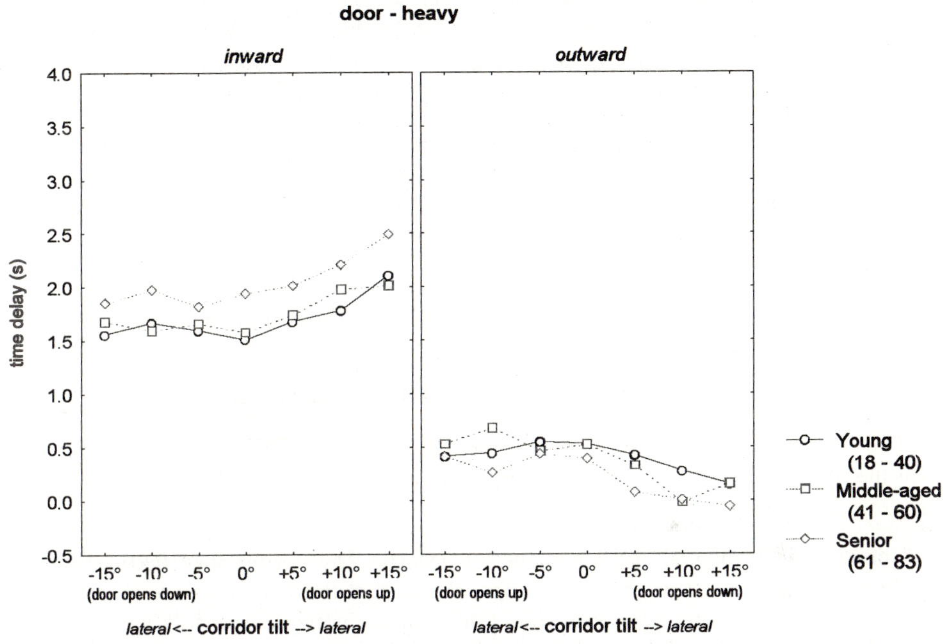

Figure 9: Time delays for opening an inward or outward swinging heavy door as a function of lateral tilt of the corridor, i.e. of the whole SMS.

Computer simulation. The current data can be used in computer simulation of (ship) evacuation. One should be aware, however, that walking impediments (canes, wheel-chairs) can have large consequences especially in narrow corridors and narrow stairs. Moreover, the build-up of fatigue will reduce the walking speed of all (e.g. climbing five stairs. This build-up is more rapid for seniors. Computer simulations based on the current data are optimistic for these reasons. One should note that:

- Healthy seniors are (already) about 10 – 15% slower than younger subjects. For ships it might be wise to use these low speeds in the calculation.

- The walking speed slows down linearly with increasing list angle.
- Subjects walk laterally listed corridors using the handrail. Without a handrail about half of the people deviates towards the underlying wall (15° list angle). It is assumed that at an angle of 20° everybody will cling to the handrail. Handrails are less important for walking up or down, but most people will prefer using a handrail at angles of 20° or beyond.
- This implies that for a listed ship specific flow should be calculated with density defined as *number of subjects per m handrail* rather than *number of subjects per m²*. Maximum flow is limited to a value of about 2, which means a considerable reduction of specific flow, especially in wide corridors and open spaces. For instance, the flow capacity per second in a 4 m wide corridor can reach as much as 6.4, assuming a density of 3 subjects/m². When tilted laterally over 20°, the flow capacity is according to our data at most 0.56 (two people per meter handrail, approximated walking speed 70% of 0.40 m/s).
- Listing makes part of the floor inaccessible because the wall leans towards the people. If people stand upright, their feet will be about 50 cm away from the wall. In narrow corridors that barely permitted two people walking side by side, listing reduces the capacity to one person at the time.
- Similar considerations hold for ship motion if the horizontal accelerations due to ship roll exceed $0.5 m/s^2$ since then subjects cannot maintain their postural control without a handrail.

4.3 Comparison of Data with IMO Interim Guidelines

The International Maritime Organization has issued interim guidelines (Table I) for the relationship between density, speed of persons and specific flow to be used for evacuation analysis (5).

Comparison of the walking speed and the specific flow for the specified densities with those for evacuation from buildings, shows that the IMO data are more conservative, they remain on the safe side. For instance, unperturbed walking speed in the IMO guidelines is 1.4 m/s, whereas for building evacuations it is 1.6 m/s (6). Compared with the data from our own study, 1.4 m/s is exactly the walking speed of our senior subjects, and 1.6 m/s the walking speed of our younger age group.

For evacuation analysis it makes sense to use the walking speed for seniors. The estimate for the specific flow in Table I is also considerably less for increasing densities compared to building evacuation guidelines (6), which seems to be more realistic, certainly in case of ship listing (see section 4.2).

In contrast to guidelines for evacuation from buildings (6), IMO provides data for descending and ascending stairs, which makes sense for the evacuation from ships. For stairs the climbing speed of 0.8 m/s for a level stair has been divided in 0.8 m/s for ascending stairs and 1.0 m/s for descending stairs. Our data suggest that people are much slower on stairs: for level stairs our seniors have an ascending velocity of only 0,7 m/s (along the climb-line), and of 0.6 m/s for descending (along climb-line), so in this case the numbers of the IMO are rather high. For the different densities on stairs, we have the same comments as for corridors: during ship listing everybody needs a handrail, so it is the number of handrails available on a stair which determines the density, and as such the specific flow.

Type of facility	Condition	Density (p/m^2)	Speed of Persons (m/s)	Specific Flow (p/ms)
Corridors	Low	< 1.9	1.40	0.76
	Optimum	1.9 to 2.7	0.70	1.30
	Moderate	2.7 to 3.2	0.39	1.10
	Crush	> 3.2	0.18	0.55
Stairs (down)	Low	< 1.9	1.0	0.54
	Optimum	1.9 to 2.7	0.50	0.94
	Moderate	2.7 to 3.2	0.28	0.77
	Crush	> 3.2	0.13	0.42
Stairs (up)	Low	< 1.9	0.80	0.43
	Optimum	1.9 to 2.7	0.40	0.75
	Moderate	2.7 to 3.2	0.22	0.62
	Crush	> 3.2	0.10	0.32

Table 1: Approximations for speed of persons and specific flow as a function of density for different facilities for the calculation of evacuation times on ships as presented by the IMO.

Recently there were comments on these interim guidelines like the one from the Working Group, Sub-Committee on Fire Protection (7). The working group suggests a further simplification like only one typical speed of persons and one specific flow, defined for corridors and stair (up and down). Since they claim to have used the same source for these data as the IMO guidelines, these values suggest a density of about 2 subjects per m^2, which means that they assume that walking is not unperturbed by other subjects anymore. For listing this makes sense, because than the densities higher then 2 p/m^2 are unlikely to occur as discussed previously (see section 4.2). It implies that direct comparison with our data is difficult, because our data refer to unobstructed walking.

Similar considerations hold for the speeds on stairs: our data refer to unobstructed walking. If we compare the given specific flow with the table as presented in the IMO Guidelines, here too a density of about 2 p/m^2 is considered. This implies for them a decrease of the climbing speed with about 50%, which means that in our view these climbing speeds are too high: 50% of our senior climbing speeds would be 0.34 m/s for descending and 0.30 m/s for ascending.

In both the IMO Guidelines (6) and the Comments (7), no attention has been paid to the strong decrease in walking speed and the capacity of open spaces due to ship listing.

References

1. W. Bles, E.L. Groen, and L.C. Boer, *Design Features: Effects of ship listing and ship motion on mustering speed,* BriteEuram project 97-4229, "MEPdesign" WP2b, (2000).

2. W. Bles, S. Nooy, and L.C. Boer, *Influence of ship listing and ship motion on walking speed,* Report TNO-TM 2001 C-15, (TNO Human Factors, Soesterberg, The Netherlands, 2001).

3. W. Bles, J.E. Bos, B. De Graaf, E.L. Groen, and A.H. Wertheim, *Motion sickness: only one provocative conflict?,* Brain Res. Bull. **47,** 481 (1998).

4. B. de Graaf and W. van Weperen, *The retention of balance: an experimental study into the limits of the acceleration which the human body can endure without a loss of postural balance,* Nederlands Tijdschrift voor Geneeskunde, 139e jaargang, **8**, 377 (1995).

5. *Interim guidelines for a simplified evacuation analysis on RO-RO passenger ship,* International Maritime Organization, MSC/Circ. 909, (1999).

6. *Brandbeveiligingsconcept Gebouwen met een publieksfunctie,* Ministerie van Binnenlandse Zaken, Directie Brandweer en Rampenbestrijding, Afdeling Preventiebeleid, Den Haag, (1995).

7. *Recommendation on evacuation analysis for passenger ships and high-speed passenger craft,* International Maritime Organization, Sub-Committee on Fire Protection, Report of the Working Group FP45/WP.6, (2001).

Druck: Strauss Offsetdruck, Mörlenbach
Verarbeitung: Schäffer, Grünstadt